THE ARMED FORCES OF INDEPENDENT INDIA: 1947–2006

THE ARMED FORCES OF INDEPENDENT INDIA: 1947–2006

KAUSHIK ROY

MANOHAR
2010

First published 2010

ISBN 978-81-7304-778-7

Published by
Ajay Kumar Jain for
Manohar Publishers & Distributors
4753/23 Ansari Road, Daryaganj
New Delhi 110 002

Typeset at
Digigrafics
New Delhi 110 049

Printed at
Salasar Imaging Systems
Delhi 110 035

To
Suhrita
again

Contents

Preface

Frederick the Great distrusted civilians writing military history. In the aftermath of the First World War, the government of Britain ordered the military officers rather than the civilian historians (though Britain had a tradition of military history being taught in the universities) to write the multi-volume official history of the war. The share of the uniformed officers in writing the British official history of the Second World War is also significant. In the Indian context, can a civilian (a comparatively young and unknown academic) who has never donned a military uniform, nor has held a gun and has never been present in the field of fire, be capable enough to write a book dealing with the aims, characteristics and functions of the armed forces? Military history is too important to be left to the military officers. This is not to show any disrespect to their writings. The officers' accounts reflect genuine emotions, passions and occasionally a worm's eye view of the frontline. Without such source materials, a historian is totally lost.

Senior historians who have whitened their hair have not dared (or cared?) to chart the historical trajectory of India's armed forces after Independence. The security analysts are interested in the present and near future. Political scientists and international relations experts will probably question the temerity of one who has been trained in the historical craft, intervening in the fields of grand strategy. Since in this book I will be wearing two hats—that of a historian and another of a security analyst—I will probably draw the ire of both groups. Despite all these challenges, and because there is no academic monograph on the history of post-1947 Indian armed forces, I have been encouraged to write one thus displaying what Napoleon termed as 'three-o'clock-in-the-morning courage'. Whether, this book will turn out to be a grand success like Austerlitz, or will be a disaster like Borodino or will remain unknown like the *Grande Armee's* campaign in Illyria, I can only wonder.

After finishing a doctorate in 2001 on the Indian Army under the British Raj, I was wondering what happened to that army after Independence. History in India stops at 1947. Contemporary history is yet to be integrated fully with the university and college syllabus. Being unemployed for one year, I started buying books dealing with independent India's armed forces. Along with the Institute of Defence Studies and Analyses Library, Central Secreteriat Library, Nehru Memorial Museum & Library, Jawaharlal Nehru University Library, the bookshops of Delhi are a treasure trove in this regard. Though the bookshops of Delhi are good, the water is bad. After finishing three chapters, I contracted jaundice. Looking back, I believe that it was a blessing in disguise. As I ate boiled papaya and rice and lay on the bed, I had the time to devour more military literature.

My trip to Europe in the summer of 2003 was fruitful because I met lot of men and women who had some sort of first hand experience with the military. Yadein, a civilian doctor told me about his experience in the Israeli Defence Force, especially when it conducted 'dirty war' against the Palestinians. Kerim and Ivana narrated their experiences as refugees from Sarajevo, while the American fighter-bombers romped over the skies of erstwhile Yugoslavia. Maya told me how horrible it was to experience the Russian T-72 tanks rolling in and shooting at the civilians in Georgia. While Irene Bernebu informed me about the role that the NGOs could play in resolving low intensity conflicts, Tumaini gave information as regards the frequent coups in Africa. At the far north of Europe, the Library of International Peace Research Institute of Oslo (PRIO) is mind boggling. The librarian kindly allowed me to photocopy free of cost as much as I could. Prof. Scott Gates was generous enough to give his time and explain his views about the rebel armies. And Sven, a researcher enlightened me about civil-military relations in post-Communist Russia. The only odd person out was the beautiful Kristine. She advised me that I would do better if I visit the nightclubs at midnight. I remain grateful to her for her friendship. The final chapters matured while I was roaming among the lakes and mountains of Norway.

Back in India, in the summer of 2004, when I was thinking of publishing the chapters as articles, Anil Chandy (an old JNUite and friend) encouraged me to write a book on independent

India's armed forces. Special thanks also to Prof. Indivar Kamtekar for encouraging me to pursue the project. This project was funded by Nehru Memorial Museum & Library, New Delhi and I am deeply thankful to them for their support. However, my wife warned me not to write any more books but rather to look for permanent jobs. She had a point. Writing a book especially on military history brings neither money nor much status in India. In contrast, a university job requires not much research but is more lucrative. Not for the last time, I overruled her advice. I only hope that she is wrong, at least this time.

Presidency College KAUSHIK ROY
Kolkata, 2008

Abbreviations

AAG	Anti-Aircraft Gun
ADS	Air Defence Ship; Indian name for indigenously manufactured aircraft carrier
AFV	Armoured Fighting Vehicle
AI	Air Interdiction
AJT	Advanced Jet Trainer
APC	Armoured Personnel Carrier
ATGMs	Anti-Tank Guided Missiles
ATV	Advance Technology Vessel; Indian name for the indigenously manufactured nuclear power submarine.
BAI	Battlefield Air Interdiction
BMD	Ballistic Missile Defence
BMP	A sort of APC
BSF	Border Security Force
CAS	Close Air Support
CDFS	Chief of the Defence Force Staff
CDS	Chief of Defence Staff
CENTO	Central Treaty Organization
CGS	Chief of General Staff
C3I	Command, Control, Communications and Intelligence
C4I2	Command, Control, Communications, Computer, Intelligence and Interoperability
C4ISR	Command, Control, Communications, Computer, Intelligence, Surveillance and Reconnaissance
CNS	Chief of Naval Staff
COAS	Chief of Army Staff
CRPF	Central Reserve Police Force
CTBT	Comprehensive Test Ban Treaty
DEW	Directed Energy Weapon
DIA	Defence Intelligence Agency
DLA	Defence Logistic Agency
DPS	Defence Planning Staff
DPSU	Defence Public Sector Undertaking

DRDO	Defence Research and Development Organization
EEZ	Exclusive Economic Zone
EPRF	*Ealam* People's Revolutionary Front
FEBA	Forward Edge of the Battle Area
FMCT	Fissile Material Cut Off Treaty
GDP	Gross Domestic Product
GNP	Gross National Product
GSLV	Geo Stationary Launch Vehicle
HAL	Hindustan Aeronautics Limited
HEAT	High Explosive Anti-Tank Ammunition
IAF	Indian Air Force
IAS	Indian Administrative Service
ICBM	Inter-Continental Ballistic Missile
IDF	Israeli Defence Force
IDR	*Indian Defence Review*
IDSAJ	*Institute for Defence Studies and Analyses Journal*
INA	Indian National Army
IPKF	Indian Peace-Keeping Force
ISI	Inter-Services Intelligence of Pakistan
ISRO	Indian Space Research Organization
JCO	Junior Commissioned Officer
JCS	Joint Chiefs of Staff
JDW	*Jane's Defence Weekly*
JICH	*Journal of Imperial and Commonwealth History*
JKLF	*Jammu and Kashmir Liberation Front*
JSS	*Journal of Strategic Studies*
JUSII	*Journal of the United Service Institution of India*
LMG	Light Machine Gun
LOC	Line of Control
LST	Landing Ship Transport
LTTE	Liberation Tigers of Tamil *Ealam*
Luftwaffe	Air Force of the Third Reich
MAS	*Modern Asian Studies*
MBT	Main Battle Tank
MDL	Mazagon Docks Limited
MoD	Ministry of Defence
MoDAR	*Ministry of Defence, Annual Report*, Government of India
NATO	North Atlantic Treaty Organization
NBC	Nuclear, Biological Chemical

NCO	Non-Commissioned Officer
NDFB	National Democratic Front of Bodoland
NEFA	North East Frontier Agency (now Arunachal Pradesh)
NLI	Northern Light Infantry of Pakistan
NMML	Nehru Memorial Museum & Library
NSAB	National Security Advisory Board
NSC	National Security Council
NSCN	National Socialist Council of Nagaland
OMG	Operational Manoeuvre Group
OODA	Observe, Orient, Decide and Act
ORBAT	Order of Battle
PAF	Pakistan Air Force
PGM	Precision Guided Munition
PLA	Peoples Liberation Army
PLAAF	Peoples Liberation Army's Air Force
PLAN	Peoples Liberation Army's Navy
POK	Pakistan Occupied Kashmir
RAF	Royal Air Force
RAPIDS	Reorganized Army Plains Infantry Divisions
RAW	Research and Analysis Wing
R&D	Research and Development
RIAF	Royal Indian Air Force, predecessor of IAF
RIN	Royal Indian Navy
RMA	Revolution in Military Affairs
RNA	Revolution in Naval Affairs
SAF	Soviet Air Force
SAM	Shoulder Armed Missile/Surface to Air Missile
SEATO	South East Asia Treaty Organization
SLOC	Sea Lines of Communication
TAC	Tactical Air Centre
TMD	Theatre Missile Defence
UAV	Unmanned Aerial Vehicle
UCAV	Unmanned Combat Aerial Vehicle
ULFA	United Liberation Front of Asom
USAF	United States Air Force
USN	United States Navy
VCO	Viceroy's Commissioned Officer
WLR	Weapons Locating Radar

Glossary

Ahimsa	Non-violence
Auftragstaktik	Mission oriented command system or decentralized command set-up of the German armed forces.
Befehlstaktik	A top down centralized command system; opposite of *Auftragstaktik*.
Bitva	Russian concept for a great battle involving lot a of bloodshed.
Blitzkrieg	Lightning attack involving tanks and infantry in armoured personnel carriers supported by self-propelled guns and tactical air force.
Bundeswehr	Army of post-1945 Germany
der totale krieg	Total War which means complete mobilization of society and economy for conducting warfare.
Chana	Parched pulse
Ealam	Autonomous homeland demanded by the Tamils in Sri Lanka.
Festung	Fortress
Fidayeen	Islamic suicide bomber
Furlough	Paid leave
Generalfeldmarschall	Field-Marshal of the German Army
Gur	Jaggery
Hawala	Informal transaction of money about which no written accounts exist. This system originated in Mughal times. People still use it to evade taxes and to conduct illegal transaction away from the eyes of the authorities.
Izzat	Prestige
Jawan	Soldier
Jihad	Religious war conducted by the Muslims against the non-believers.

Kamikaze	Literally, divine wind; suicide attack by Japanese aircraft on enemy ships.
Kesselschlacht	Literally cauldron battle, i.e. encirclement battle.
Madrasa	Religious seminary preaching the precepts of Islam.
Mandalas	Kautilyan concept of interstate relations based on circle of states.
Mansabdari system	Mughal military system in which the military chiefs were given *mansabs* (ranks) which were of various grades.
Mujahideen	Warriors conducting *jihad.*
Mukti Bahini	Liberation Force in East Pakistan.
Neta-Babu Raj	Alliance between politicians and the civilian bureaucrats.
OMG (Operational Manoeuvre Group)	Combined infantry-tank assault team of the Soviet armed forces designed to carry out deep inside the enemy front. Blitzkrieg style attacks.
Pakfront	Anti-tank defense belt manned by infantry.
Panchsheel	Literally five principles; Jawaharlal Nehru's five principles governing interstate relationship.
Pandit	Literally means a Hindu Brahmin.
Panzer	Tank or armour in German.
Rabbis	Jewish religious leaders.
Raksha Mantri	Defence Minister of India.
Rashtriya Rifles	Special units of the Indian Army detailed for conducting counter-insurgency tasks.
Realpolitik	Pragmatic politics devoid of morality.
Saathis	Fellow soldiers.
Shaheed	Martyr
Sayarot	Elite units of Israeli Defence Force which conduct probing attacks along the enemy lines.
Tanzeem	An Islamic militant outfit operating in Kashmir.
Wehrmacht	German armed forces under the Third Reich.
Yeshivots	Jewish religious seminaries.

INTRODUCTION

Indian Armed Forces and Military History

> Rational pacifism must be based on a new maxim—'if you wish for peace, understand war'. History has ample evidence of how often a move to preserve peace, or to restore it, has been paralysed by so-called 'military reasons' that were no more than a rationalization of unreasonable impulses. Hence we need to understand not only the causes but the conduct of war. This understanding can only be attained if we study war in a purely scientific spirit, with our minds freed from any pro-military or anti-military bias which impair our judgement—and thereby nullify our deductions.
>
> B.H. LIDDELL HART[1]

The above lines which Britain's foremost military thinker wrote in April 1932 is one of the justifications for writing this book. In this monograph, the aim is to understand the historical evolution of the Indian armed forces after independence. Can history teach us any important lesson? The British and the American armies rely heavily on past experiences in order to formulate their approaches to warfare.[2] John Gooch, a British professor of international history rightly says: 'A glance at the rear-view mirror of history may therefore offer some guidance to those whose task is not merely to predict the future, but to master it.'[3]

However, the Indian armed forces and their civilian analysts are yet to use history for constructing military doctrine. The Institute for Defence Studies and Analyses, the most prestigious Centre for Defence Studies and manned by military officers plus civilian scholars (especially political scientists and international relations experts) while writing about military affairs rarely go back beyond the 1980s. Their ahistorical approach is due to the conviction that rapid advance in technology has made studies of past campaigns redundant. The basic assumption behind this monograph is that history could teach us something useful. A historical approach to the study of post-independent Indian military is helpful for several reasons. The absence of a purely utilitarian motive (which remains embedded in the strategic studies framework) will hopefully result in a more rigorous analysis of the empirical data, free from the dictates of policy-making. Hence, the genesis and growth of the Indian armed forces from a historical perspective is attempted in the hope that pros and cons of our military organization will thus become evident. This will also reduce the technical bias present in current military theorization.

Some books dealing with the history of the Indian Army have been written by the men in uniform. But, the chronological narratives charting the activities of the Indian Army by retired army officers[4] fail to address any of the issues discussed in this monograph. However, there have been some exceptional studies by some civilian scholars. Stephen P. Cohen's book on the Indian Army focuses on one particular issue: professionalism of the officer corps.[5] Then, Stephen Peter Rosen by using what could be categorized as 'armed forces and society' methodology assumes a direct linkage between the social structure and the military organization.[6] This book is partly influenced by the assumption of the cultural determinists who assume that warfare and military establishments are products of particular cultural formations of the host countries.[7] Barry R. Posen's organizational analysis emphasizes that militaries are functionally specialized bureaucracies with distinct components whose organizational behaviour ought to be studied. The constituent parts and the interaction between the parts, asserts Posen, need to be analysed.[8] But, cultural influence on the militaries cannot be denied completely. In recent times organizational analysts are also taking into account the cultural predisposition of the individual military organization. Hence, we have organizational culture analysis.[9]

This book follows a more nuanced methodology. This monograph is a history of both the evolution of the military organization and the flowering of ideas about conducting warfare. It is challenging to integrate the history of ideas with the organizational culture methodology. The development of the armed forces as an institution, its nature and purposes and the formulation of theories as regards the functions of the Indian military are the central concerns of this book. About 70 per cent of the monograph addresses the issues associated with the ground forces. This is because compared to the other two services, the army has been predominant as regards manpower, funding and deployment. A history of India's air warfare is yet to be written.[10] The present monograph devotes one chapter in analysing the growth of the Indian military's air arm and another chapter on the institutional development of the Indian Navy. Besides the official histories of the Indian Navy which are rich in details but not interpretative, we have two excellent works by Rahul Roy-Chaudhury.[11] Roy-Chaudhury is more concerned in charting out the nature of sea power which suits India's present security needs rather than in tracing the developmental story of the navy after 1947. A comparative analysis of the Indian and Pakistani navies from present day security perspectives is provided by K.R. Singh.[12] In this volume, the development of the Air Force and the Navy are treated separately because joint operation and joint doctrine by the three armed services are yet to be formulated and practiced. The three services possess distinct infrastructure filled with service-specific trained personnel. The focus in this monograph remains on the institutional evolution of the three services within the armed forces, which throws light on their fighting techniques and organizational structure. Manning and utilizing the armed forces are a crucial part of statecraft. So, the study of armed forces also opens up a window regarding the character of the state structure.

A comparative methodology is developed in order to assess the level of success achieved by India in harnessing military power vis-à-vis the other developing and developed countries. A proper evaluation of the Indian armed forces is impossible without a detailed analysis of the militaries of Pakistan and China. Hence, the Pakistani and the Chinese militaries make their frequent appearance in this monograph. Comparison of the Indian armed forces with the Israeli Defence Forces' behaviour in the battlefield and its attitude towards the political system in peacetime provides crucial insights as regards the strengths and weaknesses of the Indian political and military establishments.

The author deliberately refused to use interviews of the military officers. This is because the officers while talking in private would say certain sensitive things but refused to back them formally, and did not want to be quoted on sensitive issues. In the perspective of this author, to include such sensitive stuff in a book without quoting authority is an example of bad history writing. Most of the official sources are not open to the public. Moreover, with the passage of time, memories tend to get blurred. 'Cowards' tend to develop psychological defence and depict themselves somewhat unconsciously as 'heroes'. This is known as the 'bull frog effect' and is a serious defect as regards utilizing oral history techniques for writing history.

Departmental files are not yet open to the researchers. But, some amount of important data can be extracted from the bland annual reports published by the Ministry of Defence. This book mainly depends on the open sources: accounts published by the participants in the commercial press as well as in the service journals. The Indian officer corps continues to write extensively about the various aspects of the military. Books published by military officers about various campaigns, biographies and autobiographies are indeed valuable sources. Cross checking through various books regarding a single issue establishes the degree of veracity of a particular author on that theme. Among the various journals to which the military officers contribute, the most useful ones are the *Journal of the United Service Institution of India* and the *Indian Defence Review*. For political inputs in the military establishment, published papers of Jawaharlal Nehru and unpublished oral transcripts of some of the security managers available at the Nehru Memorial Museum & Library have been used.

The historical trajectory of the armed forces is influenced by the fluctuating power politics. The twenty-first century seems to be quite different from the last century. The bipolar world disintegrated with the end of the Cold War. However, the collapse of communism did not result in the total victory of capitalism. Islamic fundamentalism and the rise of offstage extra-regional powers along the Eurasian rim are the chief characteristics of the emerging global order. How is the Indian military preparing for the challenges of the new millennium? Before assessing the security threats of India in the future, it is necessary to assess the military threats faced by the country in the past. India continues to face military challenge from its two neighbours: China and Pakistan. After the end of the Cold War, the strategic odds faced by India worsened. The first chapter analyses the changes and continuities in security threats faced by India. The discussion on nuclear strategy remains short because in India nuclear policy remains a civilian baby. The three services continue to play a marginal role in the evolution of the nuclear doctrine and nuclear policy of India.

While the first chapter charts out the broad outlines of the political canvas within which the military operated, the second chapter discusses the various mechanism evolved by the civilian ruling elite for subordinating the armed services within the polity. Since a manpower intensive army constitutes the most serious political threat to civilian supremacy in any country, particular attention is given to the interaction between the civilian rulers and the army's officer corps.

The term command culture stands for the attitudes and mechanism of command within the armed forces. The third chapter addresses this issue. The philosophy and apparatus of command of the air force, navy and the army remains distinct. Chapter 3 tackles the issue of command mechanism mostly in the army. Some attention has also been drawn to the command techniques involving air-ground operations. Command techniques of pure aerial duels and naval operations are dealt separately in Chapter 6 and 7 respectively. The command system in the Indian Army, the fourth largest in the world, was never able to conduct fast mobile campaigns. The history of past campaigns shows that India's military effort was dogged by inter-service rivalry. There is an intense debate among the Indian officer corps for setting up a joint integrated command. Even the Indian Air Force (IAF), which claims to be technologically most advanced compared to the other two sister services continues to have serious deficiencies in conducting joint warfare. The end of the Cold War and the subsequent Revolution in Military Affairs (hereafter RMA) made India's Worse Possible Scenario a reality. The RMA also known as Information Warfare is the result of Digital Revolution. The development of microchips necessitated in the reorganization of the Command, Control, Communications and Intelligence (C3I) components of the armed forces. To get inside the loop of enemy's decision-making cycle for quick response, advanced battlefield surveillance and electronic sensors for acquiring the 'real time' picture of the enemy, are all dependent on advanced software programmes. All these technologies render the industrial age armies obsolete. China accepts this implication. So, Beijing is engaged in transforming its mass peasant army into a capital-intensive force equipped with advanced C3I network. The command system of India's armed forces is yet to undergo this transformation.

Social and cultural aspects of the military are certainly important. However, the armed forces' battlefield performance cannot be negated, because after all the military exists to deter and ultimately to wage war. Chapter 4 analyses the Indian Army's performance in limited conventional and quasi-conventional warfare. The aim is not to give a detailed narrative of the actions undertaken by the various regiments of the Indian Army, but to portray the Indian Army's theoretical innovations at the strategic, operational and tactical planes and the linkages between the military officers' thought processes and the war machine's performance in the killing field.

Besides fighting and deterring external enemies, policing India remains a crucial task for the armed forces. The nature and scope of the IAF and the Indian Navy's role in counter-insurgency operations remains marginal hence treated separately. Chapter 5 discusses the Indian Army's counter-insurgency operations. Most of the infantry units have been engaged in counter-insurgency duties in the north-east and in Kashmir because of the failure of the paramilitary forces. Such tasks sap the soldiers' morale and their training suffers. In general it has been observed that internal deployments of the army erodes civilian control. How far this generalization is true as regards India is to be tested. Probably in an underdeveloped country like India, use of the army in counter-insurgency operations is essential due to the defective culture of governance and a weak civil society.

Chapter 6 portrays the evolution of the IAF. For the last fifty years, the army demanded that the IAF function as the ground force's flying artillery. In contrast, the airmen wanted to shape the air force as a strategic deterrent. This debate is still continuing and will probably never be resolved. Chapter 7 portrays the gradual emergence of the Indian Navy and its increasing role in

India's grand strategy. The Indian Navy remains the Cinderella of the armed services. Since Independence, the navy got the least funding compared to the other two services. Nevertheless, the Indian Navy played a crucial role in transforming East Pakistan into Bangladesh. And the mission of maintaining India's sea based economic lifeline is dependent on India's naval forces. Further, the task of eradicating modern piracy remains with the navy.

The Indian armed forces have to integrate the techniques for waging Information War which requires combined arms operation and further, a sophisticated weapons system. The development of the defence production base is the theme of Chapter 8 and examines why India's military industrial base is yet not self-sufficient. If India is to survive in the present state-system characterized by anarchy, then utilizing the recent technological changes is a sheer necessity and not a luxury. Proper integration of the latest technology in turn requires organizational and doctrinal transformation. What is required is the political will necessary for rapid transformation of the armed forces. Let us have a flashback to trace the origin and growth of India's armed services.

NOTES

1. B.H. Liddell Hart, *Thoughts on War,* 1944, rpt., Staplehurst: Spellmount, 1999, pp. 9–10.
2. J.J.A. Wallace, 'Maneuver Theory in Operations Other than War', *Journal of Strategic Studies* (*JSS*), vol. 19, no. 4 (1996), p. 207; John E. Jessup Jr. and Robert W. Coakley, eds., *A Guide to the Study and Use of Military History,* Washington D.C.: Centre of Military History US Army, 1982.
3. John Gooch, 'Addition, Multiplication and Step-Function: The "Arithmetic" of Military Revolution in Modern Times', in Stuart A. Cohen, ed., *Democratic Societies and Their Armed Forces: Israel in Comparative Context,* London: Frank Cass, 2000, p. 28.
4. Two examples of this genre of writing are V. Longer, *Red Coats to Olive Green: A History of the Indian Army, 1600–1974,* Bombay: Allied, 1974; Lieutenant-General S.L. Menezes, *Fidelity and Honour: The Indian Army from the Seventeenth to the Twenty-first Century,* New Delhi: Viking, 1993.

5. Stephen P. Cohen, *The Indian Army: Its Contribution to the Development of a Nation,* 1971, rpt., New Delhi: Oxford University Press, 1991.
6. Stephen Peter Rosen, *Societies and Military Power: India and Its Armies,* New Delhi: Oxford University Press, 1996. See especially Chapter 6.
7. Richard Buel Jr., 'Review Essay', *History and Theory,* vol. 34, no. 1 (1995), p. 106.
8. Barry R. Posen, *The Sources of Military Doctrine: France, Britain and Germany Between the World Wars,* Ithaca/London: Cornell University Press, 1984, p. 35.
9. Theo Farrell, 'Figuring out Fighting Organizations: The New Organizational Analysis in Strategic Studies', *JSS,* vol. 19, no. 1 (1996), p. 123.
10. George K. Tanham and Marcy Agmon, *The Indian Air Force: Trends and Prospects,* 1995, rpt., New Delhi: Vision, 1996 and Air Vice-Marshal Viney Kapila's *The Indian Air Force: A Balanced Strategic and Tactical Application,* New Delhi: Ocean Books Pvt. Ltd., 2002 are not histories of the Indian Air Force but address the role of air power in India's national security policy. The only exception is Air Marshal M.S. Chaturvedi's *History of the Indian Air Force,* New Delhi: Vikas, 1978. But, Chaturvedi's account ends in the 1970s.
11. Rahul Roy-Chaudhury, *Sea Power and Indian Security,* London/ Washington: Brassey's, 1995 and *India's Maritime Security,* New Delhi: Knowledge World, 2000.
12. K.R. Singh, *Navies of South Asia,* New Delhi: Rupa, 2002.

ONE

India's National Security Policy: The Military Dimension

In 1947, the British Indian Empire (Raj) was partitioned into India, a secular sovereign state of many nationalities, and Pakistan. In 1950, India became a Republic. Post-independent India witnessed certain far-reaching changes. The *neta-babu* combine replaced the white sahibs. The end of the twentieth century registered massive changes in the field of nuclear and information technologies. However, key aspects of geopolitical relationship remain constant. Despite political and technological changes, geo-strategic choke points remain same. This ensures continuity regarding security management in the *longue duree*.[1]

Security policy is also known as grand strategy. For Paul Kennedy, grand strategy concerns itself with long-term objectives of the state and not just the immediate issues facing it. The principal essence of grand strategy is balancing means with ends and burden sharing among the various sectors in pursuit of security objectives. The former requires net assessment of the pros and cons of any path that is to be followed.[2] The Americans term grand strategy as national security policy. India's strategic community and the government accept this concept. In 1977, Major-General D. Som Dutt defined security policy in the following words: 'National Security can be defined as a nation's endeavour to safeguard its social, political and economic institutions from attack by other independent states. The means by which this aim is achieved, constitute military policy.'[3] Military policy refers to the military component of grand strategy while military strategy is an amalgam of operational doctrine and tactical requirements including military hardware.

A British security analyst Lawrence Freedman defines grand

strategy as an amalgam of both internal threats and external security. He writes:

> The focus of strategic thinking must be the ability of a state to sustain itself. Much writing on strategy and international politics distinguishes the problem of the state in its external relations from the requirements of internal order. This is a false dichotomy. A state with problems in internal order is more vulnerable to external pressure—it is a supplicant, requiring powerful friends to put down insurgency and provide economic assistance. It is vulnerable to an unfriendly opponent stirring the pot a little.[4]

Similarly, the Ministry of Defence's (MoD) *Annual Report* for 1970-1, defines national security as: 'National security goes beyond mere defence preparedness, the deployment of forces or defence managements; it is interwoven with what we consider to be our national objectives, our foreign policy, the economic potential of the country, internal security and the policies and declared objectives of our potential adversaries.'[5] This chapter analyses the role of the armed forces in India's national security policy. Despite Freedman and the MoD's *Annual Report's* strictures, for reasons of clarity, the emphasis of this chapter will be on external security, as internal security will be addressed in Chapter 5.

In recent times it has become fashionable to analyse national security policy without referring to the armed forces. Due to the influx of post-modern methodology, the trend is to move away from state apparatus towards civil society and to focus on cultural politics, human geography and mentality of the peripheral groups. The new approach focuses on hitherto marginal groups like low castes, tribes, etc. The focus is to ensure food security, health security, etc., of the underprivileged groups. The feminist paradigm for understanding inter-state relations tends to reject what it terms as 'military-masculine' considerations in analysing national security policies.[6] It could be questioned whether such a non-statist and non-realist approach is at all helpful.

In the Indian case, the state even in the new millennium remains overtly strong. And the civil society continues to be weak. All sorts of developmental activities have been state directed. The globalization policy of Narasimha Rao's government and the successor regimes have not been able to convert the state directed mixed economy into a full-fledged market economy. The influence of independent academic centres with a dedicated band of intellectuals for shaping national security policies remains weak. All academic centres especially those concerned with defence policies like the Institute for Defence and Strategic Analyses, United Service Institution of India, etc. are at least indirectly funded and controlled by the government. The core instrument for conducting defence in independent India has been the army. The British introduced the regimental format which replaced the *mansabdari* organization. From the 1790s, the Sepoy Army was organized in regiments. Each regiment of 800–1,000 soldiers was composed of ten companies.[7] And the Indian and Pakistan armies remain organized on the same lines.[8] The post-1947 Indian Army remains a long service volunteer force numbering about one million.[9] Most of the men serve between fifteen to twenty years. The Indian Army is not only the third largest in the world but it has also been called out almost every year for maintaining law and order in several states. And it has fought four conventional wars. So, a state centric top down approach for studying national security policy focusing on the Indian military is a viable option.

States are like living organisms which compete and struggle in a world where the rule is 'might is right'. So, in the Darwinian geopolitical scenario every state for its survival tries to expand and maintain its influence and this gives rise to conflicts.[10] Preventing conflicts from breaking out as well as to win the conflicts once they have broken out require security oriented strategies. Two American international relations experts who could be categorized as representatives of the Strategic Culture paradigm[11] assert that from the dawn of civilization, due to India's social and cultural conditions, India had no strategy to speak of. George K. Tanham writes that Hinduism continues to dominate India, and because of Hindu fatalism and conservatism, the Indian elite remains inward looking. Therefore they have no consistent security policy to ward off invasions or to project power outside the subcontinent.[12] Continuing in the same vein, Stephen Peter Rosen claims, that because of its ingrained system, the Indian ruling elite is always immersed in sorting out internal communal problems. They neither have the intention nor the energy to formulate a long-term strategic defence policy vis-à-vis foreign political entities.[13] Whether India's ruling elite has pursued a clear-cut strategic agenda or not can only be answered after a rigorous historical analysis.

INDEPENDENT INDIA'S PASSIVE SECURITY POLICY

Independent India has to encounter two hostile neighbours within South Asia. The situation further worsens when the extra-regional powers interfere in the affairs of the subcontinent. When India became independent, it was left with a 'moth eaten' North-West frontier. Occupation of the strategically important cities like Multan, Uch (west of Bhawalpur), Und and Lahore had been historically necessary to prevent intrusion of the enemy into India's heartland after it had pierced the North-West frontier.[14] However, all these cities went to Pakistan after Independence. The Indian polity lacked the economic muscle and the political will to maintain an outer defensive arc for establishing India's hegemony over the states of South-East Asia and West Asia.

India's military helplessness was clear from Prime Minister Jawaharlal Nehru's speech of 13 October 1949 at Washington. He admitted that India was new to world politics and its military strength insignificant for the moment.[15] Nehru took care to reduce the role of the armed forces as far as the formulation of India's national security policy was concerned. In his speech of 20 November 1955, Nehru emphasized: 'Not being military minded, we do not appreciate the use of military phraseology or military approaches in considering the problems of today.'[16] In an attempt to reduce the role of military power within the anarchic international political order, Nehru focused on the issue of global disarmament. In a speech to the United Nation's General Assembly on 10 November 1961, Nehru declared: 'I do think that the major question and the biggest question today is this question of war and peace and disarmament. There is no conflict between those. In fact, the whole atmosphere of the world will change if disarmament comes in and these present problems go towards solution.'[17] To be fair to Nehru, it can be said that elements of *realpolitik* were embedded in his complex strategic paradigm. As early as 1937, Nehru had observed: 'We live in an abnormal world, full of wars and aggression, when international law has ceased to be and treaties and undertakings have no value. . . . The only thing to be done to protect oneself is to rely on one's strength as well as to have a policy of peace.'[18]

Nehru's policy of 'Non-Alignment', led to India refusing to align itself with any of the extra-regional powers.[19] In 1946, Nehru went against the Kautilyan *mandala* dictum and said: 'We propose as far as possible, to keep away from the power politics of groups, aligned against one another, which have led in the past to World Wars and which may again lead to disasters on an even vaster scale.'[20] On 8 March 1949, Nehru described his policy of non-alignment in the following words: 'We have stated repeatedly that our foreign policy is one to keep apart from big blocs of nations—rival blocs—and to be friendly to all countries and not become entangled in any alliances, military or other, that might drag us into any possible conflict.'[21] India missed a golden chance to establish a rapport with Burma because of Nehru's rigidity. When U. Nu requested military supplies, Nehru in a letter dated 14 April 1949 replied: 'More especially our army supplies came down to rock bottom level and because of the difficulty in getting spare parts, much that we had cannot be used. We have recently sent a special military mission to Europe and America to get supplies and spare parts. . . . But our present position is such that we just cannot find adequate supplies for you.'[22] Burma, even now, continues to remain outside India's strategic orbit.

Instead of alliance and pacts, Nehru introduced the concept of *Panchsheel*. On 17 September 1955, Nehru attempted to educate the Indian Legislative Assembly regarding his new foreign policy:

> What are these five principles (*Panchsheel*)? They are very simple. The first one is the recognition by countries of their independence and each other's independence, sovereignty and territorial integrity. The second one is non-aggression; the third is non-interference with each other; and the fourth is mutual respect and equality; and the fifth is co-existence. I think we may take some credit for spreading this conception of a peaceful settlement, and above all, of non-interference. That each country should carve out its own destiny without interfering with others is a important conception, though there is nothing new about it.[23]

Probably Nehru believed that the high moral principles of *Panchsheel* and the emphasis on the policy of disarmament would enable him to do away with the traditional power politics and display of force. Further, Nehru believed that defence spending was detrimental for economic growth,[24] and hence expenditure on the military declined drastically after Independence. Between 1872 and 1947, defence expenditure amounted to 30 per cent of the total government expenditure. In contrast, the proportion of defence expenditure to total expenditure decreased to 16 per cent between 1951 and 1960, and it was further reduced to 15.7 per cent between 1961 and 1967.[25] In contrast to India, during the 1960s, Portugal spent over 70 per cent of her Gross National Product (GNP) and 50 per cent of her annual budget on defence which has led historian Douglas Porch to categorize Portugal as a 'warfare state'.[26]

Nehru with the aid of his Defence Advisor Prof. P.M.S. Blackett charted the intellectual framework of India's grand strategy. Blackett asserted that before India could become a big power, it has to industrialize. And in the interim period, the Indian armed forces should be ready to wage 'marginal war' against small powers like Hyderabad, Goa, etc. Against Pakistan, it would be adequate for the Indian armed forces to reach a stalemate.[27] Nehru assumed that against hostile neighbours, Pakistan and China, the Himalayas in the north and the desert in the west would function as natural defence lines.[28]

Nevertheless, New Delhi threw the high principles of *Panchsheel* to the wind and attempted to establish an inner defensive arc which comprised the Indian subcontinent, an area of 16,00,000 square miles.[29] The 1949 India–Bhutan Treaty of Friendship required the latter to be guided by the government of India as regards its external relations. In accordance with the 1950 India–Sikkim Treaty of Peace, New Delhi took over the task of defending Sikkim. In the same year, India signed a treaty of peace and friendship with Nepal. By dint of this treaty, India became Nepal's principal arms supplier.[30] As a response to the Chinese move into Tibet, in 1951, the Indian government undertook the responsibility to meet Bhutan's foreign exchange needs. On 7 April 1952, a Military Mission was sent to Nepal to reorganize the Nepali Army.[31] In a note dated 6 January 1955 addressed to the Minister of Defence Organization and the Defence Secretary, Nehru said: 'So far as Nepal is concerned, there is no direct danger to us. But there is a possibility of internal trouble and our having to take some steps to meet such a situation.'[32] Nehru commented: 'Much as we stand for the independence of Nepal we cannot allow anything to go wrong in Nepal or permit that barrier (the Himalayas) to be . . . weakened, because that would be a risk to our own security.'[33] In pursuing the policy of establishing a glacis around north India, New Delhi annexed Sikkim in 1975.[34]

The logical culmination of the construction of glacis around India was the Indira doctrine which was enunciated in 1983. It clarified that India had a right to intervene in the internal affairs of the neighbouring countries if disorders threatened to spill over from the confines of the affected country. In addition, India would not tolerate any intervention by an outside power. If any external help was required for meeting any internal crisis, the affected state must look to India for aid. In a way, Indira doctrine is a sort of India's Monroe doctrine.[35] One aspect of the Indira doctrine is a return to Kautilya's *mandala* theory. According to one retired Indian Foreign Secretary, in the 1970s India perceived Afghanistan as an enemy's enemy hence India's friend.[36] The polar opposite of the Indira doctrine was the Gujral doctrine. It emphasized a policy of absolute non-intervention in the neighbouring countries. But it is the Indira doctrine, which by and large has remained in vogue.

Before the enunciation of the Indira doctrine, there were some attempts to theorize India's probable strategy by some military persons. One Brigadier Rajendra Singh in 1954 was a proponent of aggressive defence. He wrote:

> I must emphasize before proceeding further that once aggression has been committed, the destruction of the opponent by aggressive means would be a corollary. So, we start from a negative angle, that is, India would not attack the would-be aggressor, and shall take up arms only in self-defence. This negative angle is likely to induce an attitude which is most dangerous to security, because it tends to produce passivity, which in turn decries the use of arms. It has become doubly important for us to guard against this mentality, because our minds are still soaked with ideas of the efficacy of non-violence to overpower an oppressor.[37]

Rajendra Singh challenged the Gandhian paradigm of non-violence. In place of Nehruvian idealism, Singh was for injecting a strong dose of realism. He wrote: 'In plain, clear language—that the strong are tempted to swallow the weak and in the society of virile people and aggressive nations, only strength counts.'[38] Lieutenant-General Harbakhsh Singh (who commanded the Western Sector during the 1965 War with Pakistan) was also a proponent of 'Forward Defence'. He wrote: 'It is my conviction that with a leadership more mentally robust and imbued with a instinctive urge for offensive action, our achievements might have been more impressive. A defensive attitude of mind can at best avert a defeat—it cannot achieve decisive results.'[39] Such sort of thinking in the long run probably facilitated the emergence of the Indira doctrine.

Occasionally both internal political compulsions and extraneous strategic dangers forced India's overseas commitments. Besides rounding off the southern *glacis* of India, the aim of New Delhi was to prevent any foreign influence in Sri Lanka when it was riven by conflict from within. In 1987, India under Rajiv Gandhi intervened in Sri Lanka because the ruling Congress party could not overlook the public opinion in Tamil Nadu regarding the killing of Tamils by the Sinhalese. Secondly, Sri Lanka's attempt to acquire support of foreign countries for modernizing its armed forces and the possibility of the Trincomalee port being made available to the USA accelerated India's decision to send the Indian Peace Keeping Force (IPKF).[40] As part of India's Monroe doctrine, Rajiv Gandhi, the Prime Minister of India and President Jayawardene of Sri Lanka signed the India–Sri Lanka Accord in 1987 which resulted in the induction of the IPKF in Sri Lanka.[41]

Delhi's policy to establish the inner defensive arc faced obstruction from Pakistan and China. The nervous mindset of a weak Pakistan vis-à-vis a seemingly hegemonic India could be gleaned from the Pakistani dictator Field Marshal Ayub Khan's writings. In his memoirs, Ayub Khan states: 'Contrary to all agreements and principles, India forcibly occupied a major part of the State of Jammu and Kashmir and concentrated her forces there, thus posing a constant threat to our security. At the back of it all was India's ambition to absorb Pakistan or turn her into a satellite.'[42] Kashmir remains important to India for cultural and especially strategic reasons. As early as 2 November 1947, Nehru had declared: 'We were naturally concerned about this not only because of our close ties with Kashmir and her people but also because Kashmir is a frontier territory adjoining great nations and therefore we were bound to take interest in developments there.'[43] Pakistan continues to have a problem with India in general and New Delhi's policy vis-à-vis Kashmir in particular. Pakistan's ambition to incorporate Kashmir has remained an important constant in Pakistan's grand strategy from 1947 onwards.[44] Just after the British withdrawal from the subcontinent, Britain took a soft stance vis-à-vis Pakistan over the Kashmir issue. London was anxious that if Pakistan became discontented with Britain then an adverse reaction against Britain would also follow among the oil rich Muslim states of the Middle East.[45] This encouraged the Pakistani misadventure during 1947–8. According to General Roy Bucher (Commander-in-Chief of the Indian Army during 1948–9), both Nehru and the Defence Minister Baldev Singh, unlike the Pakistani Prime Minister Liaquat Ali Khan were eager to end the war with Pakistan even before the intervention by the United Nations.[46] Nehru realized that the Muslim populace of west Kashmir were eager to join Pakistan, so he was ready to stop the war once the Indian Army had gained control of the Kashmir Valley.[47]

After the first India–Pakistan War (1947–8), India decided to maintain qualitative and quantitative superiority in military equipment compared to the Pakistan Army.[48] According to Ayub Khan, Liaquat Ali was itching for war with India in 1951 and he warned Ali by telling him that the Pakistan Army was not yet ready for combat.[49] Nehru was aware that Pakistan posed a limited threat to Jammu & Kashmir and in a note to the Minister of Defence Organization and the Defence Secretary dated 6 January 1955, he wrote:

> In regard to the two specific points, it has been clearly stated previously that an attack by Pakistan on Jammu and Kashmir will be considered an act of aggression against India. We shall then be entitled to use our armed forces against Pakistan anywhere. . . . The question of minimum period of warning is almost a hypothetical one. . . . Any kind of a sudden attack on India is exceedingly unlikely for a considerable time to come. If the situation worsens, even then nothing is going to happen suddenly. . . . In view of this present position, it does not appear necessary for our forces to be kept in readiness for a ten day's warning.[50]

The interference of the USA in the subcontinent's politics further weakened India's position vis-à-vis Pakistan. The Cold War made USA look for friends and allies in Asia. The United

States perceived that an extension of the Soviet influence into the Middle East would deprive the West of oil and the principal communications routes connecting East Asia with Europe. As early as 1948, Nehru, was frustrated and irritated with American foreign policy. Nehru, conscious of his own intellectual superiority, asserted that US foreign policy was naïve and unintelligent. It was based on crude power and money,[51] and he made it quite clear that he would not join the West in any pact to check the Soviet Union.

Pakistan after 1949 initiated policies to garner foreign military aid in order to redress its weakness vis-à-vis India. During the early 1950s, Anthony Eden tried to induct India into a sort of anti-Communist bloc for checking the spread of communism in South-East Asia. However, the Nehruvian Non-Alignment policy resulted in failure on part of Britain to induct India in a sort of collective defence pact against the 'communist threat'.[52] And this pushed Britain and USA further against India vis-à-vis Pakistan who was willing to become an ally of the West in their fight against Communism. In 1954, the USA and UK formed the Central Treaty Organization (CENTO) and South East Asia Treaty Organization (SEATO) alliances for checking the spread of Soviet power.[53] The Military Defence Agreement Pact in 1954, and Pakistan's membership of SEATO in September that year, and then CENTO a year later resulted in an inflow of American military aid into Pakistan.[54] The military aid included direct grants, loans, and sales of equipment (the latter for hard or soft currency and by deferred or immediate payment). In addition, military assistance also took the form of cash subsidies to the recipient state.[55] A blueprint of the Military Defence Aid Pact was secured by India's Intelligence Bureau in 1956 and was shown both to Nehru and the Army Headquarter. It was clear from the document that with the supply of US hardware, the Pakistani military in the next five years, would be better equipped than its Indian counterpart.[56]

The continuation of American aid plus India's defeat in the hands of China in 1962 enabled Pakistan to take an aggressive stance vis-à-vis India from 1963 onwards.[57] Rising insecurity among the Pakistani military *junta* also pushed Pakistan to start the war in 1965. In 1963, at a Cabinet meeting, the Director of Inter-Services Intelligence (ISI) of Pakistan emphasized India's military build-up. He argued that India with a greater industrial base and a million strong army which was three times bigger than Pakistan, would be able to carry out a military adventure against Islamabad by 1967. The ruling class of Pakistan concluded that under the guise of preparing against China, India was doubling its air force and enhancing tank and naval construction. Pakistan represented a classic case of a small country being afraid of a bigger neighbour outstripping its military resource in the long run. So, in 1965 Ayub Khan settled for a pre-emptive strike against India. However, he miscalculated in thinking that the big neighbouring enemy whom Pakistan was attacking was a 'tottering giant'.[58]

Besides a nervous Pakistan, India had to tackle China. China's grand strategy has been to bully recalcitrant neighbours and then launch local limited wars.[59] One of the principal factors behind the India–China hostility remains the competition between Beijing and New Delhi for power, status and privileges in South-East Asia. Nehru had high hopes for India as far as South-East Asia was concerned. On 12 September 1948, Nehru declared: 'South-East Asia would include Australia and New Zealand. . . . India is the natural leader of South-East Asia if not of some other parts of Asia also. There is at present no other possible leadership in Asia, and any foreign leadership will not be tolerated.'[60]

Initially the Indian leaders underestimated Chinese power. On 2 September 1949, Nehru wrote to the Chief Ministers:

> Probably, in a reasonably short time, some kind of a Central Government will be established by the Chinese Communists and this will control a very great part of China. Hong Kong is likely to continue as a British outpost at least for the present. The island of Formosa also will remain outside the Chinese Communists' domain. There may also be other pockets in the interior.[61]

After 1950, conditions worsened for India on its northern border. From 1900 onwards Tibet had been a satellite state of the British Raj. But, this buffer state was lost to India when in October 1950 the Peoples Liberation Army (PLA) entered Tibet.[62] As a result, India had to take over the responsibility of guarding the 5,000 km long border with Tibet.[63] During the late 1950s, China violated India's territorial integrity in Ladakh, while building a highway through Aksai Chin. The 2,480 km long road linked Kashghar in Xiniang with Lhasa in Tibet.[64] However, in a note to the Minister of Defence Organization and the Defence Secretary dated 6 January 1955, Nehru underrated the Chinese threat in the following words:

> The Chiefs of Staff have referred to various frontier regions on the north-east. . . . Apart from the possibility of internal troubles in Nepal, there is no foreseeable danger in the north-east border, that is Sikkim, Bhutan and NEFA. . . . So far as these border areas are concerned, it has been proposed that we should have some kind of a border militia or something like it. There is no question of a regular invasion on that border. Nevertheless, the border has to be well protected from any possibility of infiltration by odd groups.[65]

On 3 March 1955, in a note to the Defence Minister, Nehru dragged his feet over the issue of developing the logistical infrastructure of India's northern border. Nehru pointed out that:

> Our frontier roads should be developed rapidly. The frontier of course, includes the border of Jammu & Kashmir, Punjab, Himachal Pradesh, Uttar Pradesh and Assam. Also the NEFA . . . there are other reasons also why these roads should be developed. There are defence reasons. . . . The calculations thus far made have been on the basis of wide roads over which three ton lorries can go. It might be possible to consider this matter in a more modest way, i.e. a road to be constructed for jeeps or like vehicles.[66]

In 1959, China refused to recognize the MacMahon Line and claimed 32,000 square miles in the North-East Frontier Agency (henceforth NEFA later Arunachal Pradesh) and 15,000 square miles in Ladakh.[67] In Ladakh, China demanded Aksai Chin through which lies the strategic lines of communication between Xinjiang and Tibet.[68] In 1959, following Dalai Lama's flight from Tibet to India, Chinese troops fired at the Indian patrol at Longju east of Bhutan.[69] India had tried to establish a diplomatic entente with the signing of the Trade Agreement with China in 1954,[70] but Sino–Indian relation continued to deteriorate. In the note given by the Ministry of External Affairs, New Delhi to the Embassy of China in India on 26 July 1962, it is pointed out that:

> Tension has been increasing in the past few months on the North-Western Frontier of India as a result of the recent establishment of a number of Chinese military posts and the increasing aggressive activities of the Chinese forces in that area. Details of Chinese posts recently

established in Indian territory in the Ladakh region and instances of intensive and aggressive patrolling in this area have been given in the notes presented by this Government on 14 May, 6 June, 16 June, 6, 10, 12 and 14 July 1962.[71]

In early 1962, the Chinese economy was passing through a crisis China had to strengthen its forces along the Taiwan sector because of the American aid to the latter country. Nehru therefore assumed that China would not open a front against India.[72] It turned out to be the greatest miscalculation of his career. Even after the defeat in the hands of China in October–November 1962, Nehru continued to soft pedal the Chinese menace. In a letter to Bertrand Russell dated 4 December 1962, Nehru wrote: 'Certainly we do not want this frontier war with China to continue and even more certainly we do not want it to spread and involve the nuclear powers. Also there is the danger of the military mentality spreading in India and the power of the army increasing.'[73]

Nonetheless, from the 1960s, the threat of strategic encirclement haunted the Indian ruling elite. The Indian strategists remained apprehensive of a Sino–Pakistani tie up and Pakistan launching local attacks for limited territorial gains. On 26 December 1962, Pakistan announced that it had agreed to cede 2,050 square miles of territory in Pakistan occupied Kashmir (POK) to China and this enabled China to build a road linking its Xinjiang province with Tibet.[74] The Sino–Pakistan Treaty of Friendship in 1963 was the first stage in this encirclement process.[75] Pakistan ceded 2,700 square miles of territory to China and in 1964, the construction of the Karakoram Highway linking Pakistan and China started and the project completed by 1978. Besides this road, another road connected Pakistan's northern territories with the Xinjiang province of China and the latter was linked with Tibet by another road not far from the Karakoram Pass. In February 1971, the all-weather highway between Kashgar and Gilgit was opened. The net result was the heightening of tension between India, China and Pakistan along the glaciers of the Karakoram mountain.[76]

The momentous break in India's defence policy occurred in the 1970s. The military equipment provided by the USA, allowed Pakistan in 1965 to launch a twenty-one day war with India.[77] In reaction, India moved closer to the USA's Cold War enemy—USSR. The USA's alliance with Turkey, the Arab states of the Middle East and Pakistan was an attempt to ring off the Soviets from the warm water ports of Indian Ocean. The pro-Indian approach of the Soviets was part of Moscow's policy to penetrate this Western ring in Southern Asia.[78] Already in 1955, Nikita Khruschev supported India on the Kashmir issue. Moscow was particularly incensed by the Pakistani dictator Yahya Khan's initiative to act as a mediator between China and USA. The China–USA rapprochment resultted in Henry Kissinger's visit to Beijing.[79] The Soviets and the Chinese despite being Marxists had fallen out. These two countries shared 3,125 miles long frontier and some parts of it were disputed.[80] Thanks to soft term loans and military sales on favourable rupee terms by the USSR, India's arsenal improved.[81] Especially, the light amphibious tanks supplied by the Soviets gave mobility and firepower to the Indian Army during its victorious campaign in East Pakistan during December 1971.[82] Before embarking on war with Pakistan in 1971, New Delhi signed a Treaty of Peace, Friendship and Cooperation with the Communist superpower. This treaty enjoined each party to consult each other in case of a military threat to either one and not to join any alliance hostile to either of them. In line with this policy, when in the early 1980s, the Soviets invaded Afghanistan, India did not come out with a public condemnation.[83] Friendship between India and Russia had a historical precedent. In 1907, the British Government of India signed a treaty of friendship with Czarist Russia.[84] The Soviet tilt towards India was perceived by the Western statesmen as the latest move in the long term Soviet game plan to create a chain of satellite states along its Asian frontier.[85] And this perception further motivated the USA to keep Pakistan under its own wings.

After the 1971 India–Pakistan War, China aided Pakistan to make up for the war losses inflicted by India and thus Pakistan became the biggest recipient of Chinese military aid amongst the non-Communist countries. In addition to supplying arms, China also aided Pakistan in building up arms production facilities.[86] Between 1979 and 1980, the transfer of military technology from China to Pakistan and the latter's effort to acquire nuclear technology sent shivers along the spine of India's body politic.[87]

In 1979, despite the Soviet threat along the Pakistan–Afghanistan border, about 90 per cent of Pakistani military assets were deployed along the Pakistan–India border.[88]

China also improved the existing roads and constructed new roads in Tibet along with additional bunkers.[89] Just as Czarist Russia pressurized British–India by deploying troops in Central Asia, similarly, Beijing put pressure on New Delhi by increasing troop deployment in Tibet. In addition, China increased patrolling along the India–Tibet border. To cap it all, China lent moral and material support to the various insurgent movements in north-east India.[90]

In the east, New Delhi's relationship with Dacca deteriorated when the pro-Indian President Mujibur Rehman was assassinated. The military cabal that came to power pursued anti-Hindu policy which resulted in the influx of large numbers of that community into India.[91] The Bangladeshi influx in turn alienated the people of north-east India from Delhi. The frustration of the people of the north-east against Delhi was one of the principal causes behind the rise of insurgencies. During the 1980s, India's military posture could be summed up as ignoring Bangladesh, maintaining a status quo against China and launching a counterstrike in case of a war with Pakistan. For this purpose India maintained a slight edge over Pakistan in conventional weaponry. In contrast to Pakistan's doctrine of pre-emptive strike, India adhered to a passive strategy. However, India's military strategy included the provision of escalation which meant limited thrusts by the air–land forces against Pakistan.[92] The Indian military analysts planned to launch a local attack in the Sind–Bhawalpur–Rajasthan sector, because the Indian generals deduced that operations in the mountainous terrain of Kashmir would not only be slow and inconclusive, but also prove to be costly. India deployed about 3,00,000 soldiers in Kashmir to check both the Pakistani sponsored insurgents as well as to deter the Pakistani regulars from crossing the Line of Control.[93]

From the economic perspective, the armed forces did not gain much even under the most aggressive Prime Minister that India ever had, i.e. Indira Gandhi. During 1973-4, while India spent 15 per cent of the combined centre and state's budget on defence, Pakistan's military budget was over 50 per cent of the national budget and 10 per cent of the GNP.[94] In the 1970s, Egypt and Israel were putting 30 per cent and 32 per cent of their GNP respectively on defence. In the same decade China devoted 10 per cent of its GNP to defence, whereas India spent only 3 per cent of its GNP. In the 1970s, the military lobby within India argued that New Delhi should raise its defence spending to about 5 per cent of the GNP. Their logic was that increased spending on defence by India would widen the 'military gap' between India and Pakistan. In the long-run this would be economical as Pakistan would be convinced of the futility of its sterile policy of confrontation with India and the Pakistani ruling elite would then be induced to cooperate instead of confronting India as regards the regional issues.[95] Nevertheless, the military lobby's recommendations had no effect in policy formulation. In 1983, India's defence expenditure as percentage of Gross Domestic Product (GDP) was 3.5 while that of Pakistan was 5.8.[96] Compared to India's military expenditure of $10 per capita in 1988, Pakistan spent $26, Egypt $61, South Korea $147, Taiwan $192, Iraq $433, Israel $481 and USA $1,061.[97] And this was despite the fact that India was the tenth largest economy in the world.[98] Not only the fund given to the Indian military was in real terms comparatively insignificant, but most of the military budget went towards maintaining manpower. After 1973, despite the Pay Commission's recommendations and additional Dearness Allowance, the rise in the prices of food, clothing and oil, left little to the military budget for modernization. In 1973–4, pay, pension allowances, petrol, oil, lubricants, rations and transport accounted for 55.9 per cent of the defence budget. And during 1974–5, this percentage went upto 66.9 per cent.[99]

THE POST-COLD WAR ERA

The end of the twentieth century witnessed the worsening of India's strategic scenario. The emergence of a unipolar world with the collapse of Soviet Union and the rise of a powerful China adversely affected India's strategic balance.[100] The Soviet disintegration proved to be a boon for Pakistan. From the 1990s, India's prospect of launching a limited strike against Pakistan has declined. Several factors in the post-Cold War era are obstructing India's conventional build up which in turn has put severe constraints on India's security planning. Most of the Indian weapon systems are of Soviet origin and dependent on the Russian military industries for spare parts. And Russia, requires cash payments from India for the military equipment which it once provided at highly subsidized rates.[101] The biggest weakness of the Indian armed forces from the last decade of the twentieth century has been the failure to integrate the technology associated with the Communication Revolution with its weapon platforms. The spread of precision weapon systems and rapid advancement in microcircuitry necessitates rapid response. Hence, use of latest technology associated with information processing is becoming a dire necessity,[102] and Russia is quite backward in this field. Secondly, the Indian economy is not in a position to acquire sophisticated Western weaponry as the Western firms operate on 'cash and carry' principle. Even after opening up its economy, India lags behind China as far as external investment is concerned. This is partly because India's bloated bureaucracy and internal political turmoil scare away foreign investors.[103] Until 2001, India had only attracted $3 billion in foreign investment compared to China's $40 billion.[104] Despite liberalization in the age of globalization, India's share in world trade in 2001 was a mere 1 per cent.[105] And China's $100 billion trade with USA prevents India from playing Washington against Beijing.[106]

Pakistan despite possessing a weak economic base is able to sustain and expand its conventional forces because of international power politics. Pakistan's external debt is more than 100 per cent of its GDP.[107] Islamabad maintains its arsenal with financial aid from USA and its economy is dependent on the loans from the US directed International Monetary Fund.[108] As a reward for allowing the US to use Pakistan as a frontline state during Operation Enduring Freedom,[109] and for allowing over flights and bases, the Bush administration is pouring money into Pakistani coffers. In October 2001, the US agreed to release $600 million to Pakistan. Even after the collapse of the Taliban, Washington needs Pakistan in order to project its power in the Eurasian heartland. At least one key concern of the USA is Central Asia's oil and natural gas resources[110] and Pakistan provides a short route to the Indian Ocean for the projected oil and gas pipeline from Central Asia.[111]

Eurasia accounts for 75 per cent of the world's population plus resources and 60 per cent of the world's GNP. Collectively Eurasia's power overshadows even that of USA. Hence, USA's interest in Eurasia is to prevent the emergence of any continental bloc which in future could challenge Washington's hegemony,[112] a strategy the US is following in the post-Cold War era.[113] In April 1999, the North Atlantic Treaty Organization (NATO) approved a new strategic concept of military missions in the volatile regions outside its borders.[114] The NATO has almost become the vehicle for the US for the expansion of political and military power in Eurasia. The point to be noted is that the USA provides more than 50 per cent of NATO's annual operating budget. At the Helsinki Summit in 1999, the big powers of Europe decided to set up a 60,000 strong rapid reaction force for quick deployment to enforce peace.[115] The eastward expansion of the NATO would pose a threat not only to Russia but also to India and both face the prospect of Western intervention in internal affairs under the garb of human rights issue.[116] Today NATO is intervening in Bosnia and Afghanistan, tomorrow it might intervene in Chechnya and Kashmir. There is a possibility that to counteract the rising power of NATO, the Bear might move closer to the Dragon.[117] Further, Moscow has trade interest to keep Beijing in good humour. By 1995, China had become the world's largest importer of Russian weapons. China is buying $1 billion worth of weapons from Russia every year,[118] a situation which will deter Russia from allying strongly with India against China. In December 1998, the Russian Prime Minister Y. Primakov proposed the scheme of India–China–Russia triangle to counter American dominance in the unipolar world. However, because of India–China border tension this type of strategic partnership will be difficult.[119]

The reduction of military threat along its northern border enables China to shift focus to Southern Asia and the Asia-Pacific rim. The dismemberment of the Soviet Union has placed China in a position to transfer massive military assets from its northern frontier to Tibet in case of any probable tension with India. To keep the pressure on India's northern border, China continues to enhance infrastructural facilities in Tibet by building roads, railways, airfields, missile sites and communication facilities.[120]

India's satellite technology is seriously lagging behind China, and China is willing to transfer such technology to Pakistan. In 2002, China had three meteorological satellites in sun-synchronous polar orbit and geosynchronous orbit. These satellites give China an added advantage of having accurate weather inputs, which is an important variable in planning an amphibious strike or even concerted air and missile attacks.[121] Chinese military-technical aid to Pakistan continues, because Beijing's aim is to prevent Indian 'expansionism' beyond the subcontinent. To prevent any competition from India, China is indirectly engaging India by strengthening Pakistan. China's policy of expanding its tentacles in South-East Asia is to encircle India. In 2000, Indonesia and China, moved towards defence and security cooperation in a newly concluded memorandum on bilateral cooperation that covered finance, trade, culture, and research and development. Indonesia's difficulties in obtaining military equipment and spare parts from the traditional suppliers like the UK and the USA explains Jakarta's orientation towards Beijing. Beijing has also signed a strategic partnership deal with Thailand.[122]

Further deterioration of the strategic environment for India started occurring when the missile race started. India has failed to evolve any sort of missile superiority over Pakistan and China. By 1971, China had developed medium range ballistic missiles like the DF-21 with a range of 3,000 km. Its accuracy is improved through global positioning system. This missile is fitted with a radio frequency explosive warhead, i.e. an electromagnetic pulse bomb in a new nose cone. These missiles are capable of reaching India from Tibet. China at present possesses the DF-31 which is a mobile, solid-fuel missile fitted with a nuclear warhead and a range of 8,000 km and is in the process of developing an inter-continental ballistic missile (ICBM) with a range of up to 10,000 km.[123] In 1983, India started the Integrated Guided Missile Development Programme. In the last decade of the twentieth century, India developed Prithvi battlefield tactical missile with a range of 250 km and a circular error probability of only 250 metres.[124] In response, Pakistan developed and deployed tactical missiles like Haft-I and II and Chinese M-7,[125] the latter two being equivalent to Prithvi.[126] To avert US restrictions regarding supply of completely built units of missiles, Beijing is supplying M-11 missiles in semi-knockdown state to Islamabad.[127]

China spends about 5 per cent of its military budget on ballistic missiles[128] while India's investment in defence technology has been comparatively less. Between 1984 and 1990, India's defence expenditure on an average hovered around 3.5 per cent of the GNP.[129] Between 1987–8 and 2000–1, India's defence expenditure as a proportion of its GDP declined from 3.85 to 2.5 per cent. In 2001–2, defence expenditure reached 2.53 per cent of the GDP.[130] During the 1990s, Israel spent between 13–25 per cent of its budget on the military.[131] China compared to India could afford to spend more (both in absolute and in percentage terms) because of its average growth rate of 8 per cent in the GDP.[132] In 1999, China's defence budget registered an increase of 15.01 per cent. Independent analysts say that the real defence budget is three times higher than the official figure of $12.6 billion. The rate of inflation in China in that period was 3 per cent. Even then, since 1994, in real terms the Chinese defence budget has risen by 10 per cent.[133]

Pakistan and China pose a direct nuclear threat to India. In 1955, Nehru had declared complete helplessness of India in case of an attack by an enemy equipped with thermonuclear weapons.[134] China tested its first nuclear device in October 1964. Then in 1974, China carried out the Lop Nor nuclear test in Sinkiang province.[135] After the humiliation at the hands of China during the 1962 War and the explosion of Chinese nuclear device in 1964, India continued to follow a non-nuclear weaponization policy partly because of its idealism and partly because of indecision.[136] Nehru's nuclear programme faced strong opposition from C. Rajagopalachari and Morarji Desai.[137]

In 1972, Z.A. Bhutto decided to pursue a nuclear programme.[138] After the 1998 nuclear tests (which used highly enriched uranium), Islamabad is trying to use plutonium in a weapon that is small and light enough to be carried on a ballistic missile. Pakistan began operating its reactor at Khushab from 1998. This reactor supplied the plutonium bearing spent fuel that is needed for reprocessing and making it into a weapon core.[139] Pakistan views its nuclear capability as an equalizer against India's conventional military superiority.[140] The defence analysts of Pakistan favour countervalue targets (cities which are concentrations of industry and populace) rather than counterforce targets (nuclear weapons and military assets of the enemy) because of the inaccuracy of its long range missiles. Islamabad neither possesses the economic muscle to maintain a vast nuclear arsenal, nor has the space to disperse the weapons warheads and the delivery vehicles against a possible enemy nuclear strike. In other words Pakistan cannot maintain a second strike capability, and hence, is against 'no first strike' policy and focuses on countervalue targets.[141] By the 1990s, Pakistan possessed Chinese M-11 missiles as well as Haft-I and II nuclear delivery vehicles.[142] The nuclear weapons acquired by Pakistan provide Islamabad the confidence to go on supporting insurgency within India with impunity and India's conventional forces cannot pursue the insurgents inside Pakistan for fear of the conflict escalating into a nuclear exchange between the two countries.[143] The acquisition of nukes by both Pakistan and India has established a balance of terror between these two states. New Delhi lacks any significant nuclear superiority over Pakistan. And this prevents India from launching deep conventional operations. The nuclear parity encouraged Pakistan during 1999 to launch a limited conventional strike in Kargil. In the near future, Pakistan's hostility against India over Kashmir will not decrease because of the rise of Islamic parties within Pakistan which pursue an aggressive policy vis-à-vis India.[144] India assumes that, though hostile, a stable Pakistan is a lesser threat compared to an unstable disintegrating Pakistan.[145]

As regards the nukes, the strategic theorists of India have always been divided roughly into two camps: nuclear weaponization *versus* nuclear disarmament lobby. And these two camps cut across the traditional civilian and military divide. Field Marshal Sam Manekshaw under whom the Indian Army got its biggest victory in 1971, was against nuclear weaponization. He fails to conceptualize any scope for the nukes in a strategic war fighting scenario.[146] The anti-nuclear lobby attempts to take a broader view of national security policy. S.R. Valluri writes that since minimum deterrence is useless against China, only a credible deterrence makes sense as regards security matters. But, warns Valluri, such an attempt would bankrupt the Indian state.[147]

In a similar tune, the retired Admiral L. Ramdas continues that nuclear weapons are over-expensive. He warns that human security could never be separated from human development. And these two are important constituents of national security.[148] While retired Air Marshal Brijesh D. Jayal is for nuclear disarmament, he writes that Pakistan has already decided to use nukes as battlefield weapons. So, nuclear weaponization vis-à-vis an unstable state like Pakistan might result in an arms race which not only would prove to be economically ruinous for India but might result in a nuclear war.[149] Ramdas assumes that all sorts of theorizing about deployment and use of nukes is useless because in case of a nuclear conflagration there will be neither winners nor losers.[150]

During the early 1990s, General K. Sundarji believed that Pakistan had nuclear weapons and in case of an Indian conventional attack deep inside Pakistani territory, Islamabad would use nukes. In fact, argued Sundarji, the very fear that Pakistan might use nuclear weapons against the advancing Indian ground forces would prevent the Indian military leaders from concentrating adequate mass of weapons and troops for conducting a breakthrough battle.[151] Sumit Ganguly writes that crisis stability between India and Pakistan will emerge when the nuclear forces of both these countries possess a secure second strike capability.[152] Raju G.C. Thomas says that in a psychological war of nerves in the nuclear context, minimum deterrence has no credibility and leads to political capitulation.[153] According to Thomas and Amit Gupta, minimum deterrence in the Indian context means the minimum capability for a retaliatory strike by New Delhi against India's two potentially hostile neighbours. They argue that it might grow into credible deterrence which is based on ICBMs with global reach.[154] To deter the USA, asserts Bharat Karnad, India needs city-busting megaton warheads fitted on ICBMs.[155] For Bharat Karnad, a soft state is categorized by the ruling class' lack of 'will to power', and he categorizes India on the basis of its de-mated and de-alerted nuclear posture as a soft state.[156]

Most of the Indian analysts belonging to the pro-nuclear lobby are against India possessing tactical nukes. Tactical nuclear weapons have short range (below 100 miles) and are low yield weapons designed for combat use in the battlefield/sea.[157] In recent times, the retired Air Commodore Jasjit Singh has asserted that nuclear weapons are for deterrence only. On no account is he willing to consider the nukes as tools for fighting a nuclear war.[158] And Jasjit Singh is backed up by a civilian analyst Matin Zuberi who concurs that nukes are not fighting tools but political instruments for deterrence.[159] According to an expert on India's nuclear policy, India is capable of making tactical nuclear devices suitable for artillery shells.[160] Whether the Indian Army would be given the option to handle such devices remains to be decided. However, India's self-imposed ban on further nuclear testing because of American pressure is preventing the engineers and scientists from manufacturing miniaturized and multiple nuclear warheads. A significant chunk of the Indian strategic analysts opposed Atal Behari Vajpayee government's decision during 1999–2001, of not testing any more nuclear bombs. They argued that the self-imposed moratorium on further nuclear testing would prevent development of sophisticated nuclear warheads.[161]

China is carrying on research and development on missile interception capability. China has fairly advanced research programme on lasers and the ability to use them as interceptors in outer space.[162] In recent times India has attempted to curry favour with the USA by supporting the latter's National Missile Defence programme also known as the Anti-Ballistic Missile Defence.[163] In an attempt to bolster its own nuclear defence, India since 1995 has been negotiating with Russia for buying the mobile anti-ballistic missile system, the S-300V. India is trying to buy six Russian batteries that would cost $1 billion. This weapon system will be supplemented by India's indigenous missile Akash with the Rajendra radar array.[164] However, this seems to be a waste of expenditure because like the bombers, nuclear missiles will always get through. And the penetration of even one nuclear tipped missile would cause horrendous casualties.

Thomas points out that India is wary of possible Western 'humanitarian interventions' in the near future, and Indians understand as said in the *Panchatantra,* that friendship can only be among the equals.[165] Sundarji accepted that in the near future, a superpower might resort to arm-twisting policies to force India to accept certain programmes that might be detrimental to the latter's sovereignty. And if India resists then a Desert Storm like operation might unfold. He argued in the book published posthumously that India will have to resist such an intrusion through an amalgam of conventional and nuclear weapons. The first line of defence would include the Indian Navy to prevent concentration of enemy ships (deployed for launching amphibious operation) in the northern Indian Ocean. The second line of defence would consist of adequate air–land capability to prevent the lodgement of US Marines and back up amphibious forces in India's coastal enclaves. The conflict might escalate to a limited nuclear exchange, with the enemy hitting first with nukes. So, India required to possess a second strike nuclear capability.[166] Thus, for the first time, nuclear weapons are enmeshed in the war strategy of the Indian armed forces at least theoretically.

Most of the Indian military budget goes to the conventional arms to equip the three services rather than for the creation of a strategic nuclear force. During 1984–5, the shares of the army, navy and air force in defence expenditure were 67.1 per cent, 8.6 per cent, and 24.3 per cent respectively.[167] In 1990, the army consumed about 58 per cent of the military budget, the Indian Air Force about 30 per cent, and the Navy came last with only 12 per cent.[168] Thus, the lion's share of India's defence budget went to the army. Karnad rightly asserts that for the present and near future, possession of a strategic arsenal is more vital than continuous upgradations of conventional military apparatus especially the armoured elements. He points out that the funds earmarked for upgradations of the conventional weapons should be used for building up the strategic triad. For instance between 2000–20, the Indian Army plans to spend Rs. 9,00,000 crore on the armoured and mechanized components. And the cost of building and maintaining a strategic nuclear arsenal will be about Rs. 74,352 crore ($14.9 billion). Karnad continues that India could generate further resources for a full-fledged nuclear deterrent by cutting the subsidies from the budget which amounts to 14 per cent of the GDP.[169]

CONCLUSION

India after 1947 became an autonomous regional power and the evolution of the nuclear doctrine from the 1990s probably represents the greatest change in India's defence policy. A credible nuclear deterrent is necessary because an arrogant China is unwilling to discuss confidence-building measures with India on the basis that the Nuclear Non-Proliferation Treaty does not recognize India as a nuclear power.[170] The issue of interfacing nuclear deterrence with conventional operations is the greatest challenge for the Indian armed forces in the near future.[171]

Till 2003, all sorts of power projection by independent India were confined within the geographical environs of South Asia. The only exception was the Maldives operation under Rajiv Gandhi. Internal insurgencies continue to be provoked by Pakistan and China. Military expenditure remains marginal. One could argue that Nehru was right in drastically reducing military expenditure while pursuing a policy of sustainable development of an underdeveloped agrarian economy. But, then Nehru should have realized that with a weak military, India stands no chance in any confrontation with a dynamic power like Red China. The low military expenditure in the aftermath of independence is partly due to the reduced status and power of the military within the post-independent polity. The role of uniformed men was drastically reduced after 1947, a phenomenon that will be analysed in the next chapter.

NOTES

1. Colin S. Gray, 'History for Strategists: British Seapower as a Relevant Past', *JSS*, vol. 17, no. 1 (1994), p. 7.
2. Paul Kennedy, 'Grand Strategies and Less-than-Grand Strategies: A Twentieth Century Critique', in Lawrence Freedman, Paul Hayes and Robert O'Neill, eds., *War, Strategy and International Politics: Essays in Honour of Michael Howard,* Oxford: Clarendon Press, 1992, pp. 228–31.
3. *Recruitment into the Officer Corps of the Armed Forces,* Report of a Seminar held at the United Service Institution of India on 14 February 1977, New Delhi: United Service Institution, 1977, p. 11.
4. Lawrence Freedman, 'Strategic Studies and the Problem of Power', in Freedman et al., eds., *War, Strategy and International Politics,* p. 293.
5. *Ministry of Defence Government of India, Annual Report: 1970–71,* (*MODAR*), p. 1.
6. Anuradha M. Chenoy, *Militarism and Women in South Asia,* New Delhi: Kali, 2002.

7. *The Army in India and Its Evolution Including an Account of the Establishment of the Royal Air Force in India,* 1924, rpt., Delhi: Anmol Publications, 1985, pp. 9–10.
8. For the continuity of the regimental structures of the colonial Sepoy Army and the Indian and Pakistan armies see John Gaylor, *Sons of John Company: The Indian and Pakistan Armies, 1903–91,* 1992, rpt., New Delhi: Lancer International, 1993.
9. *MODAR: 1975–76*, p. 17.
10. For the Organic theory of state see G. Pope-Atkins, 'Redefining Security in Inter-American Relations: Implications for Democracy and Civil-Military Relations', in John P. Lovell and David E. Albright, eds., *To Sheathe the Sword: Civil-Military Relations in the Quest for Democracy,* Westport: Greenwood, 1997, p. 168.
11. The Strategic Culture approach assumes that the world-view of the ruling elite is shaped by their cultural ethos.
12. George K. Tanham, 'Indian Strategic Thought: An Interpretative Essay'; and idem, 'Indian Strategy in Flux', in Kanti P. Bajpai and Amitabh Mattoo, eds., *Securing India: Strategic Thought and Practice,* New Delhi: Manohar, 1996, pp. 28–139.
13. Stephen Peter Rosen, *Societies and Military Power: India and its Armies,* New Delhi: Oxford University Press, 1996, pp. 33–60.
14. Jagadish Narayan Sarkar, 'Aspects of Military Policy in Medieval India', *Proceedings of the Indian History Congress,* 36th Session, Aligarh (1975), p. 243.
15. S. Gopal and Uma Iyengar, eds., *The Essential Writings of Jawaharlal Nehru,* vol. 2, New Delhi: Oxford University Press, 2003, p. 263.
16. Ibid., p. 293.
17. Ibid., p. 484.
18. S. Gopal and Uma Iyengar, eds., *The Essential Writings of Jawaharlal Nehru,* vol. 1, New Delhi: Oxford University Press, 2003, p. 41.
19. Stephen P. Cohen, *Emerging Power India,* 2001, rpt., New Delhi: Oxford University Press, 2003, p. 130.
20. Gopal and Iyengar, eds., *Essential Writings of Nehru,* vol. 1, p. 323.
21. Ibid., vol. 2, p. 247.
22. Ibid., p. 249.
23. Ibid., p. 163.
24. Cohen, *India,* p. 128.
25. K. Reddy, 'India's Defence Expenditure: 1872–1967', *Indian Economic and Social History Review,* vol. 7 (1970), p. 474.
26. Douglas Porch, *The Portuguese Armed Forces and the Revolution* London: Croom Helm, 1977, p. 12.
27. Transcript of Oral History Interview with Prof. P.M.S. Blackett, Manuscript Section, Accession no. 284, Nehru Memorial Museum & Library (NMML), New Delhi, pp. 1, 5–6.

28. Gopal and Iyengar, eds., *Essential Writings of Nehru*, vol. 1, p. 41.
29. T.A. Heathcote, *The Indian Army: The Garrison of British Imperial India, 1822–1922*, Newton Abbot: David & Charles, 1974, p. 13.
30. Cohen, *India*, p. 130.
31. Lorne J. Kavic, *India's Quest for Security: Defence Policies, 1947–65*, Berkeley/Los Angeles: University of California Press, 1967, pp. 54, 56–7.
32. Ravinder Kumar and H.Y. Sharada Prasad, eds., *Selected Works of Jawaharlal Nehru*, Second Series, vol. 27 (1 October 1954–31 January 1955), New Delhi: Jawaharlal Nehru Memorial Fund, 2000, p. 494.
33. Quoted in Sandy Gordon, 'Domestic Foundations of India's Security Policy', in idem and Ross Babbage, eds., *India's Strategic Future: Regional State or Global Power?*, New Delhi: Oxford University Press, 1992, pp. 7–8.
34. Gary Klintworth, 'Chinese Perspectives on India as a Great Power', in Babbage and Gordon, eds., *India's Strategic Future*, p. 97.
35. Pervaiz Iqbal Cheema, 'Arms Procurement in Pakistan: Balancing the Needs for Quality, Self-Reliance and Diversity of Supply', in Eric Arnett, ed., *Military Capacity and the Risk of War: China, India, Pakistan and Iran*, Oxford: Oxford University Press, 1997, p. 151.
36. Jagat S. Mehta, 'The Geopolitical and Socio-Economic Implications of USA's Involvement in Afghanistan', in Salman Haidar, ed., *The Afghan War and Its Geopolitical Implications for India*, New Delhi: Manohar, 2004, pp. 71, 78.
37. Brigadier Rajendra Singh, *Soldier and Soldiering in India: Twelve Essays*, 1954, rpt., New Delhi: Army Educational Stores, 1964, p. 2.
38. Ibid., p. 3.
39. Lieutenant-General Harbakhsh Singh, *War Despatches: Indo-Pak Conflict, 1965*, New Delhi: Lancer International, 1991, p. 68.
40. Hugh Tinker, 'From British Indian Army to Indian Army', *Indo-British Review*, vol. 16, no. 1 (1989), p. 15; S.P. Sinha, 'Prabhakaran as Leader of the LTTE', *Journal of the United Service Institution of India* (*JUSII*), vol. 131, no. 544 (2001), pp. 205–6.
41. Sumona Dasgupta, 'Militarization of the Indian State Since the 1980s', in Maroof Raza, ed., *Generals and Governments in India and Pakistan*, New Delhi: Har-Anand, 2001, pp. 61, 63.
42. Mohammad Ayub Khan, *Friends not Masters: A Political Autobiography*, Lahore: Oxford University Press, 1967, p. 115.
43. Gopal and Iyengar, eds., *Essential Writings of Nehru*, vol. 2, p. 326.
44. Sanjay Badri-Maharaj, *The Armageddon Factor: Nuclear Weapons in the India-Pakistan Context*, New Delhi: Lancer Publishers & Distributors, 2000, p. 32.
45. C. Dasgupta, *War and Diplomacy in Kashmir: 1947–48*, New Delhi: Sage, 2002, p. 53.

46. Transcript of Oral History Interview with General Roy Bucher, 2 March 1970, Manuscript Section, Accession no. 59, NMML, p. 6.
47. Prem Shankar Jha, *Kashmir 1947: The Origins of a Dispute*, New Delhi: Oxford University Press, 2003, pp. 174–5.
48. B.N. Mullik, *My Years with Nehru: The Chinese Betrayal*, Bombay: Allied Publishers, 1971, p. 538.
49. Khan, *Friends not Masters*, p. 40.
50. Kumar and Prasad, eds., *Selected Works of Nehru*, Second Series, vol. 27, p. 495.
51. Sarvepalli Gopal, *Jawaharlal Nehru: A Biography, 1947–56*, vol. 2, 1979, rpt., New Delhi: Oxford University Press, 1988, p. 44.
52. Anita Inder Singh, 'Containment through Diplomacy: Britain, India and the Cold War in Indo-China, 1954–56', in Ian Nish, ed., *South Asia in International Affairs: 1947–56, International Studies*, London: London School of Economics and Political Science, 1987, pp. 3–4, 11.
53. Ian Talbot, 'Does the Army Shape Pakistan's Foreign Policy?', in Christophe Jaffrelot, ed., *Pakistan: Nationalism Without a Nation?* New Delhi: Manohar, 2002, p. 316; Sultana Afroz, 'Pakistan and the 1951 Middle East Defence Plan: The US and UK Positions', *Asian Affairs*, vol. 19 (1988), pp. 171, 177.
54. Lieutenant-Colonel Gautam Sharma, *Path of Glory: Exploits of 11 Gorkha Rifles*, Ahmedabad: Allied, 1988, p. 49.
55. Stephen P. Cohen, 'US Weapons and South Asia: A Policy Analysis', *Pacific Affairs*, vol. 49, no. 1 (1976), p. 49.
56. Mullik, *My Years with Nehru*, pp. 53–40.
57. Harbakhsh Singh, *War Despatches*, pp. 5–6.
58. Mohammad Asghar Khan, *The First Round: Indo-Pakistan War 1965*, Ghaziabad: Vikas, 1979, p. 7.
59. Paul H.B. Godwin, 'Military Technology and Doctrine in Chinese Military Planning: Compensating for Obsolescence', in Arnett, ed., *Military Capacity*, p. 41.
60. Gopal and Iyengar, eds., *Essential Writings of Nehru*, vol. 2, pp. 236–7.
61. Ibid., vol. 1, p. 712.
62. Kavic, *India's Quest for Security*, p. 38.
63. General V.N. Sharma, *India's Defence Forces: Building the Sinews of a Nation*, United Service Institution National Security Paper, no. 13, New Delhi: United Service Institution of India, 1994, p. 8.
64. *The Indian Army*, New Delhi: Lancer & Army Head Quarter, 1990, p. 71; Brigadier Ashok Malhotra, *Trishul: Ladakh and Kargil, 1947–93*, New Delhi: Lancer Publishers & Distributors, 2003, p. 32.
65. Kumar and Prasad, eds., *Selected Works of Nehru*, Second Series, vol. 27, p. 494.
66. Ravinder Kumar and H.Y. Sharada Prasad, eds., *Selected Works of Jawaharlal Nehru*, Second Series, vol. 28, New Delhi: Jawaharlal Nehru Memorial Fund, 2001, pp. 522-3.

67. Kavic, *India's Quest for Security*, p. 64.
68. Ashley J. Tellis, *India's Emerging Nuclear Posture: Between Recessed Deterrent and Ready Arsenal*, New Delhi: Oxford University Press, 2001, p. 757.
69. Premen Addy, 'Historical Problems in Sino-Indian Relations', in Nish, ed., *South Asia in International Affairs: 1947–56*, p. 18.
70. Mullik, *My Years with Nehru*, p. 538.
71. *Notes, Memoranda and Letters Exchanged between the Governments of India and China, July 1962–October 1962, White Paper No. VII, Ministry of External Affairs Government of India*, New Delhi: General Manager Government of India Press, 1962, p. 3.
72. Yaacov Vertzberger, 'India's Strategic Posture and the Border War Defeat of 1962: A Case Study in Miscalculation', *JSS*, vol. 5, no. 3 (1982), pp. 378–82.
73. Gopal and Iyengar, eds., *Essential Writings of Nehru*, vol. 2, p. 313.
74. Sumit Ganguly, *Conflict Unending: India-Pakistan Tensions since 1947*, New Delhi: Oxford University Press, 2002, p. 33.
75. Shankar Roychowdhury, *Officially at Peace: Reflections on the Army and its Role in Troubled Times*, New Delhi: Viking, 2002, p. 279.
76. Lieutenant-General V.R. Raghavan, *Siachen: Conflict Without End*, New Delhi: Viking, 2002, pp. xii, 25; Robert Jackson, 'The Strategic Outlook for the Indian Subcontinent: Some Lessons of Recent History', *Asian Affairs*, vol. 3 (1972), p. 264.
77. General J.N. Chaudhuri, *An Autobiography* (as narrated to B.K. Narayan), New Delhi: Vikas, 1978, p. 177.
78. Amin Saikal, 'The Future of India and Southwest Asia', in Babbage and Gordon, eds., *India's Strategic Future*, p. 126.
79. M.S. Dahiya, 'Indo-Soviet Relations under Kosygin', *JUSII*, vol. CII, no. 427 (1972), pp. 119, 127.
80. Edgar O'Ballance, *Afghan Wars, Battles in a Hostile Land: 1839 to the Present*, 2002, rpt., Karachi: Oxford University Press, 2003, p. 112.
81. Gregory Austin, 'Soviet Perspectives on India's Developing Security Posture', in Babbage and Gordon, eds., *India's Strategic Future*, p. 144.
82. Chaudhuri, *Autobiography*, p. 184.
83. A. Hasnan Habib, 'Southeast Asian Perceptions of India's Strategic Development: An Indonesian View', in Babbage and Gordon, eds., *India's Strategic Future*, p. 108.
84. M.A. Yapp, 'British Perceptions of the Russian Threat to India', *Modern Asian Studies*, vol. 24, no. 4 (1987), p. 664:
85. Geoffrey Wheeler, 'The Indian Ocean Area: Soviet Aims and Interests', *Asian Affairs*, vol. 3 (1972), p. 272.
86. *MODAR: 1975–76*, p. 3.
87. *MODAR: 1979–80*, p. 2.
88. Michael McKinley, 'Indian Naval Developments and Australian Strategy in the Indian Ocean', in Robert H. Bruce, ed., *The Modern Indian Navy*

and the Indian Ocean, Perth: Centre for Indian Ocean Regional Studies, 1989, p. 145.

89. *MODAR: 1970–71,* p. 1.
90. *MODAR: 1975–76,* pp. 3–4.
91. Ibid., p. 4.
92. Eric Arnett, 'Beyond Threat Perception: Assessing Military Capacity and Reducing the Risk of War in Southern Asia', in idem, ed., *Military Capacity,* p. 20.
93. Badri-Maharaj, *Armageddon Factor,* p. 34.
94. *MODAR: 1974–75,* p. 10.
95. *The Imposition of a Manpower Ceiling on the Army: Report of a Seminar at the United Service Institution of India,* Chairman General M.L. Thapan, 12 January 1976, New Delhi: United Service Institution of India, 1977, p. 3.
96. Clive Dewey, 'Some Consequences of Military Expenditure in British India: The Case of the Upper Sind Sagar Doab, 1849–1947', in idem, ed., *Arrested Development in India: The Historical Dimension,* New Delhi: Manohar, 1988, p. 159.
97. Raju G.C. Thomas, 'The Growth of Indian Military Power: From Sufficient Defence to Nuclear Deterrence', in Babbage and Gordon, eds., *India's Strategic Future,* p. 36.
98. Kanti P. Bajpai, 'State, Society, Strategy', in idem and Mattoo, eds., *Securing India,* p. 140.
99. *MODAR: 1974–75,* p. 9.
100. Satish Nambiar, 'India and the World Strategic Environment', in J. Baranwal (Editor-in-Chief), *SP's Military Yearbook: 1998–99,* New Delhi: Guide Publications, 1998, p. 5.
101. Amit Gupta, 'Indian Security Planning in the 1990s: Learning to Live in a New World Order', in Marvin G. Weinbaum and Chetan Kumar, eds., *South Asia Approaches the Millennium: Reexamining National Security,* Boulder: Westview, 1995, pp. 184–6.
102. *Joint Warfare of the Armed Forces of the United States, Joint Pub-1,* 10 January 1995, I–3.
103. Gupta, 'Indian Security Planning in the 1990s', in Weinbaum and Kumar, eds., *South Asia Approaches the Millennium,* pp. 199, 202.
104. R.H. Tahiliani, 'Corruption in Arms Trade', *JUSII,* vol. 131, no. 544, 2001, p. 185.
105. 'India's Role 1% in World Trade', *Times of India,* 2 November 2001, p. 3.
106. Vice-Admirals K.K. Nayyar, R.B. Suri, Air Marshal B.D. Jayal, Lieutenant-General V.K. Singh and Major-General Afsir Karim, *National Security: Military Aspects,* New Delhi: Rupa, 2003, p. 11.
107. Shri Prakash, 'USA's Involvement in Afghanistan: Third World Perceptions with Special Reference to South Asian Countries and

Implications for the Future', in Salman Haidar, ed., *The Afghan War and its Geopolitical Implications for India*, p. 178.

108. Ayesha Siddiqa-Agha, *Pakistan's Arms Procurement and Military Buildup, 1979-1999: In Search of a Policy,* Houndsmill and Basingstoke: Palgrave, 2001, pp. 108, 197.
109. For a detailed analysis of the military aspects of US war on Afghanistan see Kaushik Roy, 'Limitations of Western Warfare: American Military Operations in Afghanistan, 2001', in *The Afghanistan Crisis: Problems and Perspectives,* New Delhi: Nehru Memorial Museum & Library, 2002, pp. 91–112.
110. 'It Pays to Wage War', *Outlook*, vol. XLI, no. 39, 8 October 2001, p. 47; K.K. Katyal, 'Big Brother Comes Calling', *The Hindu*, 11 February 2002, p. 8; Kanti Bajpai, 'The War in Afghanistan and US Policy Towards India and Pakistan', in Haidar, ed., *The Afghan War and Its Geopolitical Implications for India*, p. 31.
111. Major-General Afsir Karim, 'War on Terrorism: Strategic Diversion', *Aakrosh*, vol. 6, no. 18 (2003), p. 11.
112. Zbigniew Brzezinski, 'A Geostrategy for Eurasia', *Foreign Affairs,* September–October 1997, pp. 50–3.
113. Jerry Harris, 'The US Military in the Era of Globalization', *Race and Class*, vol. 44, no. 2 (2002), p. 1.
114. Colonel Gurmeet Kanwal, 'NATO's Kosovo Intervention: Threat to Stable World Order', *Indian Defence Review* (*IDR*), vol. 14, no. 2 (1999), p. 50.
115. Tom Lansford, *All for One: Terrorism, NATO and the United States,* Aldershot: Ashgate, 2002, pp. 17, 66–7.
116. V.K. Shrivastava, 'Indian Army: The Challenge Ahead', *Strategic Analysis*, vol. 25, no. 4 (2001), p. 493.
117. Zdzislaw Lachowski, 'Conventional Arms Control', *SIPRI Yearbook 1998: Armaments, Disarmament and International Security,* Oxford: Oxford University Press, 1998, pp. 507, 526–7.
118. Dennis Woodward, 'The People's Liberation Army: A Threat to India?', *Contemporary South Asia*, vol. 12, no. 2 (2003), p. 233.
119. Kanwal, 'NATO's Kosovo Intervention', p. 51.
120. V.K. Shrivastava, 'Indian Army 2020: A Vision Statement on Strategy and Capability', *Strategic Analysis*, vol. 25, no. 6 (2001), p. 755.
121. A.V. Lele, 'China as a Space Power', *Strategic Analysis*, vol. 26, no. 2 (2002), pp. 260, 262–3.
122. John Haseman, 'Indonesia, China Expand Cooperation', *Jane's Defence Weekly* (*JDW*), vol. 33, no. 21, 24 May 2000, p. 4.
123. Paul Beaver, 'China Prepares to Field New Missile', *JDW*, vol. 31, no. 8, 24 February 1999, p. 3; *MODAR: 1970–71*, p. 2.
124. Sango Panwar, 'The Shape of India's Emerging Missile Shield', *Strategic Analysis*, vol. 13, no. 3 (1990), pp. 315, 317.

125. Yezid Sayigh, 'Arms Production in Pakistan and Iran: The Limits of Self-Reliance', in Arnett, ed., *Military Capacity*, p. 176.
126. Raju G.C. Thomas, 'Arms Procurement in India: Military Self-Reliance Versus Technological Self-Sufficiency', in Arnett, ed., *Military Capacity*, pp. 124-5.
127. Siddiqa-Agha, *Pakistan's Arms Procurement*, p. 108.
128. P.K. Ghosh, 'Economic Dimension of the Strategic Nuclear Triad', *Strategic Analysis*, vol. 26, no. 2 (2002), p. 287.
129. Sreedhar, 'Defence Expenditure During the Rajiv Gandhi Years', *Strategic Analysis*, vol. 13, no. 3 (1990), p. 227.
130. *Eleventh Report, Ministry of Defence, Demands for Grants, Standing Committee on Defence, Thirteenth Lok Sabha*, New Delhi: Lok Sabha Secretariat, 2001, p. 8.
131. Stuart A. Cohen, 'The Israel Defence Forces (IDF): From a "People's Army" to a "Professional Military"—Causes and Implications', *Armed Forces and Society*, vol. 21, no. 2 (1995), p. 24.
132. Vinod Anand, 'Warfare in Transition and the Indian Subcontinent', *Strategic Analysis*, vol. 23, no. 5 (1999), p. 713.
133. Robert Karniol, 'China's Defence Budget is Increased Again', *JDW*, vol. 31, no. 11, 17 March 1999, p. 15.
134. Kumar and Prasad, eds., *Selected Works of Nehru*, Second Series, vol. 27, p. 494.
135. *MODAR: 1974–75*, p. 5.
136. General K. Sundarji, *Blind Men of Hindoostan: Indo-Pak Nuclear War*, 1993, rpt., New Delhi: UBSPD, 1996, p. xiii.
137. Rear-Admiral Raja Menon, 'Reflections on India's Nuclear Doctrine and Command and Control', *JUSII*, vol. CXXXIII, no. 552 (2003), p. 196.
138. Smruti S. Pattanaik, 'Pakistan's Nuclear Strategy', *Strategic Analysis*, vol. 27, no. 1 (2003), p. 95.
139. 'Fear over Nuclear Tests Preparation in Pakistan', *JDW*, vol. 33, no. 22, 31 May 2000, p. 6.
140. Ejaz Haider, 'Managing Nuclear Weapons in South Asia: In Search of a Model', in M.V. Ramana and C. Rammanohar Reddy, eds., *Prisoners of the Nuclear Dream*, New Delhi: Orient Longman, 2003, p. 136.
141. Pattanaik, 'Pakistan's Nuclear Strategy', p. 102; Rajesh Rajagopalan, *Second Strike: Arguments about Nuclear War in South Asia*, New Delhi: Penguin, 2005, pp. 52–3.
142. Sundarji, *Blind Men of Hindoostan*, p. 4.
143. Jasjit Singh, 'Pakistan's Fourth War', *Strategic Analysis*, vol. 23, no. 5 (1999), p. 690.
144. Sumita Kumar, 'The Role of Islamic Parties in Pakistani Politics', *Strategic Analysis*, vol. 25, no. 2 (2001), p. 271.

145. Bharat Verma, 'Pakistan: The Counter Strategy', *IDR*, vol. 14, no. 2 (1999), p. 6.
146. Lieutenant-General Depinder Singh, *Field Marshal Sam Manekshaw: Soldiering with Dignity*, Dehra Dun: Natraj, 2002, pp. 80–1.
147. S.R. Valluri, 'Lest We Forget: The Futility and Irrelevance of Nuclear Weapons for India', in Raju G.C. Thomas and Amit Gupta, eds., *India's Nuclear Security*, New Delhi: Vistaar Publications, 2000 pp. 266–7.
148. Admiral L. Ramdas, 'Nuclear Weapons and National Security', in Ramana and Reddy, eds., *Nuclear Dream*, p. 57.
149. Brijesh D. Jayal, 'Look Beyond the Madness', *The Telegraph*, 18 February 2003, p. 12.
150. Ramdas, 'Nuclear Weapons', in Ramana and Reddy, eds., *Nuclear Dream*, p. 73.
151. Sundarji, *Blind Men of Hindoostan*, pp. 44–5.
152. Sumit Ganguly, 'Explaining the Indian Nuclear Tests of 1998', in Thomas and Gupta, eds., *India's Nuclear Security*, p. 57.
153. Raju G.C. Thomas, 'India's Nuclear and Missile Programme: Strategy, Intentions, Capabilities', in Thomas and Gupta, eds., *India's Nuclear Security*, p. 103.
154. Raju G.C. Thomas and Amit Gupta, 'Introduction', in Thomas and Gupta, eds., *India's Nuclear Security*, p. 8.
155. Bharat Karnad, *Nuclear Weapons and Indian Security: The Realist Foundations of Strategy*, New Delhi: Macmillan, 2002, p. 690.
156. Ibid., p. 701.
157. M.S. Mamik, 'Tactical Nuclear Weapons at Sea: Are They Essential?' *Strategic Analysis*, vol. 13, no. 3 (1990), pp. 246–7.
158. Jasjit Singh, 'Nuclear Command and Control', *Strategic Analysis*, vol. 25, no. 2 (2001), p. 147.
159. Matin Zuberi, 'The Proposed Indian Nuclear Doctrine', *Contemporary India*, vol. 1, no. 1 (2002), p. 62.
160. Badri-Maharaj, *Armageddon Factor*, p. 79.
161. Nayyar et al., eds., *National Security*, pp. 50–1.
162. M.V. Rappai, 'China's Nuclear Arsenal and Missile Defence', *Strategic Analysis*, vol. 26, no. 1 (2002), p. 74.
163. Nayyar et al., eds., *National Security*, pp. 33, 51.
164. Andreas Katsouris and Daniel Goure, 'Strategic Crossroads in South Asia: The Potential Roles for Missile Defence', *Comparative Strategy: An International Journal*, vol. 18, no. 2 (1999), p. 180.
165. Thomas, 'India's Nuclear and Missile Programme', in Thomas and Gupta, eds., *India's Nuclear Security*, pp. 100–1.
166. General K. Sundarji, *Vision 2100: A Strategy for the Twenty-First Century*, New Delhi: Konark, 2003, p. 200.

167. Sreedhar, 'Defence Expenditure during the Rajiv Gandhi Years', p. 223.
168. Vice-Admiral Mihir K. Roy, 'Asymmetry of India's Defence Forces: A Sailor's View', *Strategic Analysis*, vol. 13, no. 3 (1990), p. 233.
169. Karnad, *Nuclear Weapons and Indian Security*, pp. 688-9, 692.
170. Pran Krishan Pahwa, 'The Nuclear Dimension', in Rear Admiral Raja Menon, ed., *Weapons of Mass Destruction*, New Delhi: Sage, 2004, p. 23.
171. Shrivastava, 'Indian Army 2020', p. 758.

TWO

Frock Coats versus *Brass Hats*: *Political Control over the Indian Army*

The politicians might be able to formulate a national security policy but whether the army would accept it or not remains questionable. The uniformed men might be unwilling to embrace the fabric of grand strategy chalked out by their political masters and unenthusiastic to carry out the functions allotted to them. This has proved to be especially true in the Latin American as well as in most of the Afro–Asian countries during the post-World War II era. However, the situation is different in India. Before undertaking a historical analysis of the political supremacy over the military in India, let us have a glance at the various theorists who have conceptualized the interaction between the politicians and the 'specialists on violence'.

THEORIES OF CIVIL–MILITARY RELATIONS

The various theories dealing with civil–military interaction could broadly be divided into two groups: Conflict and Cooperation theories. While Cooperation theory points out that the civilian rulers and the military generally complement each other, Conflict theory focuses on competition and resultant tension between the civilian government and its military component. The foremost proponent of the Conflict theory is Samuel P. Huntington for whom the core issue of civilian control over the armed forces is minimizing the military's power within the government.[1] In a similar tone, Eva Etzioni-Halevy asserts that when the political elite fails to ensure clear separation between the government and the military spheres, the functioning of democracy is undermined.[2] A most extreme proponent of the Conflict theory is Eliot A. Cohen. He asserts that the distinction between civilian policy-making and military strategy is actually a red herring. Politics and the conduct of warfare intermingle with each other, and a successful statesman especially in wartime has no other option but to meddle in matters military. The net result is tension between the civilian leadership and the military professionals.[3]

According to the Conflict theorists, the social origins of the officer corps are irrelevant. Rather, the strength of the political institutions and organizational interests of the bureaucratic military machine determine the officers' behaviour and action.[4] The Conflict theory argues that the armed forces play an active role in public affairs when the political institutions fail to function properly and lose legitimacy.[5] For instance, the chance of a military coup increases if the ruling party forms the government after rigging an election.[6] Jimi Peters states that military intervention occurs when the government fails to maintain law and order and the country is threatened by the spectre of a civil war. The army fears that the ensuing chaos might engulf itself in the long run, so it steps in to stem the rot.[7] Astride R. Zolberg comments that too frequent use of the army to aid in civil tasks also encourages the military officers to substitute the politicians with themselves as rulers.[8] Some spokesmen of the Conflict theory assert that when the civil administration fails to modernize the society, the military steps in as a modernizing agency.[9] In the same vein, Constantine P. Danopoulos and Adem Chopani assert that when the political system is in the throes of a transformation, then the army concerned too suffers from the temptation to intervene in an attempt to introduce good governance.[10] Douglas Porch writes that when there is political vacuum and the army is humiliated after being defeated in a war, it tries to capture political power.[11] Besides the institutional factors emphasized by the above mentioned scholars, Stephen P. Cohen brings in the personality factor to explain the civil–military clash. He argues that even in the absence of any substantive difference over policy issues, aggressive and overblown ego of the civilian masters and the generals might contain the seeds of conflict.[12]

On the otherhand, the Cooperation theorists argue that there need not be continuous tussle between the civilians and the military. One retired general of the United States Army, Fred

Woerner claims that inter-relationship between the civil government and the military is not a zero-sum game. In other words, the loss of power on part of the civilian authority does not mean gain of authority on part of the army or vice versa. Rather, a harmonious civil–military relationship might result in strengthening of both the political and military sectors.[13]

The Soviets considered military policy as composed of two parts. The socio-political component shaped by the party/civilian authority encompassed all aspects of economic and social issues related with the political goals of war. The military technical component was determined by the armed forces' high command and it included force structure, technology, training, doctrine, etc. And this functional distinction despite having some grey area with overlapping jurisdiction was accepted by both the civilian ruling elite as well as the military officers.[14] David E. Albright writes that the Soviet military accepted this scenario because the Communist Party bought off the military elite by giving them a large stake in the existing system. During the 1970s, the military officers occupied a significant number of seats in the Communist Party's Central Committee and even sat at the party's top decision-making body, the Politburo. And major military representation also existed in the government's formal parliamentary bodies—the national, regional and local Soviets.[15] Andrew Scobell shows that the Communist Party of China maintains a civil–military consensus by giving the crucial ministerial posts of foreign affairs and internal security to the retired military generals and the marshals.[16]

In a similar tone, Stephen P. Cohen in one of his articles and Apurba Kundu in his monograph claim that the post-independent Indian state by giving the senior officers of the Indian Army a large share in running the country has generated a civil–military consensus.[17] While Sumona Dasgupta asserts that India from the 1980s is witnessing militarization of politics. This means that the civilian leadership is increasing the extent of military participation in the fabrication and execution of national policies. So, militarization of politics is the effect of the civilian ruling group's policy of encouraging the military to participate more vigorously in the formulation and implementation of national policies. Militarization of politics is opposite of politicization of the military. In the latter case, the military bosses themselves decide how and to what extent the military should participate in the political process.[18]

In recent times, there has been a shift within the supporters of Cooperation theorists from focusing on tangible political structures to intangible cultural values. Rebecca L. Schiff gives a new name to the Cooperation theory which she calls the theory of Concordance. She writes that instead of pointing only on the power-sharing aspects within the institutions, one must also analyse the cultural values that occasionally result in civil–military concordance.[19] The assumption is that culture shapes the matrix of power-politics. A dialectical relationship exists between the culture of governance and the languages of power. Instead of overt control, she emphasizes on partnership and dialogue between the civilian bosses and the military officers. A true concordance represents a sort of cultural agreement between three partners: political elites, military and the citizenry at large.[20] Sanjay Dasgupta claims that an apolitical army in the sense that it is politically naive is undesirable. In contrast to Huntington, he conceptualizes a professional military force as an army thoroughly imbued with civilian ethos and directed by a politically mature officer corps.[21] Sanjay Dasgupta is probably influenced by M. Janowitz who asserts that a professional officer corps requires political orientation in order to relate harmoniously with the civilian society.[22] So the proponents of the Cooperation theory harp on the absence of contradictory values between the army and the civilian society.

Probably the Concordance theory's emphasis on cultural convergence is of some use in the case of the communist one-party system where the military officers are also socialized in the values of Marxism–Leninism.[23] The Cooperation theory is applicable to some extent in countries like the erstwhile USSR and China. In such countries in contrast to the democracies, the polities resorted to conscription and indoctrination of the recruits in the values of communist ideology before enlisting them into the armies.[24] In erstwhile USSR, the Communist Party ideologically indoctrinated both the civilians and the military officers in the principles of Marxism–Leninism. About 93 per cent of the officers were members of the Communist Party or the *Komsomol*.[25] So, in the Communist countries, ideological consensus was actively inculcated through certain institutions.

In such cases communist ideology legitimized the social arrangements as chalked out by the party.

The Concordance theory could also be applied to societies, where despite the absence of an institution inculcating a set of uniform secular values, a prevailing system of shared norms exist which inform the behavioural patterns of the society. Such values are backed by ethno-religious imperatives.[26] For instance in Pakistan, many educated persons believe that the concept of democracy is totally foreign to the ideology of Islam.[27] This sort of attitude presumably prevents the democratic values from gaining deep roots within the Pakistani psyche. And such an attitude on part of the civilian society in Pakistan partly facilitates occasional military takeovers. However, cultural values whether secular or religious, being intangible, are difficult to document and assess.

Besides emphasizing the political and cultural dimensions, some scholars also point to the economic aspects of the Cooperation theory. According to Volker R. Berghahn, industrialized societies are characterized by militarism. The features of militarism are expansion of military values in society and undue preponderance of the military in the national decision-making process. One feature of the cooperation between certain crucial segments of the civil society and the military is the handshake between the politicians, military officers and big business which give rise to the military–industrial complexes even in the democratic polities. The foremost example of this case is the USA. The expansion of the military–industrial complex enables the military to participate significantly in the making of national security policy and often the outcome is an extension of a militaristic ethos in society.[28]

Veena Kukreja in the context of modern India rightly says that no clear causal linkage between economic development or lack of it and assumption of political role by the army could be established.[29] Finally, due to the low level of industrialization in India and operation of a sort of 'command economy' till the 1990s, a viable military–industrial complex did not emerge here. In Chapter 8 we will see that the military industrial infrastructure remains in the hands of the civilian bureaucrats and the private industrialists and the managers of the armed forces continue to play a marginal role in the production process.

The problem with the Cooperation theory is that it fails to accept that rivalry is in-built within the structure of civil–military interaction. Share in the administration is a finite cake. The political elite has to decide what share should go to which occupational and social groups. Not only is the army's share in administration marginal within India but it is also minimal when compared to the scenario in other countries. For instance in India, the Indian Administrative Service (IAS) officers both during their service and after retirements enjoy a larger share of the administrative cake compared to the uniformed men. To cite an example, at any given time more retired IAS officers are appointed as Governors compared to the military personnel. But in Israel, unlike in India, large numbers of retired army officers are given crucial ministerial posts.[30]

The presence of specific features in the civil–military relations of various countries point to the necessity of undertaking historical case studies. It would be simplistic to categorize all the democratic states into a single category as far as the civilian domination of the military is concerned. This is because the degree of subordination of the militaries by the civilian authorities in the various democracies differs a lot and even in a particular country the contours of civil–military relations change with time. A democracy could be categorized as 'minimalist democracy' where the concerned political system accepts conditions imposed by the military on candidates, parties and election procedures.[31] Unlike in India, during the 1991 elections in Russia, many military figures stood as candidates.[32] The nature of military power also varies within a mature democratic system. For instance both the USA and India are democracies. But the armed forces in USA have always enjoyed a lot more power in the body politic vis-à-vis their counterparts in India. Unlike India, in USA several military generals like George Washington, Andrew Jackson, U.S. Grant and Dwight Eisenhower became Presidents after retiring from service.[33] In contrast, not a single general has either been appointed President or has been Prime Minister in independent India. Again, assuming that cooperation exists between the military officers and the political leaders, the power of the 'men on horseback' depends on the scope, size and mission for the armed forces within the paradigm of grand strategy as chalked out by the politicians. Since the US armed forces play a significant role in American foreign policy, the armed forces enjoy disproportionate power within the polity.[34] However, in India's grand strategy, the role of the army remains more or less on the borderline. A word of caution is necessary before proclaiming that India is militaristic since it possesses a million-strong army, for a small army does not necessarily means a loyal instrument. Most of the sub-Saharan African states like Botswana, Burundi, Chad, etc., possess small armies, but these forces pose threat to their respective political systems.[35]

One could not deny the role of state organization in subordinating the army to the political establishment and forcing it to cooperate with the civilian security managers. The Soviet system and the countries in eastern Europe which followed the Soviet model included several organizational mechanisms for subordinating the army to the party. One was the commissar system under which party workers were appointed commissars and kept a watch over the military commanders. The reports of the political commissars were sent to the central party organ.[36] Another mechanism was altering the social base of the officer corps. In the post-World War II Polish Army, the government deliberately recruited workers' sons in larger numbers compared to the bourgeoisie in the belief that since the former social group was given an opportunity for upward mobility by the socialist system, they would remain loyal to it. Added with this was also the belief that they would be more receptive of the working-class ideology of the Communist Party.[37] However, such an institutional fabric is absent in the democracies and the democratic polities like India could not afford to follow a policy of planned 'proleterization' like the communist countries.[38] After the collapse of the Soviet system with its attendant institutional apparatus, Russia is finding it difficult to control its army. A large chunk of the Russian Army officers believe that in case of war, the army should take control of policy or at least should have a major say.[39]

In order to study the control mechanism of a democratic polity, i.e. India over its army, we will mesh Organizational theory with the cultural aspects of the Concordance theory. This is necessary because even if we accept that the army constitutes a bureaucracy alienated from the society, that organization still has its own ethos and cultural code of conduct. One scholar has rightly said that military profession gives birth to a sort of homogeneous professional ideology.[40] So the approach followed in this book could be categorized as an Organizational Cultural approach. This methodology will enable us to study the 'military mind'.

THE EVOLUTION OF CIVILIAN DOMINATION OVER THE MILITARY IN INDIA

The attitude of the officer corps towards their political masters is a crucial component of the civil–military relations. The Indian Army's officers express themselves in the premier service journal known as the *Journal of the United Service Institution of India*. Analysis of their writings opens up a window to their mind. When Independence was imminent and inevitable, both the white and brown officers expressed doubt about the Indian politicians' capacity for maintaining a healthy relationship with the Indian officers in the near future.

In 1946, Major A.J. Wilson warned the Indian officer corps in the following words:

> Closely allied to training in the responsibilities of a citizen lies the relationship of the Army officer to politics. The young Indian is both by upbringing and inclination a political being—and in the present nature of things it is both healthy and desirable that he should be so inclined. It is, however, difficult yet vital to persuade him that a soldier must, in his public life at least, be utterly detached from and uninfluenced by political considerations. His task is to serve the Government of the day without reference to its political complexion or his own opinions.[41]

On the eve of Independence, the editorial committee of the *JUSII* emphasized that the newly independent countries expect loyalty and devotion of the officer corps. The editorial board pointed out the duty of the servicemen:

> It is an aspect which any Servicemen worth his salt will understand, and is summed up in two words: loyalty and discipline. Individual feelings, individual careers, are naturally apt to predominate in peoples' minds just now, but a sailor, soldier, or airman has but one duty—loyalty to the Government of his country, and a determination to assist in maintaining law and order by adopting a tolerant attitude in his conversation and privately expressed thoughts.[42]

The editorial board of the *JUSII* also warned the Indian officers that regardless of their communal affiliations they should remain loyal to their regiments and to the newly emerging states of Pakistan and India. Instead of perceiving themselves as Sikh, Hindu, Muslim or Christian, the officer cadre, urged the editorial board, should conceive themselves as nationalists.[43] Even the British Commander-in-Chief of the Indian Army, in line with the *JUSII's* editorial committee hastened to emphasize the apolitical attitude of the army and its loyalty towards the new government. On 28 May 1948, General Roy Bucher in his address to the staff and cadets of the Indian Military Academy at Dehra Dun said:

> Remember that the Army, and indeed all the Services, are the servants of the Government in power at the time, and the political complexion of a particular Government makes not the slightest difference to this fact. As soldiers you are not concerned with politics. There is nothing wrong in your having political opinions and in expressing them with moderation in private conversation, but that is a very different matter to expressing political opinions in public or allowing such opinions to influence your action in any way. No Army which concerns itself with politics is ever of any value. . . . It follows, therefore, that the Army has never the slightest right to question the policy of Government. Implicit obedience to the orders issued by Government is essential, and only in this manner will the interests of the country be fully served.[44]

Some British officers doubted the ability of Indian political leadership to control the army. When India became independent, Major-General L.G. Whistler expressed himself in a very patronizing tone in an article written in the *JUSII.* Whistler said that the British soldiers and white officers as well as the Indian soldiers under British leadership had civilized India. Now, claimed Whistler, it was upto the Indians to maintain the polity that the British were leaving behind in the subcontinent. He seemed doubtful whether the Indian leaders would be able to display the necessary statecraft.[45] And a year after India's Independence, Lieutenant-Colonel J. Wilson Stephens expressed doubt about the post-colonial government being able to maintain its control over the Gurkhas.[46]

Just before Independence, some Indian officers in their writings pointed out the importance of the army in general and the officer cadre in particular for maintaining the stability and integrity of independent India. They also cautioned the politicians that in the newly independent India the politicians should not interfere with the valued apolitical cultural traditions of the Indian Army. In 1946, Lieutenant-Colonel Rajendra Singh wrote:

If freedom to India comes through peaceful means, India has to thank the Indian Army, which made it possible. A military soldier is not judged by his political fervour, personal motives or party leanings but by his military qualities. In India we have first-class fighting material and its proper use in future will depend on the attitude of the political leaders of to-day. The germs of indiscipline once sown are difficult to eradicate. They keep on spreading. To day mutiny may be for freedom, to-morrow that freedom may take the shape of Pakistan and later on the achievement of a particular "ISM". . . . When our system is overhauled or reconstructed we must not forget the value of tradition. The soldier must have something to fall back upon. . . . Every unit is proud of a gallant record and this should be carried forward in the new form.[47]

Lieutenant S.G. Chaphekar warned the Indian politicians that they should neither view the Indian Army as a band of mercenaries nor attempt to politicize the soldiers. He wrote in 1947:

Some leading public figures have condemned it as a mercenary army, but in recent speeches we have our political leaders praising it. Yet the army has not changed overnight; it is the same army, unchanged in spirit and fighting ability. . . . If a country's soldiers are dragged into politics, then we can bid goodbye to democracy and civil liberties. In short, the Army of a nation has no concern with party politics or with changes in Government; its duty is to defend the freedom of its country.[48]

A small group of Indian officers proved to be politically naïve. On the eve of Independence, these officers were anxious that the diffusion of political consciousness in the country might infect the officer cadre negatively. They argued that independent India should continue to recruit from the apolitical 'martial races' in accordance with the British sponsored Martial Race theory.[49] One such officer was Major Gurbachan Singh, an instructor of the Officer Training School. If the required number of politically inert Indians could not be recruited from the 'martial races', wrote Gurbachan Singh, then India must rely on the British officers. In 1946, Singh wrote:

Turning to politics, to have a politically minded army is to head for a national disaster. Politics, unfortunately, colour the tenor of all life in India today, but if our future Army is to carry out its normal military functions efficiently, politics should be taboo. A cadet therefore who comes from a politically active class is most undesirable in the officer ranks of the Army. . . . The defence of India will be a joint Indo-British responsibility for many a years to come. . . . India would want the very best type of British officers in her Armies.[50]

The attitude of many Indian officers towards their new political masters remained critical even after Independence. This is evident from the conversation that Brigadier J.N. Chaudhuri (at that time posted in the Planning Section of the Army Headquarter and became Chief of the Army Staff [COAS] on 20 November 1962) had with Major-General Rajender Sinhji (commanding Delhi Area in December 1947 and COAS from 15 January 1953–14 May 1955) in the last month of 1947. The latter told Chaudhuri:

I want to brief you about all these new people now in charge here whom I have met but you have not. Your very direct ways of dealing with all and sundry won't work in this set up. You will have to be a little more circumspect in your method of dealing with the politicians of your own country. You can't be as blunt with them as you could be with politicians of other countries. . . . Our present leaders are Indians. They do not behave like the British. Their techniques are different. They behave more like the Princes. You have to be more of a courtier.[51]

However, on 8 October 1947 the senior most Indian military officer Lieutenant-General K.M. Cariappa in a speech to the Indian Army officers emphasized their duty and loyalty to the new nation state in the following words:

What are our duties as officers to our country? As soldiers we are the servants of the state, and as such it is our duty to serve our Government loyally at all times. We must refrain from expressing our personal views on the wisdom or otherwise of Government policy. Our duty is to contribute our best to the maintenance of a well-disciplined, well-trained and happy Army—an Army which can effectively and efficiently support our Government in whatever manner the Government may wish to use us.[52]

The political elite of India accepted Rajendra Singh's dictum as regards maintaining cultural continuity with the Sepoy Army for the sake of maintaining regimental *esprit de corps*. The net result is that the post-colonial Indian Army still nurtures several colonial customs and traditions. The Madras Regiment's elephant crest was won under Arthur Wellesley in 1803 against the Maratha Army at the Battle of Assaye and this symbol is still in use with the Madras Regiment of the Indian Army. Further, old British tunes like '*Abide with Me*' and the Scottish soldiers' favourite '*Auld Lang Syne*' are still played by the military band along with '*Saare Jehaan se Achchha.*'[53]

Nevertheless, the government of India was unwilling to rely wholly on the fragile goodwill of the army's officer corps. After independence, the government deliberately initiated several organizational changes that resulted in reduction of the military's power drastically within the polity. Prime Minister Jawaharlal Nehru took certain steps which kept the Indian Army sanitized from the broader influence of the society. Nehru was influenced in his actions by the advice of Admiral Lord Louis Mountbatten. In 1946, Nehru visited Malaya (later named Malaysia) and met Mountbatten, then commander of the South-East Asia Command. For making immediate political gains, Nehru was thinking of inducting the Indian National Army (INA) personnel in the regular army of independent India. But, Mountbatten warned that whatever were the motives of INA, they had broken their oath of allegiance. Mountbatten asserted that India after becoming independent would require a completely apolitical army. Mountbatten said to Nehru:

> Doubtless you think that the INA are all like Subhas Chandra Bose, devoted patriots, who have sacrificed everything for their own country. In fact that is not how any of us, who have been in the war, regard them. I can give you evidence, if you wish, to show that the INA was recruited from that type of Indian soldier who was not prepared to face the rigours of Japanese captivity. . . . When you get your independence and have your own army, the people you want are those who remain loyal to their oath, and who will stay with you, and not the people who just change according to political opportunism. Otherwise you will find that if you become unpopular in your own country one day, the opposition party will call on the army to turn you out. You want people who will stick to their oath.[54]

The Indian officers of the Sepoy Army backed Mountbatten's view regarding the INA. During the Mountbatten–Nehru meeting, two Indian officers (Major Duggie Sawhney of the 16th Light Cavalry and J.N. Chaudhuri then a Brigadier in-charge of administration in Malaya) were also present. In Chaudhuri's own words: 'Sawhney and I noticed. . . . Panditji's concern about the future of the INA. He asked us a few questions about them but we were non-committal in our replies. We did not give the impression that we had been impressed by them for, quite frankly speaking, we had not.'[55] The Indian officers of the Sepoy Army were against the INA because they feared that the induction of the INA might result in increased competition and restrict their career advancement in the post-war army, which was going to be much smaller and would be resource-starved. Those Indian officers who remained loyal to the Raj till 1947, tried to legitimize their position in the post-Independence era by portraying themselves as above party politics. In the long run, the non-induction of the INA personnel into the Indian Army was a blessing in disguise as it prevented politicization of independent India's Army.

In British India, the civil–military relation was loaded heavily in favour of the army. This aided excessive expenditure on matters military. The head of the Military Department was a serving officer, a member of the Viceroy's Executive Council, and was responsible for sanctioning military expenditure.[56] The tussle between Kitchener, Commander-in-Chief of India (1902–9) and Viceroy Curzon, ended in favour of the former.[57] This further strengthened the hold of the military within the administrative apparatus of the Raj. Under the Raj, the Commanders-in-Chief held the defence portfolio[58] but after independence, Nehru began crafting an institutional mechanism for subordinating the military. Even in 1947 each of the three services had their own Commander-in-Chief. With the passage of time, the military's position was downgraded. In 1954, the President became the Supreme Commander-in-Chief of all the three services. And the Commanders-in-Chief of the three services were renamed as the COAS, Chief of Naval Staff and Chief of Air Staff respectively.[59] The Government of India deliberately lowered the *izzat* of the officer corps by downgrading the position of the military officers in the order of precedence vis-à-vis the civilian bureaucrats and the police officers.[60] The civilian secretaries of the central government who had previously ranked lower than the Lieutenant-Generals became equivalent in status to full Generals.[61]

Independent India's government drastically increased the power of the civilian bureaucrats, i.e. IAS officers especially in the sphere of forging national security policies. Fabricating the national security policy became the responsibility of the Cabinet

Committee on Political Affairs, and in this committee, the presence of the service chiefs was not mandatory.[62] The Cabinet Committee on Political Affairs was initially known as the Defence Committee of the Cabinet came into existence when Pakistan invaded Kashmir in 1947.[63] Below it came the Defence Minister's (*Raksha Mantri*) Committee. The *Raksha Mantri* is supposed to hold weekly meetings with the three service chiefs along with the secretaries of the Defence Ministry. An inter-ministerial standing forum designated as the Committee for Defence Planning is convened by the Cabinet Secretary. This committee includes the three service chiefs, Principal Secretary of the Prime Minister, plus secretaries from the departments of Foreign Ministry, Defence, Finance and Defence Production. In this committee not only a civilian bureaucrat occupies the post of chairman but the civilian bureaucrats also dominate over the uniformed men by sheer numbers. This committee undertakes regular assessments of defence policy and advises the *Raksha Mantri*. In addition, this committee also reviews the implementation of the annual plan and schemes related to the defence programme.[64]

During January 1951, in response to a request from the Assam government, General Cariappa had a platoon of paras dropped to deal with the Chinese incursion near Rima, north of Walong. To forestall any further Chinese moves, Cariappa proposed that the entire border with Tibet be placed under the army. Nehru flared up and told Cariappa that it was not the task of the Commander-in-Chief to tell the Prime Minister who the adversary was going to be. Nehru's vision at that time was clouded by the romantic notion of *Hindi–Chini bhai bhai* and he asked Cariappa to confine himself with the threat from Pakistan especially in Kashmir.[65]

In contrast, the Pakistan Army developed its own view of foreign policy even when Liaquat Ali was the Premier. In 1953, when Khwaja Nazimuddin reduced the military's budget by one-third, the Commander-in-Chief of the Pakistan Army General Ayub Khan joined hands with Governor-General Ghulam Muhammad and dismissed Nazimuddin's government. In the new Cabinet which came into existence in 1954, Ayub Khan assumed the post of the Defence Minister.[66] In Pakistan, the corps commanders continue to enjoy overt power and pose a threat to the political authority.[67]

In India, within the Ministry of Defence the military was subordinated to the *neta–babu* raj which means politician–bureaucrat combine. Civilianization of the MoD started long before Independence. Till 1920, an uniformed officer usually of Major-General rank occupied the post of Secretary to the Army Department. Only in 1921, a civilian was appointed as Secretary. In 1936, the Army Department was renamed as Defence Department. Independent India carried on the imperial tradition of further civilianizing the MoD.[68] The 'Iron Man of India' Sardar Patel maintained a tight control over the military officers. Rajender Sinhji warned J.N. Chaudhuri: 'Panditji is number one, but Panditji is extremely difficult to get hold of. . . . He is not particularly forthcoming in his replies. He tends to be indecisive at times. . . . The man today who is really looking after the internal security and the problem in Kashmir is Sardar Patel. You watch him. He is the strong man and if you fall foul of him, you are finished.'[69] Sardar Patel till his death in 1950 maintained his hold over the military through H.M. Patel, the Defence Secretary (1947–53). During the regime of two Patels, Cariappa was forced to accept a new pay code whereby Indian commissioned officers' salaries were reduced in relation to that of the civil servants.[70]

Politicization of the army due to excessive political intervention in the military sphere resulted in an army coup in Bangladesh during 1975. Sheikh Mujibur Rehman, the founding father of Bangladesh preferred the leaders of the Mukti Bahini (Liberation Army) compared to those East Pakistani officers who were repatriated from West Pakistan after the 1971 War. The Mukti Bahini included deserters from the East Bengal Regiments, East Pakistani Rifles, police as well as guerrilla fighters owing allegiance to various politicians. In 1973, about 28,000 defence personnel were repatriated from Pakistan. The regular officers on repatriation found out that they had been superseded in ranks by Mujib's men. These regular soldiers were in no mood to accept the supremacy of the guerrilla leaders of the Mukti Bahini. Mujib followed such a discriminatory policy because he doubted the loyalty of the officers of the regular army.[71] Probably, Mujib also tried to reward his political supporters by giving them high posts in the regular army. The net result of Mujib's insecure interventionist policy was to push a suspicious army into open rebellion.

But, too little political intervention might also create problems in the civil–military relationship. This was the case in Pakistan. Field Marshal Ayub Khan writes in his memoirs:

> There was considerable speculation about the prospects of a Pakistani being appointed Commander-in-Chief at the end of General Gracey's term . . . long before the appointment of the new Commander-in-Chief, Prime Minister Liaquat Ali Khan came to Rawalpindi. There was a Divisional Commanders' Conference and I had come from East Pakistan to attend. The Prime Minister sent for senior Pakistani officers and addressed them at the Circuit House. Towards the end of his address he said that it had been decided to appoint a Pakistani as the next Commander-in-Chief of the Pakistan Army. He mentioned that the government had not yet selected the person but that it was possible that the appointment would not go to the most senior officer. He wanted to know the likely repurcussions among senior officers in case a person lower in the order of seniority was chosen. He asked several officers and finally turned to me. I was sitting at the end. He said, 'General Ayub Khan, would you like to say something?' I said, 'Sir, may I say with great respect that this question should never have been asked.[72]

The Pakistani government erred by showing extra deference to the army officers, which encouraged the Pakistani officer cadre to think that they could easily topple the supine politicians in the future. And that was what exactly happened after a few years.

A balance is required between the level of civilian intervention and the sphere of military autonomy, a balance which was disturbed in India during the late 1950s. After the demise of Sardar Patel, the situation was such that the *Raksha Mantri* could create problems for the COAS when the former enjoyed the backing of the Prime Minister. Krishna Menon, the Defence Minister with Nehru's backing started intervening in promotion matters in order to fill the forces with 'yes men'. This trend continued till the 1962 disaster.[73]

Krishna Menon and General K.S. Thimayya (COAS from 8 May 1957 to 7 May 1961) developed personal animosity against each other.[74] It was a perfect case of Cohen's personality clash scenario. The result was increasing friction between the civilian and military components of the Defence Department. Krishna

Menon, tall, lean and humourless with a penchant for detail had the habit of calling the Chiefs of Staff for consultation at all hours of the day and night. Often they had to wait in his anteroom while Menon was engaged with other visitors. The service chiefs resented this sort of behaviour from Menon and their sense of dignity was offended.[75] Menon started to disregard the COAS's view even on purely military matters. Thimayya nicknamed Timmy demanded replacement of the 25-pounders by 105-mm guns and the replacement of the 5.5 inch medium guns with 155-mm guns and 75-mm mountain guns before the Indian Army could hope to meet the Chinese challenge. But, the Defence Minister overruled the general.[76] In retaliation, the Army Headquarter refused to cooperate with Menon. Whenever Menon asked for information regarding various schemes, the Army Headquarter dragged its foot. As a result, Menon was forced to bypass the Army Headquarter by communicating directly with the junior officers in charge of the various schemes, to which the Army Headquarter protested that Menon was going against the code of discipline.[77]

The situation did not improve when Timmy was succeeded by General P.N. Thapar (COAS from 8 May 1961 to 20 November 1962). One senior naval officer, Admiral S.M. Nanda in his memoirs portrays the ham-handed behaviour of Menon:

> As Deputy Chief of Naval Staff, I attended many meetings whenever the Chief of Naval Staff was out of station. General P.N. Thapar was the Chief of Army Staff. He used to brief the Defence Minister's Committee on the ground situation showing the position of Chinese troops and our troops on a map. Defence Minister Krishna Menon often told him where and how Indian troops should be moved and deployed. Somehow, General Thapar was not able to assert himself sufficiently to counter the Defence Minister. He had deployed his troops effectively but they were ill equipped.[78]

It seems that Menon like President Lyndon Johnson and his National Security Advisor Macnamara attempted to micro-manage the army. During the Vietnam War, Johnson selected the targets that were going to be bombed by the USAF.

Taking cue from the Defence Minister, even junior bureaucrats started taking pot shots at the COAS. In September 1962, when Thapar was against following the policy of evicting the Chinese from disputed territory, his view was overridden. When Thapar asked for written directive about government policy, a mere Joint Secretary of the MoD issued the following order: 'The decision throughout has been as discussed at previous meetings, that the army should prepare and throw out the Chinese as soon as possible. The Chief of the Army Staff was accordingly directed to take action for the eviction of the Chinese in the Kameng Frontier Division of North East Frontier Agency as soon as he is ready.'[79]

During the 1962 War, the COAS lost autonomy even in the operational sphere. Despite the COAS's plea, the politicians did not allow the use of the Indian Air Force against the Chinese in NEFA. The political elite feared that the use of air force by India would trigger a more intense and dangerous reaction on part of the Chinese and this might result in the loss of the whole of eastern India. The civilian bosses did not take into account the massive logistical problems which the Chinese would have faced due to their tenuous and long communication lines across Tibet in case the Peoples Liberation Army (PLA) attempted a large scale invasion of east India.[80] To be fair to the politicians it must be stressed that the attitude of the senior IAF officers in joining the war was ambivalent. This is explained in greater detail in Chapter 6.

The Indian politicians learnt something from the 1962 disaster. In 1965, Pakistan made a major thrust into the Chamb region of Jammu & Kashmir in support of a massive infiltration of armed tribal irregulars. The COAS General J.N. Chaudhuri demanded an Indian counter-offensive through Punjab. Several senior politicians and bureaucrats feared that such a military offensive by opening up a new front would result in widening the conflict and bring pressure from the UN and US. However, Prime Minister Lal Bahadur Shastri supported Chaudhuri against the wishes of his Cabinet.[81]

In general, the Indian politicians gave autonomy to the army in the spheres of tactics and operations. Even under Indira Gandhi, a dynamic Prime Minister, the army enjoyed considerable space as far as the planning of military operations was concerned. To give an example, Indira Gandhi wanted to start the war with Pakistan in April 1971. She wanted to get rid of the political and economic problems caused by mass immigration of the Bangladeshis into eastern India as quickly as possible.

However, in the end, she accepted General Sam Manekshaw's (COAS from 8 June 1969 to 14 January 1973; Field Marshal after 1 January 1973) proposal to begin the war during late 1971, when the Himalayan passes would be closed due to snowfall and Chinese intervention would be an impossibility.[82]

Nevertheless, unlike the Israeli Defence Force (IDF), the Indian Army did not enjoy any freedom in the strategic sphere. In 1982, Defence Minister Ariel Sharon started the Lebanon War without the full knowledge of the Israeli Cabinet and the Parliament.[83] Such a scenario is unthinkable in India. In 2002, the Indian government removed Lieutenant-General Kapil Vij, Commander of the II Corps because he misread the political situation. Vij believed that India would soon attack Pakistan and on his own initiative to get a favourable position before the onset of war, Vij started to move his troops forward. When Washington with the aid of satellite imagery brought this movement to the notice of the government of India, the latter removed Vij from his command.[84]

Independent India like British India faced insurgencies throughout the length and breadth of the subcontinent. A regime that uses the army for suppressing internal disorders in the long-run also suffers from military influences in the civilian sphere. In the Soviet Union, the focus of the military was on external security threats, a principal factor that prevented any military coups. However, the dissolution of the Soviet Union and consequent use of the Russian Army in internal disturbances increased the army's political power. President Boris Yeltsin's use of the army in combating organized crime as well as in tax collection resulted in the expansion of influence of the Russian military. All these developments transformed the army into a dangerous and unpredictable force.[85]

Instead of becoming over-dependent on the army for maintaining law and order, the government of India not only retained the various para-military forces inherited from the colonial state but also raised several new paramilitary forces under civilian control with separate administrative structures. The vast number of paramilitary formations function as a buffer between the army and the civilian government. The Assam Rifles which came into existence during 1835, had by 1950, five battalions. During the 1960s, the Assam Rifles and units of the

Indian Army carried out joint operations against the guerrillas in Nagaland.[86] Towards the end of the 1970s, the size of the paramilitary forces amounted to one-third of the strength of the country's armed forces.[87] During the 1980s, the central government possessed the following paramilitary forces: 1,20,000 in the Central Reserve Police Force, 90,000 in the Border Security Force (BSF), 40,000 in the Assam Rifles (organized into 32 battalions), 14,000 in the Indo-Tibetan Border Police, 70,000 in the Central Industrial Security Force, 70,000 in the Railway Protection Force and 2,50,000 in of the States Armed Police.[88] The Water Wing of the BSF guards the Creek Area of the Rann of Kutch in Gujarat[89] and additionally, the BSF has 20 artillery batteries.[90] From Table 2.1 it is evident that with the passage of time, the size of the paramilitary forces increased.

Pakistan lacks such large number of paramilitary formations under the operational and administrative control of police officers. In order to establish civilian authority over the military, Z.A. Bhutto tried to set up the Federal Security Force, a paramilitary force equipped with the latest weapons, under his direct control.[91] But, his plan came to a naught, since before he could secure political dominance over the 'men in uniform', the army moved against him. At present, Pakistan maintains several paramilitary formations like the *Mujahid* Force, *Janbaz* Force, etc. But most of the Junior Commissioned Officers and the Non-Commissioned Officers of these formations are from the army and the senior commanders are Brigadier level officers retired from the army.[92] China depends on the People's Armed Police for maintaining internal security which is staffed by men decommissioned from the PLA.[93] In India, army officers on deputation are appointed in the middle ranks of the BSF and the top slots remain the domain of the Indian Police Service officers.[94]

From the economic perspective, the Indian armed forces remain totally dependent on the Parliament and the Indian military has no autonomous economic base. The state-owned military industries in India have always been completely under civilian control. In contrast, the PLA owns a substantial business empire and, the military industries owned by the PLA also go for massive production of civilian goods like motorcycles and cameras.[95] In 1998 though President Jiang ordered the PLA to

TABLE 2.1: GROWTH OF PARAMILITARY FORCES IN INDIA

Year	National Security Guards	Indo-Tibetan Border Police	Central Reserve Police Force	Border Security Force	Assam Rifles	Central Industrial Security Force
1988	7,482	23,419	1,20,979	1,35,544	52,067	66,102
1989	7,482	25,482	1,21,206	1,49,568	52,460	71,818
1990	7,482	29,488	1,31,260	1,71,168	52,460	74,334
1991	7,482	29,504	1,59,091	1,71,363	52,460	79,620
1992	7,485	29,504	1,58,097	1,71,501	52,482	84,611
1993	7,485	29,504	1,58,693	1,71,735	52,504	87,337
1994	7,512	30,297	1,65,334	1,71,735	52,504	88,603
1995	7,360	30,293	1,65,408	1,81,269	52,223	91,212
1996	7,360	30,369	1,66,198	1,81,403	52,223	96,567
1997	7,360	29,275	1,67,383	1,82,675	52,269	96,534
1998	7,360	30,367	1,67,329	1,82,732	52,223	95,863
1999	7,357	30,367	1,67,637	1,83,790	51,985	94,598
2000	7,357	30,356	1,71,443	1,81,839	51,056	95,000

Source: Lieutenant-General R.K. Jasbir Singh, ed., *Indian Defence Yearbook: 2003*, Dehra Dun: Natraj Publishers, 2003, p. 251.

close down their commercial enterprises, it is yet to be seen how far Jiang's order will be obeyed by the PLA.[96] The Pakistan Army, too, has its finger in the economic pie of the nation. The Inter-Services Intelligence (ISI) started the Army Welfare Trust in the 1970s. It has become one of the most active business group in Pakistan. It controls a sugar factory, a shoe factory, a woollen garment factory, a glass factory, land development scheme, and housing projects. The ISI also controls the *Fauji* Foundation which has set up factories for the manufacture of fertilizers and chemicals as well as the Shaheen Foundation which controls the privately owned Shaheen Airlines. It has also been permitted by the government to start a satellite TV channel.[97]

Finally, the Indian armed forces remain out of the nuclear decision-making loop. The Prime Ministers rarely consulted the service chiefs regarding nuclear policies. Only one engineer regiment was involved in digging shafts at Pokhran for the nuclear tests of mid-1998.[98] General Shankar Roychowdhury (COAS from November 1994 to September 1997) during an interview in 1998 claimed that he was never consulted by the Indian government about any nuclear policy decision.[99] Shankar Roychowdhury's view is that the nuclear forces should be under the direct control of the Chiefs of Staff Committee or a tri-service Strategic Command.[100] The government is yet to set up such a command. The nuclear programme remains under the control of the Directorate of Atomic Energy and Defence Research and Development Organization, with the former reporting directly to the Prime Minister.[101] In contrast, the nuclear programme in Pakistan has always been completely under the control of the army, which enjoys greater prestige than the Indian Army.[102]

In a study conducted by Jerzy J. Wiatr in the 1960s, it was found that among the seven countries (West Germany, USA, Denmark, Indonesia, Philippines, Poland and Pakistan), the army in Pakistan enjoyed the greatest amount of social prestige.[103] It has been very difficult for militaristic values to spread in Indian society because the armed forces though a million strong are quite small compared to the vast population of India. So, a low military participation ratio[104] and absence of any conscription as in China has enabled the Indian state to segregate the military from the civilian society. However, the civilian rulers of China use Confucian values for attaching a low status to the military men.[105] Unlike the Indian Army which is composed of long-service professional volunteers, the IDF is composed of mostly reservists and conscripts who outnumber the long-service professionals. So, the IDF unlike the Indian Army is a 'nation in arms'.[106] In addition due to the organizational measures discussed above, the army is marginalized within India's public life. But, the IDF is the embodiment of the new Jewish history and the most respected public institution in Israel.[107] David Ben-Gurion, Israel's first Premier and Defence Minister (1948–53 and 1955–63) was the man principally responsible for creating the IDF. Ben-Gurion intended the military to become an instrument of the new Jewish 'nation building' process and a symbolic focus of national sentiment. He envisioned the IDF as a bonding institution within whose framework Israel's otherwise fractured society could be welded into a homogeneous whole. A system of universal and mandatory military service was deemed essential for this purpose. Thus, the IDF is not a segregated professional institution but a people's army, the representative of the nation. Military service remains a civic right as well as a public obligation. This situation helps to preserve the societal esteem of the men in uniform within Israel.[108]

DEFECTS IN INDIA'S CIVIL–MILITARY RELATION

Writing about twenty-three years after independence, a retired Lieutenant-Colonel, C.L. Proudfoot raised doubt about the continued loyalty of the *jawans* and their officers towards the Indian government. He asserted that political interference in the internal affairs of the military service and vulnerability of the new generation officers to the corrupting influence of wider society might result in disintegration of the fragile civil–military bonds.[109] Proudfoot might be over-reacting, but there are some elements of truth as regards the corrosion of the civilian control over the military.

The biggest deficiency as regards civil–military interaction in India remains the absence of any institution for long-range strategic planning and for coherently implementing a strategic plan during war. Field Marshal Claude Auchinleck, the last Commander-in-Chief of British India in his address to the Staff College at Quetta on 1 November 1945 proposed an organization for directing military affairs of independent India. In Auchinleck's own words:

> I will now refer to our plans for the future organization of the higher command in this country. We are at present examining in detail certain definite proposals for the setting up in India of an integrated command, on which all three Services will be equally represented, under a Supreme Commander who might be a sailor, a soldier or an airman. Under him he would have a planning and advisory staff, headed by a Chief of the Operations Staff, and a Chief of the Administrative Staff, who might be drawn from any of the three Services. They would have under them fully integrated staffs of officers of all three Services. The navy, army and air force would each have their own Flag Officer Commanding, with their own Service staffs to give executive orders which would be framed so as to give effect to the policy laid down by the Supreme Commander. I think that this is what India needs for the proper co-ordination of her defence, and I hope she may achieve it.[110]

An account of how the MoD after 1947 functioned can be glimpsed from Prof. P.M.S. Blackett's account. He says:

> As I was out to advise on the scientific set-up to the Ministry, I said, 'You cannot have a scientific set-up till you tell me what weapons you are going to use. And you cannot tell me that until you tell me what sort of war you are preparing for. You must have a model for that.' So I

asked, 'What are your plans?'. Sardar Baldev Singh [Defence Minister] asked the three Chiefs of Staff: 'Well, produce your plans.' And the next morning I had all the three plans on my desk. It added up twice the national budget, the federal budget. They had never added them together.[111]

Just after Independence, the government of India could be forgiven for the absence of clear-cut strategic objectives and for inadequate coordination between the three services and the policy makers. However, even after fifty years of Independence India is yet to achieve an organization with an autonomous secretariat for coordination and conducting future planning.[112] One positive reaction to the Chinese aggression of 1962 was the realization among the ruling class that systematic planning was necessary for ensuring defence preparedness, and hence, the introduction of the Five Years Defence Plan in 1964. However, the Defence Plan did not provide for long-term strategic planning (i.e. looking at twenty years hence), nor did it have the assurance of resources to support it.[113]

Power-projection over long distances amidst varied types of environment in pursuit of foreign policy objectives, requires cooperation and concensus among the three services.[114] For higher direction of war, the Indian military has a system in which the senior most service chief functions as the Chairman of the Chiefs of Staff. In the Chiefs of Staff Committee, three equal and independent service chiefs act in concert. As the position of the Chairman is occupied on a rotational basis, the Chairman is only an equal among the others and has virtually no power over the other two service chiefs.[115] Cooperation among three equal and independent service chiefs acting in concert within the Chiefs of Staff Committee with its annually rotating Chairman is not inducive to unified planning. Thus, systematic planning and implementation of any joint plan is impossible and this prevents close coordination of the three services in the midst of a war. The creation of a Chief of Defence Staff (CDS) might solve the problem. In Britain, the post of the CDS was created in 1956.[116] The US armed forces had a Chairman of the Joint Chiefs of Staff (JCS) right from the Second World War, and the Australian military has a Chief of the Defence Force Staff (CDFS). The CDFS acts as the government's principal military adviser with direct access to the Defence Minister.[117] In general the Indian Defence Ministers remain preoccupied with domestic issues and rarely completes two years in office, making it impossible for a politician to give close attention to defence affairs continuously.[118]

The option for India is the creation of a CDS with a permanent secretariat for ensuring tri-service integration. The CDS should ideally be the inter-service professional adviser and coordinator for the optimum use of the available resources. But the scheme is hampered by the bureaucrats who are quick to point out to their political masters that the CDS might carry out coups.[119] A fear that is baseless because the CDS would not have any armed force under its direct command and have a purely advisory function.[120]

The hostility of the bureaucrats towards joint military planning was evident in the aftermath of the Chinese invasion. General Chaudhuri writes:

> For the future there is an urgent requirement to have a Chief of Defence Staff or Permanent Chairman of the Chiefs of Staff Committee, with the responsibility of integrating plans, integrating intelligence, initiating research, advising the Minister and, where necessary, the Cabinet plus thinking well into the future. The need is accepted by most people but implementation is held back because some political groups probably fear that a strong CDS may become too powerful. Also, the civil servants are not in favour of the CDS set up as they feel some of their authority will diminish.[121]

The Indian military's sad lack of higher direction during 1971 is spelt out by Lieutenant-General J. Jacob (who was the Chief of Staff of the Eastern Command during the 1971 War):

> At the time of the Bangladesh War, no institution of the Indian Army taught or studied strategy. Our military institutions taught tactics at the unit, brigade and to some extent divisional level. Consequently, no realistic, overall estimate of war situations by the Army Headquarters was made. There was, in fact, no strategic or political definition of policy, nor an appropriate higher command organization to plan or direct the war. Political objectives were rarely spelt out clearly. As a result, planning at Service Headquarters lacked depth.[122]

Back in the 1970s, several officers clamoured for rectifying the defects in the higher echelons of the defence command. In 1972, Major-General S.N. Antia argued that the JCS with a permanent Chairman is a necessity for formulating long-term policy.[123] Brigadier N.B. Grant demanded the introduction of a CDS who, instead of the defence secretary would arbitrate in case of conflicting claims by the service chiefs.[124] In 2000, Colonel P.K. Vasudeva demanded that the CDS should be the senior most officer in the armed services and aided by a Vice-CDS. However, this scheme is facing flak from the bureaucrats who are unwilling to lose control over decisions made in the higher defence echelon. The Navy and the IAF are also hesitant about the CDS scheme fearing that the post of CDS will always go to the army because of its numerical superiority over the other two services.[125] The two sister Services must realize that India is bound to follow mostly a continental strategy because of its live land frontiers where the army will have to play the major role, and that the absence of a CDS enables the government to encourage rivalry among the three service chiefs which in turn ensures the dominance of the civil servants.[126]

The bureaucrats at the MoD have monopolized power to such an extent that they will not tolerate widening of the decision-making apparatus. During the crisis of November 1962, a Military Affairs Committee was set up to provide the defence minister with technical advice. However, as soon as the crisis with China was over, this committee became defunct.[127] In 1976, a group of military officers pointed out the necessity of constituting a body to initiate strategic planning—planning which takes into account what is going to happen in the next fifteen years—but it fell on the government's deaf ears.[128]

Since the bureaucrats and the politicians in India dominate the decision-making process, the military has been demanding the formation of a National Security Council (NSC) with a hope to get some share of the power, a situation completely in contrast to Israel where the IDF mostly formulates the national security doctrine. So, the politicians in Israel attempted to establish a NSC in order to acquire greater leverage over the higher defence apparatus.[129] Finally in 1999, the NSC in Israel came into existence. And it is headed by a retired major-general.[130] In 1999, the Bharatiya Janata Party government set up a NSC and to advise this body, a National Security Advisory Board (NSAB) has been set up. The aim is to tap the experts through this mechanism for better functioning of the government's security apparatus. But, retired bureaucrats dominate the NSAB and worse still, the NSAB is denied classified information by the government.[131] Despite the creation of the new apparatus the predominance of civilian bureaucrats have not been broken.

India has no institutional mechanism for fusing the operational, technological, logistical and economic components of grand strategy.[132] As a result, the Indian armed forces remain incapable of not only planning and waging pre-emptive long-distance strikes but also have no consistent policy to procure hardware for such contingencies.[133] Instead of a proper civilian control, what India has is rigid bureaucratic control: a system in which the generalist babus indulge in micro-managing military affairs.

The Indian government is unable to check the expanding role of the army in civilian life. During the late 1960s, the democratic government of Tanzania attempted to transform the army into a sort of developmental militia. The aim was to use the army in nation building tasks.[134] M.K. Gandhi had the idea of using the army for social welfare activities. In 1946, Gandhi in *Harijan* declared that after Independence, the army personnel should be used in agriculture and in sanitary works in the village.[135] Luckily, Gandhi's views were not followed by either Nehru or Sardar Patel. However, the army is used in various non-military roles which might result in weakening of the democratic institutions in the long-run.

The government depends on the army during emergencies (both man made and natural) when the civilian administration fails to discharge its duties properly. For instance, use of the army's manpower and infrastructure during natural disasters is a case in point. In 1982, during the monsoon season, severe floods occurred in north India. Some 30,00,000 people in 40,000 villages of Uttar Pradesh and Bihar especially were affected. The army evacuated 14,000, rendered medical aid to 17,000 and distributed 3,37,000 kilograms of food.[136] In October 1999, a cyclone originating from the Bay of Bengal caused severe devastation in the coastal areas of Andhra Pradesh and Orissa. More than 5,000 army personnel were deployed in the affected areas. The army evacuated 22,288 civilians, and medically treated 33,722 persons. In addition, the army distributed 4,259 tons of food and 2,48,000 litres of drinking water along with 371.382 kilolitres of kerosene.[137] Again, following the strike threat by Uttar Pradesh

Electricity Board Workers Union, the army took up the task of maintenance of essential services. In 1977, the army undertook the task of maintaining electric services at Allahabad between 7 to 11 November. Then at the request of the Government of Maharashtra in the wake of strike by civil employees engaged in certain essential services, the army's technical personnel manned the installations of milk supply at Bombay from 16 December 1977 to 8 February 1978, and at Pune from 21 December 1977 to 8 February 1978.[138] In August–September 1991, when the Himachal Pradesh State Electricity Board Employees went an strike, the army had to use their technical personnel for running the essential services.[139] Numerous such cases could be cited.

When the paramilitary formations fail in their duties the army is called in. In 1987, the Minister of State for Defence Arun Singh accepted the necessity of the army raising a permanent force for countering militancy. For tackling insurgency, General B.C. Joshi raised the Rashtriya Rifles in 1994. Within nine months a total of thirty battalions with ten Sector Headquarters comprising of 40,000 troops were raised by milking the existing army formations. Most of the Rashtriya Rifles personnel are from the army who are deputed for two to two and half years. Each Rashtriya Rifle battalion has 1,147 troops in six rifle companies consisting of soldiers drawn from the infantry. From 1995 onwards, the Rashtriya Rifles has been deployed in Srinagar and Doda.[140] The net result is lowering prestige of the civilian administration in the eyes of the people and the belief among military personnel that they could run the administration better. Such an attitude both on part of the civil society and the uniformed personnel might increase the probability of military coups in the long run.

In neighbouring Bangladesh, the antipathy between the military and the civilian rulers saw resulting in the former's rise to power accelerate in 1974, when the government used uniformed men in aid to civil tasks. Mujib deployed the army to stop hoarding of rice and jute. The army officers discovered that most of the Awami League politicians and their supporters were hand in gloves with the hoarders. When the army decided to take action against these corrupt persons, it was recalled to the barracks and the concerned officers cashiered. These frustrated and angry officers played a very important role in activating the coup of 15 August 1975 which resulted in the death of President Mujib.[141]

Throughout history, the political elite have maintained their control over the military machine by pursuing a policy of ethnic balancing. An ethnic group can be conceptualized on the basis of common origin and distinguished by skin colour, language or religion.[142] Cynthia H. Enloe rightly says that despite increasing internationalization, modernization and professionalization of the militaries, ethnic manipulations by the state's elites for furthering state security will continue.[143]

A serious defect of both the Indian and Pakistani armies is that these two organizations continue to remain over-dependent on the 'martial races', a colonial legacy. The British heavily recruited Punjabi Muslims and Sikhs from west and central Punjab respectively. The Punjabis constituted 60 per cent of the military personnel under the Raj.[144] After 1947, while the Indian Army remained over-dependent on the Sikhs, the Pakistan Army relied on the Punjabi Muslims from Attock, Rawalpindi and Jhelum districts of the Punjab.[145] More than 70 per cent of the personnel of the Pakistan Army hail from the Potowar region (Jhelum, Cambelpur and Rawalpindi districts of the Punjab and Kohat and Marolan districts of the North-West Frontier Province).[146] In 1979, the officer corps of the Pakistan Army was 70 per cent Punjabi. The North-West Frontier Province contributed 14 per cent, Sind 9 per cent, Baluchistan 3 per cent and *Azad* Kashmir 1.3 per cent personnel of the officer corps respectively.[147] Till 1971, though the East Pakistanis constituted 56 per cent of united Pakistan's population, in the Pakistan Army they numbered less than 7 per cent.[148] And this was one of the principal grievances of the East Pakistanis. At present, 60 per cent of the soldiers and officers of the Pakistan Army are from Punjab. The Pakistani politicians' attempt to raise the share of the Baluchis and the Sindhis have been obstructed by the Punjabi Muslim lobby within the army.[149] A tacit understanding between the civil bureaucracy and the army in Pakistan from the 1950s have further hamstrung the political establishment in Pakistan, an understanding facilitated by ethnic homogeneity, i.e. the domination of Punjabi Muslims of both these institutions.[150]

In 1981, though the Punjabis comprised only 2.45 per cent of India's population, about 15 per cent of the Indian Army's personnel came from the Punjab. The Sikhs are also over-represented in the officer cadre—in 1962, 40 per cent of the brigadiers and 40 per cent of the major-generals were Sikhs, and in 1991, the Sikhs constituted about 25 per cent of the officer cadre.[151] The proportion of the Sikhs is also much higher in the combatant branches—towards the end of the 1960s, the Sikhs in the armoured corps constituted 48 per cent of the personnel.[152]

Aid to civil tasks generally has negative impact on an ethnically unbalanced military. The most dangerous aid to civil tasks which threatened to tear apart the civil–military relations in India was Operation Blue Star. On 5 June 1984, Operation Blue Star was launched by the Indian Army to flush out the Sikh terrorists from the Golden Temple complex at Amritsar. The resultant tragic drama brought to the fore the dangers of the Indian Army remaining over-dependent on one community—the Sikhs, who constituted 10 per cent of the Indian Army at that time. Major-General Brar with commandos, infantry, mechanized infantry in armoured personnel carriers and tanks fought a fierce battle with Jarnail Singh Bhindranwale's henchmen inside the temple complex. After the news that the army had moved inside the holy shrine of the Sikhs broke out, the Indian Army was rocked by mutinies. On 7 June, 600 soldiers of the 9th Sikh Battalion mutinied at Ganganagar. Another mutiny occurred at the Sikh Regimental Centre at Ramnagar. There were also mutinies in one battalion of the Sikh Regiment deployed in the Jammu region and among the Sikhs serving in the Punjab Regiment in Pune.[153]

Maroof Raza rightly asserts that the more diverse an army is, the less chance of its political intervention.[154] Stephen P. Cohen says that new regiments should be raised from new materials for balancing the 'martial' classes. Lower caste Sikhs like the Mazhabis and the Ramdasias are now joining the army in larger numbers and they are organized into separate regiments like the Sikh Light Regiment in order to establish their own identity vis-à-vis the Jat Sikhs. Other low castes who could be considered are the Mahars of Maharashtra and the Chamars of Uttar Pradesh.[155] The number of such units must be increased, but whenever the government tried to follow such a policy, it faced obstruction from the Sikhs within the officer corps and the politicians hailing from Punjab. Some steps have been taken to widen the field of recruitment but they have proved to be largely insignificant. In 1971, the Indian government took the decision to raise a Naga regiment. Besides recruits from Nagaland, this unit also enlists men from other hill regions of India.[156] The number of such regiments ought to be increased gradually so the army is not dependent on any one community.

The emergence of militant Hinduism in civil society might also threaten the civil–military infrastructure in the near future. In other societies, rise of religious fundamentalism has resulted in the clerics calling upon the soldiers to disobey the normal chain of command on certain issues that are wrapped up in religious and national sentiments which has weakened the state's control over the army. For instance, in recent years, films and posters have appeared exhorting Israeli soldiers to disobey the government's orders to remove Jewish settlers from the occupied territories. In March 1994, three prominent *rabbis* published an edict calling on the Israeli troops to disobey orders to evacuate the Jewish settlers and alarmingly, there are ultra nationalist Israeli officers who might disobey orders to relinquish the territories to the Palestinian Muslims.[157] From the *yeshivots* (religious seminaries), Jewish religious leaders provide ideological justification for Israel's occupation of the West Bank. Another negative development is the rise of ultra orthodox parties in Israel.[158] The question is whether the national-religious troops could be trusted to remain loyal to the conventional secular military chain of command. And the spectre of religious mutiny is haunting the IDF.[159] It is to be noted that many Pakistani Army officers are religiously motivated to support Islamic insurgency in Kashmir.[160]

CONCLUSION

> If the situation is one of victory but the sovereign has issued orders not to engage, the general may decide to fight. If the situation is such that he cannot win, but the sovereign has issued orders to engage, he need not do so.
>
> SUN TZU[161]

Sun Tzu argues that the rulers (read politicians) should allow autonomy to the military persons operating as field commanders.

After the 1962 disaster, the politicians are not willing to interfere in the Indian Army's internal organization and operational plans, nor will the uniformed men tolerate such interference in case the people's representatives attempt to do so.[162] However, the dominance of the *neta–babu* raj at the strategic sphere is complete. The Vij affair shows how technology could be used by the politicians to tighten strategic control over the battlefield commanders. Civilian domination of the military in India is a shining example of the fact that apolitical military professionalism is not exclusive to Western Civilization.

However, in India, in the name of civilian control what exists today is bureaucratic control. The IAS officers have usurped the power of the politicians as far as day to day running of the defence ministry is concerned. The greatest hurdle to change in India is of course the bureaucracy. It is difficult even for powerful politicians to punish incompetent civil servants. After the 1962 debacle, the defence minister, the COAS and all the corps commanders resigned,[163] but not a single bureaucrat resigned or was forcibly pensioned off. For raising combat effectiveness of the Indian armed forces, the Indian government needs to transform the subjective civilian control with objective civilian control.[164] Some necessary steps for this transformation require setting up of a NSC with constitutional power and permanent members, and an organization of the CDS or JCS with an independent secretariat. These measures will to a significant degree result in the replacement of bureaucratic stranglehold over the military with proper civilian control. Instead of relying on IAS officers, Indian politicians should hire civilians from the technical branches and managers from the corporate sector for manning the MoD.

As a general principle, one can state that the use of the military in non-military civilian tasks is a sure path to disaster. One could argue that periodic deployment of the army for aid to civil tasks and their withdrawal reflects the strength of India's civil–military relationship. But, the Indian state requires to shore up professionalism of the Indian Army by preventing its frequent deployment for internal security duties. Despite the stultifying political control, it needs to be seen how far the army's professional officer cadre has been able to cope with command challenges in the battlefield.

NOTES

1. Samuel P. Huntington, *The Soldier and the State: The Theory and Politics of Civil-Military Relations*, 1957, rpt., Cambridge, Massachusetts: The Belknap Press, 1981, p. 80.
2. Eva Etzioni-Halevy, 'Civil-Military Relations and Democracy: The Case of the Military-Political Elites' Connection in Israel', *Armed Forces and Society*, vol. 22, no. 3 (1996), p. 401.
3. Eliot A. Cohen, *Supreme Command: Soldiers, Statesman and Leadership in Wartime*, London: Simon & Schuster, 2003, pp. 242–89.
4. Dario Canton, 'Military Interventions in Argentina: 1900–1966', in Jacques Van Doorn, ed., *Military Profession and Military Regimes: Commitments and Conflicts*, The Hague/Paris: Mouton, 1969, pp. 261–2.
5. John D. Martz, 'Contrasting Military Roles in Democratization: Colombia and Venezuela', in John P. Lovell and David E. Albright, eds., *To Sheathe the Sword: Civil-Military Relations in the Quest for Democracy*, Westport: Greenwood Press, 1997, p. 15.
6. B.J. Dudley, 'The Military and Politics in Nigeria', in Doorn, ed., *Military Profession and Military Regimes*, p. 211.
7. Jimi Peters, *The Nigerian Military and the State*, London/New York: I.B. Tauris, 1997, p. 9.
8. Astride R. Zolberg, 'Military Rule and Political Development in Tropical Africa', in Doorn, ed., *Military Profession and Military Regimes*, p. 178.
9. William Gutteridge, *Military Institutions and Power in the New States*, New York: Federick A. Praeger Publishers, 1965.
10. Constantine P. Danopoulos and Adem Chopani, 'Departyizing and Democratizing Civil-Military Relations in Albania', in Lovell and Albright, eds., *To Sheathe the Sword*, pp. 77–8.
11. Douglas Porch, *The Portuguese Armed Forces and the Revolution*, London: Croom Helm, 1977, pp. 22, 32.
12. Stephen P. Cohen, 'Issue, Role, and Personality: The Kitchener-Curzon Dispute', *Comparative Studies in Society and History*, vol. 10 (1967–8), pp. 337–55.
13. General Fred Woerner, 'Foreword', in Lovell and Albright, eds., *To Sheathe the Sword*, p. viii.
14. Brian A. Davenport, 'The Ogarkov Ouster: The Development of Soviet Military Doctrine and Civil/Military Relations in the 1980s', *JSS*, vol. 14, no. 2 (1991), p. 145.
15. David E. Albright, 'Democratization and Civil-Military Relations in Russia and Ukraine', in Lovell and idem, eds., *To Sheathe the Sword*, p. 35.
16. Andrew Scobell, *China's Use of Military Force: Beyond the Great Wall and the Long March*, Cambridge: Cambridge University Press, 2003, p. 53.

17. Apurba Kundu, *Militarism in India: The Army and Civil Society in Consensus,* New Delhi: Viva Books, 1998; Stephen P. Cohen, 'The Military and Indian Democracy', in Atul Kohli, ed., *India's Democracy: An Analysis of Changing State-Society Relations,* 1988, rpt., New Delhi: Orient Longman, 1991, pp. 120–3, 142–3.
18. Sumona Dasgupta, 'Militarization of the Indian State', in M. Raza, ed., *Generals and Governments in India and Pakistan,* New Delhi: Har-Anand, 2001, p. 50.
19. Rebecca L. Schiff, 'The Indian Military and Nation Building: Institutional and Cultural Concordance', in Lovell and Albright, eds., *To Sheathe the Sword,* pp. 119–30.
20. Rebecca L. Schiff, 'Concordance Theory: A Response to Recent Criticism', *Armed Forces and Society,* vol. 23, no. 2 (1996), pp. 277, 279.
21. Sanjay Dasgupta, 'Of Military Men and Politicians', in Raza, ed., *Generals and Governments in India and Pakistan,* pp. 127, 129.
22. M. Janowitz, 'The Military Professional', in Lawrence Freedman, ed., *War,* Oxford: Oxford University Press, 1994, p. 125.
23. Elizabeth P. Coughlan, 'The Changing Structure and Values of the Polish Military', in Lovell and Albright, eds., *To Sheathe the Sword,* p. 55.
24. Gutteridge, *Military Institutions and Power,* pp. 82–3.
25. P. Zhilin, 'The Armed Forces of the Soviet State: Fifty Years of Experience in Military Construction', in Doorn, ed., *Military Profession and Military Regimes,* pp. 167, 173–4.
26. Eric Carlton, *Militarism: Rule Without Law,* Aldershot: Ashgate, 2001, pp. 1, 4–5.
27. Stephen P. Cohen, *The Pakistan Army,* 1984, rpt., Karachi: Oxford University Press, 1993, p. 109.
28. Volker R. Berghahn, *Militarism: The History of an International Debate: 1861–1979,* Leamington: Berg, 1981, pp. 109, 119–20, 123.
29. Veena Kukreja, *Civil-Military Relations in South Asia: Pakistan, Bangladesh and India,* New Delhi: Sage, 1991, p. 27.
30. Etzioni-Halevy, 'Civil-Military Relations and Democracy', pp. 410–11.
31. Brian Loveman, '"Protected Democracy" in Latin America', in Lovell and Albright, eds., *To Sheathe the Sword,* p. 135.
32. Sven Gunnar Simonsen, 'Marching to a Different Drum? Political Orientations and Nationalism in Russia's Armed Forces', *The Journal of Communist Studies and Transition Politics,* vol. 17, no. 1 (2001), p. 45.
33. Vincent Davis, 'Foreword', in Lovell and Albright, eds., *To Sheathe the Sword,* p. xii.
34. Peter Karsten, 'The Coup d' Etat and Civilian Control of the Military in Competitive Democracies', in Lovell and Albright, eds., *To Sheathe the Sword,* pp. 157–9.

35. Michel Louis Martin, 'Operational Weakness and Political Activism: The Military in Sub-Saharan Africa', in Lovell and Albright, eds., *To Sheathe the Sword*, pp. 88, 96.
36. Jae Souk Sohn, 'Factionalism and Party Control of the Military in Communist North Korea', in Doorn, ed., *Military Profession and Military Regimes*, p. 284.
37. Jozef Graczyk, 'Social Promotion in the Polish People's Army', in Doorn, ed., *Military Profession and Military Regimes*, pp. 82–92.
38. Jacques Van Doorn, 'Political Changes and the Control of the Military: Some General Remarks', in idem, ed., *Military Profession and Military Regimes*, p. 17.
39. Pavel K. Baev, *The Russian Army in a Time of Troubles*, London: Sage, 1996, p. 22.
40. Bengt Abrahamsson, 'Military Professionalization and Estimates on the Probability of War', in Doorn, ed., *Military Profession and Military Regimes*, pp. 46–7, 50.
41. Major A.J. Wilson, 'Officer Production and Nationalization', *JUSII*, vol. LXXVI, no. 324 (1946), p. 451.
42. 'Matters of Moment', *JUSII*, vol. LXXVIII, no. 328 (1947), p. 391.
43. Ibid., no. 329 (1947), p. 598.
44. General Roy Bucher, 'Advice to Gentlemen Cadets', *JUSII*, vol. LXXVIII, no. 333 (1948), p. 331.
45. Major-General L.G. Whistler, 'A Story of Honourable Dealing', *JUSII*, vol. LXXVIII, no. 329 (1947), pp. 605–7
46. Lieutenant-Colonel J. Wilson Stephens, 'The Question', *JUSII*, vol. LXXVIII, no. 330 (1948), pp. 35–8.
47. Lieutenant-Colonel Rajendra Singh, 'Is Scientific Selection Successful?', *JUSII*, vol. LXXVI, no. 325 (1946), p. 336.
48. Lieutenant S.G. Chaphekar, 'A Frank Survey of India's Defence Problems-II', *JUSII*, vol. LXVIII, no. 328 (1947), p. 453.
49. The British-sponsored Martial Race theory laid down that only certain rural communities possessed martial traditions and being uneducated they were also apolitical. They constituted the best material for soldiers. The Jats, Sikhs, Punjabi Muslims along with the Gurkhas were martial classes for the British. For the Indian officer corps, the Raj during the Second World War had to depend especially on the university educated urban middle class who possessed political consciousness. David Omissi, *The Sepoy and the Raj: The Indian Army, 1860–1940*, Houndmills, Basingstoke: Macmillan, 1994, pp. 23–46; Lieutenant-Colonel Gautam Sharma, *Nationalization of the Indian Army: 1885–1947*, New Delhi: Allied Publishers Ltd., 1996, pp. 173–99.
50. Major Gurbachan Singh, 'The Right Type and Some Thoughts on Indianization', *JUSII*, vol. LXXVI, no. 324 (1946), p. 444.

51. Transcript of Oral History Interview with General J.N. Chaudhuri, New Delhi, 23 March 1973, by B.R. Nanda, Accession no. 426, NMML, New Delhi, pp. 16–17.
52. Lieutenant-General K.M. Cariappa, 'A Call to Duty', *JUSII*, vol. LXXVIII, no. 330 (1948), p. 6.
53. R. Prasannan, 'Colonial Hangover', *The Week*, 28 January 2001, p. 22.
54. Transcript of Oral History Interview with Lord Mountbatten, London, 26 July 1967, by B.R. Nanda, Accession no. 351, NMML p. 4.
55. Transcript of Chaudhuri, pp. 12–13.
56. Brian Robson, *The Road to Kabul: The Second Afghan War, 1878–81*, London: Arms and Armour, 1986, p. 55.
57. Stephen P. Cohen, *The Indian Army: Its Contribution to the Development of a Nation*, 1971, rpt., New Delhi: Oxford University Press, 1991, pp. 22–7.
58. General J.N. Chaudhuri, 'A Chief of Defence Staff?', in Ravi Kaul, ed., *The Chanakya Defence Annual: 1973–74*, Allahabad: Chanakya Publishing House, n.d., p. 26.
59. Chaudhuri, 'A Chief of Defence Staff?', in Kaul, ed., *Chanakya Defence Annual*, pp. 27, 29.
60. *Recruitment into the Officer Corps of the Armed Forces, Report of a Seminar held at the United Service Institution of India on 14 Feb. 1977*, New Delhi: United Service Institution, 1977, p. 5.
61. Daljit Singh and Katherine Singh, 'Decentralization and Diffusion of Power in the Army Structures of India and Pakistan', *JUSII*, vol. CI, no. 425 (1971), p. 316.
62. M.L. Sethi, 'Higher Defence Organization in India', *JUSII*, vol. CXX, no. 499 (1990), pp. 21–2.
63. Rathy Sawhny, 'Decision-Making for Defence', *Institute of Defence Studies and Analyses Journal* (*IDSAJ*), vol. 2 (1969–70), p. 59.
64. *MODAR: 1979–80*, pp. 4–5.
65. Brigadier H.S. Sodhi, *Top Brass: A Critical Appraisal of the Indian Military Leadership*, Noida: Trishul Publications, 1993, p. 22.
66. Smruti S. Pattanaik, 'Military Decision-Making in Pakistan', in M. Raza, ed., *Generals and Governments in India and Pakistan*, pp. 77, 80.
67. Major Karim, 'War on Terrorism: Strategic Diversion', *Aakrosh*, vol. 6, no. 18 (2003), p. 10.
68. P.R. Chari, 'Civil-Military Relations in India', *Armed Forces and Society*, vol. 4, no. 1 (1977), p. 6.
69. Transcript of Chaudhuri, pp. 17–18.
70. Major-General Sukhwant Singh, *General Trends: India's War Since Independence*, vol. 3, 1982, rpt., New Delhi: Lancer Publishers, 1998, p. 6.
71. The East Pakistani Rifles was a paramilitary unit whose duty like the Border Security Force of India was to guard the borders. Some 9,000 men from the East Pakistan Rifles joined the Mukti Bahini to fight the

occupying Pakistani regular soldiers. After the 1971 India–Pakistan War, the East Pakistani Rifles was reorganized and renamed the Bangladesh Rifles. P.B. Sinha, *Armed Forces of Bangladesh*, Occasional Paper, no. 1, New Delhi: Institute of Defence Studies and Analyses, 1979, pp. 1–4, 7, 33.

72. Mohammad Ayub Khan, *Friends not Masters: A Political Autobiography*, Lahore: Oxford University Press, 1967, p. 34.
73. General V.N. Sharma, *India's Defence Forces: Building the Sinews of a Nation*, United Service Institution National Security Paper, no. 13, New Delhi: United Service Institution of India, 1994, pp. 8–9.
74. Chari, 'Civil-Military Relations', p. 17.
75. B.N. Mullik, *My Years with Nehru: The Chinese Betrayal*, Bombay: Allied Publishers, 1971, p. 543.
76. Major-General Pratap Narain, *Indian Arms Bazaar*, Delhi: Shipra Publications, 1994, p. 63.
77. Mullik, *My Years with Nehru*, pp. 546–7.
78. Admiral S.M. Nanda, *The Man Who Bombed Karachi: A Memoir*, New Delhi: HarperCollins, 2004, pp. 118–9.
79. Sodhi, *Top Brass*, p. 24.
80. Sharma, *India's Defence Forces*, p. 10.
81. Ibid., p. 11.
82. Rakesh Datta, 'From Henderson to Subrahmanyam: Army to be Blamed. And Political Leaders?', *IDR*, vol. 15, no. 1 (January 2000), p. 21.
83. Chris C. Demchak, 'Numbers of Networks: Social Constructions of Technology and Organizational Dilemmas in IDF Modernization', *Armed Forces and Society*, vol. 23, no. 2 (1996), p. 186.
84. S. Kalyanaraman, 'Operation Parakram: An Indian Exercise in Coercive Diplomacy', *Strategic Analysis*, vol. 26, no. 4 (2002), p. 485.
85. Baev, *Russian Army in a Time of Troubles*, pp. 21, 24.
86. Joe, 'The Assam Rifles: An Appraisal', *JUSII*, vol. LXXXXV, no. 398 (1965), pp. 27–9.
87. K.P. Misra, 'Paramilitary Forces in India', *Armed Forces and Society*, vol. 6, no. 3 (1980), p. 372.
88. Rajesh Kadian, *India and Its Army*, New Delhi: Vision Books, 1990, p. 173; Lieutenant-General R.K. Jasbir Singh, ed., *Indian Defence Yearbook: 2003*, Dehra Dun: Natraj, 2003, p. 249.
89. E.N. Rammohan, 'Disquiet on the Western Front', *JUSII*, vol. CXXXIII, no. 552 (2003), p. 253.
90. R.K. Jasbir Singh, ed., *Indian Defence Yearbook: 2003*, p. 250.
91. Shirin Tahir-Kheli, 'The Military in Contemporary Pakistan', *Armed Forces and Society*, vol. 6, no. 4 (1980), p. 645.
92. Lieutenant-General M.L. Chibber, *Para Military Forces*, United Service Institution Paper, no. 4, New Delhi: United Service Institution of India, 1979, p. 17.
93. Scobell, *China Use of Military Force*, p. 36.

94. Personal communication with certain Border Security Force officers in Delhi in 2001.
95. Deba R. Mohanty, 'Defence Industry Conversion in China', *Strategic Analysis*, vol. 22, no. 12 (1999), pp. 1846, 1854.
96. Scobell, *China's Use of Military Force*, p. 67.
97. B. Raman, 'The Kosovo Tragedy and NATO's Psywar', *IDR*, vol. 14, no. 2 (1999), pp. 66–7.
98. Colonel Gurmeet Kanwal, 'Safety and Security of India's N-Weapons', *Strategic Analysis*, vol. 25, no. 1 (2001), p. 6.
99. Sanjay Dasgupta, 'Command and Control in the Nuclear Era', in Raza, ed., *Generals and Governments in India and Pakistan*, p. 117.
100. Shankar Roychowdhury, *Officially at Peace: Reflections on the Army and its Role in Troubled Times*, New Delhi: Viking, 2002, p. 276.
101. Kanwal, 'Safety and Security of India's N-Weapons', p. 9.
102. Dasgupta, 'Command and Control in the Nuclear Era', in Raza, ed., *Generals and Governments in India and Pakistan*, p. 121.
103. Jerzy J. Wiatr, 'Social Prestige of the Military: A Comparative Approach', in Doorn, ed., *Military Profession and Military Regimes*, pp. 75–8.
104. The Military Participation Ratio is the proportion of personnel in the military compared to the total population. For an elucidation of this concept see Stanislav Andreski, *Military Organization and Society*, 1954, rpt., Berkeley/Los Angeles: University of California Press, 1968, p. 33.
105. Scobell, *China's Use of Military Force*, p. 52.
106. Cohen, 'The Israel Defence Forces (IDF) From a "People's Army" to a "Professional Military"—Causes and Implications', *Armed Forces and Society*, vol. 21, no. 2 (1995), p. 237.
107. Gad Barzilai and Efraim Inbar, 'The Use of Force: Israeli Public Opinion on Military Options', *Armed Forces and Society*, vol. 23, no. 1 (1996), p. 69.
108. Stuart A. Cohen, *Towards a New Portrait of the (New) Israeli Soldier*, Ramat Gan: Bar Ilan University, BESA Centre, 1997, pp. 88, 98.
109. Lieutenant-Colonel C.L. Proudfoot, 'The Indian Soldier: Cornerstone of Democracy', *JUSII*, vol. C, no. 421 (1970), p. 349.
110. General Claude Auchinleck, 'The Future of India's Armed Forces', *JUSII*, vol. LXXVI, no. 322 (1946), p. 7.
111. Transcript of Oral History Interview with Prof. P.M.S. Blackett, NMML, Manuscript Section, Accession no. 284, p. 5.
112. K. Subrahmanyam, 'Decision Making in Defence', *IDSAJ*, vol. 2 (1969–70), p. 441.
113. *MODAR: 1977–78*, p. 4.
114. *Joint Warfare of the Armed Forces of the United States*, Joint Publ, 10 January 1995, p. 2.

115. Chaudhuri, 'Chief of Defence Staff?', in Kaul, ed., *Chanakya Defence Annual*, p. 32.
116. Rajesh Kadian, *India and Its Army*, New Delhi: Vision Books, 1990, p. 152.
117. Jeffrey Grey, *A Military History of Australia*, Cambridge: Cambridge University Press, 1990, p. 248.
118. Air Commodore Jasjit Singh, *India's Defence Spending: Assessing Future Needs*, New Delhi: Knowledge World, 2000, p. 72.
119. P.K. Vasudeva, 'The Role of the Chief of Defence Staff', *JUSII*, vol. CXXX, no. 542 (2000), pp. 565–7.
120. Kadian, *India and Its Army*, p. 150.
121. General J.N. Chaudhuri, *Arms, Aims and Aspects*, Bombay: Manaktalas, 1966, p. 30.
122. Lieutenant-General J.F.R. Jacob, *Surrender at Dacca: Birth of a Nation*, New Delhi: Manohar, 1997, p. 59.
123. Major-General S.N. Antia, 'Indo-Pak War 1971: Some Reflections', *JUSII*, vol. CII, no. 427 (1972), p. 111.
124. Brigadier N.B. Grant, 'Inter-Services Cooperation and All That: A Top Defence Management Myth', *JUSII*, vol. CVI, no. 455 (1976), pp. 309, 311.
125. Vasudeva, 'The Role of the Chief of Defence Staff', pp. 563–5.
126. Chaudhuri, 'Chief of Defence Staff?', in Kaul, ed., *Chankya Defence Annual*, pp. 33–4.
127. P.V.R. Rao, 'Governmental Machinery for the Evolution of National Defence Policy and the Higher Direction of War', *IDSAJ*, vol. 1 (1969), p. 9.
128. The military officers concerned were as follows: Lieutenant-General M.L. Thapan, Lieutenant-General Jaswant Singh, Deputy Chief of the Army Staff, Lieutenant-General M.M.L. Chabbra, Director Electrical and Mechanical Engineers, Army Headquarter, Brigadier Inder Sethi, Director of Organization, Army Headquarter, Rear Admiral M.S. Grewal, Assistant Chief of Personnel, Naval Headquarter, etc. *The Imposition of a Manpower Ceiling on the Army: Report of a Seminar at the United Service Institution of India*, New Delhi: United Service Institution of India, 1977, p. 5.
129. Baruch Kimmerling, 'The Social Construction of Israel's "National Security"', in Stuart A. Cohen, ed., *Democratic Societies and Their Armed Forces: Israel in Comparative Context*, London: Frank Cass, 2000, p. 218.
130. Mark A. Heller, *Continuity and Change in Israeli Security Policy*, Adelphi Papers, Oxford: Oxford University Press, 2000, pp. 52–3.
131. D. Shyam Babu, 'India's National Security Council: Stuck in the Cradle?', *Security Dialogue*, vol. 34, no. 2 (2003), pp. 216, 226.

132. Amitabh Mattoo, 'Raison *d'etat* or Adhocism', in idem and Kanti Bajpai, eds., *Securing India: Strategic Thought and Practice: Essays by George K. Tanham with Commentaries,* New Delhi: Manohar, 1996, p. 203.
133. Grant, 'Inter-Services Cooperation and All That', p. 311.
134. Ali A. Mazrui, 'Anti-Militarism and Political Militancy in Tanzania', in Doorn, ed., *Military Profession and Military Regimes,* pp. 226–7.
135. Sharma, *India's Defence Forces,* p. 7.
136. Jerrold F. Elkin and W. Andrew Ritzel, 'Military Role Expansion in India', *Armed Forces and Society,* vol. 11, no. 4 (1985), pp. 493–4.
137. *MODAR: 1999–2000,* p. 113.
138. *MODAR: 1977–78,* p. 135.
139. J. Baranwal, ed., *SP's Military Yearbook: 1992–93,* New Delhi: Guide Publications, 1992, p. 607.
140. Pravin Sawhney, *The Defence Makeover: 10 Myths that Shape India's Image,* New Delhi: Sage, 2002, pp. 149–52.
141. Sinha, *Armed Forces of Bangladesh,* pp. 14–15.
142. Marta Reynal-Querol, 'Ethnicity, Political Systems, and Civil Wars', *Journal of Conflict Resolution,* vol. 46, no. 1 (2002), p. 32.
143. Cynthia H. Enloe, *Ethnic Soldiers: State Security in Divided Societies,* Harmondsworth, Middlesex: Penguin, 1980, p. 234.
144. Cynthia H. Enloe, 'Ethnicity in the Evolution of Asia's Armed Bureaucracies', in idem. and Ellinwood, eds., *Ethnicity and the Military in Asia,* New Brunswick/London: Transaction Books, 1981, p. 9.
145. Ian Talbot, 'The Punjabization of Pakistan: Myth or Reality?', in Christophe Jaffrelot, ed., *Pakistan: Nationalism Without a Nation,* New Delhi: Manohar, 2002, p. 59.
146. Kaushik Roy, 'Good Governance Versus Bad Governance in South Asia: Civil-Military Relations in India and Pakistan (1947–2000)', *Asian Studies,* vol. 18, nos. 1–2 (2000), p. 92.
147. Cohen, *Pakistan Army,* p. 52.
148. Mohammad Asghar Khan, *Generals in Politics: Pakistan 1958–1982,* London/Canberra: Croom Helm, 1983, p. 18.
149. Maroof Raza, 'Generals and Government in India and Pakistan', in idem, ed., *Generals and Governments in India and Pakistan,* p. 18.
150. Ayesha Jalal, *The State of Martial Rule: The Origins of Pakistan's Political Economy of Defence,* 1990, rpt., New Delhi: Foundation Books, 1992, p. 298.
151. Apurba Kundu, 'The Indian Armed Forces' Sikh and non-Sikh Officers' Opinion of Operation Blue Star', *Pacific Affairs,* vol. 67, no. 1 (1994), pp. 48–9.
152. Margaret Macmillan, 'The Indian Army Since Independence', *South Asian Review,* vol. 3, no. 1 (1969), p. 53.

153. Mark Tully and Satish Jacob, *Amritsar: Mrs Gandhi's Last Battle*, Calcutta: Rupa, 1985, pp. 155–72, 192–7.
154. Raza, 'Generals and Government in India and Pakistan', in idem, ed., *Generals and Governments in India and Pakistan*, p. 19.
155. Stephen P. Cohen, 'The Indian Military and Social Change', *IDSAJ*, vol. 2 (1969–70), pp. 14, 22; A. Bopegamage, 'Caste, Class and the Indian Military: A Study of the Social Origins of Indian Army Personnel', in Doorn, ed., *Military Profession and Military Regimes*, p. 151.
156. *MODAR: 1970–71*, p. 16.
157. Steven R. David, 'Democracy, Internal War and Israeli Security', in Cohen ed., *Democratic Societies and Their Armed Forces*, pp. 282–3.
158. Heller, *Continuity and Change*, pp. 36, 45.
159. Stuart A. Cohen, 'The Scroll or the Sword? Tensions Between Judaism and Military Service in Israel', in idem, ed., *Democratic Societies and Their Armed Forces*, p. 258.
160. Pramod K. Kantha, 'Transition from Authoritarianism in South Asia: Pakistan's Renewed Hope for Democracy', in Lovell and Albright, eds., *To Sheathe the Sword*, p. 114.
161. Quoted from Mark McNeilly, *Sun Tzu and the Art of Modern Warfare*, Oxford/New York: Oxford University Press, 2001, p. 273.
162. Sharma, *India's Defence Forces*, p. 1.
163. Lieutenant-General E.A. Vas, 'National Security Council: Its Role and Functions', in *Field-Marshal K.M. Cariappa Memorial Lectures: 1995–2000*, New Delhi: Directorate General of Infantry in association with Lancer Publishers & Distrbutors, 2001, p. 142.
164. Subjective civilian control results in maximization of power of certain civilian groups vis-à-vis the military. In the Indian case the concerned civilian group is the Indian Administrative Service officers. Objective civilian control on the other hand allows space to the armed forces for their professional development. For a discussion on the concepts regarding these two types of civilian control, see Huntington, *The Soldier and the State*, pp. 80–3.

THREE

The Culture of Command in the Indian Armed Forces

Historical case studies show that in the battlefields, ideological beliefs tend to lose their motivational effects on the soldiers.[1] The combat effectiveness of a military machine is mostly a function of the leadership capabilities of the command mechanism.[2] The command apparatus is responsible for generating cohesion within the units. Sam C. Sarkesian, a military sociologist defines cohesion as 'the attitudes and commitments of individual soldiers to the integrity of the unit, the "will" to fight. . . . Unit cohesion, in the most simple terms, is *esprit de corps*.'[3] An Israeli military officer turned anthropologist Eyal Ben-Ari claims that unit cohesion is generated by the leadership emphasizing primordial bonds and a selected version of shared past.[4] Besides generating *esprit de corps*, command also involves managerial functions: directing and coordinating the military assets effectively. Management also includes assessment and dissemination of information for directing the military force.[5]

Thus, command culture involves a set of ideas, attitudes and institutions. The culture of command in the Indian Army remains a *terra incognita*. Stephen Rosen's categorization of the post-1947 Indian Army's command culture as sterile due to the banal influence of the caste system, is inadequate.[6] The focus of this chapter is on the Indian armed forces' command performance in conventional wars. The aim is not to deal with each campaign separately, but to analyse two aspects of command. The first section analyses the human factor in command-dealing with fighting troops and not the logistical apparatus. And the next section attempts to evaluate the elements of command required for mobile warfare in a high technology context warfare. In most of the wars fought by India against its neighbours, the Indian

Army played the most important part. Neither strategic bombing nor amphibious operations were crucial in these wars. The Indian Air Force (IAF) shall be referred to only when it impinged on ground operations. This chapter concentrates on military leadership both at the tactical and operational levels. The ambit of this study is quite broad and oscillates between the air-conditioned offices of the South Block to the foxholes in the snowy wastes of Siachen.

COMMAND: THE HUMAN DIMENSION

Carl Von Clausewitz focuses on intangible elements of warfare like courage, bravery, generalship, leadership capabilities, etc., that escapes quantitative analysis. All these intangible elements are connected, says Clausewitz, with human creativity. The human agency in general and the personality of the commander in particular are allowed decisive roles in Clausewitz's paradigm of warfare.[7] Modern warfare remains a conglomeration of small unit actions and in such scattered small combat, along with the regimental commanders, the platoon and the section leaders play a vital role. Man management is all the more vital in the armies of the democratic countries because unlike totalitarian countries like the erstwhile USSR or China, the military leaders of the democracies cannot display an indifferent attitude towards the lives of the military personnel under their command during battles.[8] Nor can the commanders of armies composed of volunteers use brute force in ensuring obedience among their men.[9]

In the history of Indian armed forces, the successful commanders have always recognized the worth of human factor in warfare. Lieutenant-General Harbakhsh Singh, the aggressive commander at North East Frontier Agency during the 1962 India–China War and commandant of the Western Command during the 1965 War with Pakistan, writes in his memoirs: 'Manpower is militarily our greatest asset.'[10] In a similar tune, Squadron-Leader K.N. Parik asserted in 1972, that whether it is guerrilla, conventional or nuclear warfare, man is the central figure.[11]

Proper command requires professionalism on part of the officer corps which means expertise in training and education for commanding the soldiers.[12] The professionalism of the Indian commissioned and non-commissioned officers (NCOs) is derived mostly from the professional ethos of British officers. Even after Independence in 1947, 300 British officers were retained for training the Indian officer cadre.[13] Just after the Partition, the Director of Military Training was Major-General Wilkinson.[14] This was so because before Independence, no Indian officer had risen above the rank of brigadier.[15] In 1947, only ninety-three officers had over thirteen years of service. According to Lieutenant-General S.K. Sinha, the Indian Army's command between 1947 to 1949 passed through a crisis.[16] Most Indian officers had served for only three to seven years and in order to fill the higher ranks many majors were promoted to major-generals within a year. The Indian Army got its first Indian Commander-in-Chief in 1949 with the appointment of K.M. Cariappa.[17] But shortage of staff officers forced dependence on the British officers and even after Independence, for seven years, Major-General W.D.A. Lentaigne commanded the Indian Staff College in the Nilgiris. He was assisted by a number of British instructors who trained officers for Grade 2 Staff appointments in order to make them fit for higher staff duties.[18] The staff officers' duty is to draft and transmit orders from the commander to their subordinates.[19] Staff officers are desk bound and deal with paper work; hence they lack the glamour of the fiery field commanders. Behind every successful commander, rightly says a British military historian, stands an energetic chief of staff.[20] Even in the 1970s, the Indian Army's link with the British Army continued. Many Indian officers of the rank of brigadier attended the Imperial Defence College Course to gain further competence in man management at the higher level.[21]

According to the British armoured officer-turned-scholar Richard E. Simpkin, the Anglo–American officers conceive war as a team game and the members playing the game (both officers and the subordinates) are bound by a code of written and unwritten rules.[22] How do the officers enforce such rules? Moral authority is required in order to make the men throw themselves against enemy fire. By displaying courage and bravery in the field of fire, the officers legitimize their command to lead the men in the combat zones. Soldiers, writes John Keegan, who are supposed to fight, must not feel that they might die alone in the battlefield. The commanders must share the risks and privations of the soldiers in order to retain the 'Mask of Command'.[23] Keegan goes on to say that the essential characteristics of military command should be inspirational leadership and complete disregard for the commanders' personal safety.[24]

The British majors and captains by leading the companies and the platoons[25] laid down the paradigm of 'heroic command'. Heroic leadership on part of the British officers in the Sepoy Army (British officered Indian Army of the pre-1947 era) involved risking their lives and limbs for the Indian subordinates. Thus, the British officers were able to establish bonds of loyalty with the Indian subordinates. Let us take a flashback. The scene is Italy in 1943. At Hangman's Hill near the town of Cassino, B Company of the 1st Battalion of the 9th Gurkha Rifles attacked the German trenches. A German shell hit Jemadar Kashiram Khatri. Despite heavy shelling and sniping by the *Wehrmacht* soldiers, Lieutenant Miles dug him out and saved the jemadar's life at personal risk.[26] Heroic leadership was not unique to the British officers. The German concept of frontline command as evolved during the early 1940 emphasized that the troop commander was supposed to lead the company by remaining within the battle group.[27] During the Second World War, the Japanese infantry fought tenaciously because their officers lead the attacks with their swords drawn, shouting *banzai*.[28]

Just before Independence, the editorial of the journal most widely read by the Sepoy Army's officers emphasized that the leaders should exercise authority by displaying heroic acts.[29] The Indian Army's officer corps absorbed the techniques of heroic command as practised by the British officers. The young and middle-aged Indian commissioned officers were urged to display gallantry in order to maintain their 'Mask of Command'. Field Marshal Sam Manekshaw who led the Indian Army to victory in the 1971 War with Pakistan writes regarding leadership: 'But to be frightened is one thing and to show fear is something different'.[30] General Shankar Roychowdhury, a retired Chief of Army Staff (COAS) points out that the Indian Army's manual on leadership states: 'Courage means being present at the point of maximum danger. The officers are supposed to show this sort of courage and not to exhibit fear.'[31] Lieutenant-General S.K. Sinha wrote: 'With my little experience of war, I am convinced that calmness and self-confidence during stress and strain are very important assets for a good leader. Self-confidence in a leader is contagious, it soon spreads among the led, and a self-confident fighting team is a battle winning factor.'[32] Lieutenant-Colonel S.C. Bhattacharya wrote in 1963 that 'Leadership must be earned, during training and in action.'[33] The situation is not much different in the IAF. Air Marshal Arjan Singh says: 'I would define an Air Force leader as a man whom his subordinates would follow with confidence and are prepared to undertake the most hazardous operations even at the risk of supreme sacrifice. . . . It is of utmost importance to display personal leadership by exposing oneself to danger first of all.'[34]

The Indian Army's privates fought with courage and determination because of the heroic leadership displayed by the officers. And statistics proves that the Indian officers were willing to fight and die. In the wars that were fought by the Indian Army after Independence, the proportion of casualties amongst officers had been much higher than the *jawans*—the proportion being 5:3. An officer corps which is willing to die while leading the men into battle, writes M.L. Chibber, a retired major-general of the Indian Army, enjoys the confidence, affection and respect of the troops.[35] Lieutenant-General P.S. Bhagat calculated that during the 1965 War with Pakistan, the Indian Army lost 80 commissioned officers and 1,253 privates. The proportion came to about 1:15, a high proportion considering that normal officer–privates ratio was 1:60 in the battlefield.[36] One of the principal factors behind the superb performance of the Israeli Defence Force during the Sinai Campaign and the 1967 Six Days War was the courageous attitude of its officer corps which suffered heavy casualties. [37]

Some examples can be given of commanders of the Indian armed forces who were able to extract loyalty from their subordinate officers and the privates through their personal heroism. Even very senior officers at times exhibited themselves in the frontline during emergencies to prevent collapse of the war machine. On 26 May 1948, a lone Dakota flown by Air Commodore Mehar Singh and carrying General K.S. Thimayya, General Officer Commanding Srinagar Division landed at the small landing strip of Leh.[38] The objective of these two senior commanders was to display their solidarity with the privates stationed in the snowy wastes. The junior-most commissioned officer also displayed courage. During 1971, the 4th Kumaon Regiment was deployed in the Sylhet district. Between 24 and 28 November, they clashed with the Pakistanis. When Second-Lieutenant R.S. Sandhu commanding Number 4 Platoon of the B Company saw that an enemy light machine gun in a bunker was holding up advance, he crawled towards it. Though injured in the chest, he threw a hand grenade that finally silenced the enemy gun. Seeing this spectacle of heroism, Subedar Shib Charan rushed forward and blew up another enemy automatic with grenades. Risk-taking attitude of the officers created bonds of affection between them and the privates. On 10 December 1971, 14th Kumaon Battalion engaged the Pakistanis at the strongpoint of Laksham. In the morning when a Pakistani havildar belonging to 25th Frontier Force Regiment suddenly came out of his hideout and aimed his sten gun towards Major Jesudas, *jawan* Harak Singh risking his own life jumped in front of the Pakistani to cover the major and shot the enemy dead.[39] During the 1999 Kargil War, many young and mid-level officers led from the front and suffered casualties. For instance Major M. Saravanan of the Bihar Regiment commanded a platoon of thirty men in the Batalik sector. On 29 May 1999, he led an attack and was moving ahead of the privates when he fell victim to the machine gun fire of the intruders.[40]

Besides showing gallantry in order to encourage the men in the field of fire, what other techniques did the officers have up their sleeves? At times, the officers appeal to their subordinates' ego in order to get difficult tasks done. In August 1965, the 31st Light Regiment (Artillery) was deployed at Naushera. News came that some Pakistani soldiers had infiltrated along the Naushera–Rajauri highway. Since no infantry was available, men from the regimental headquarter of the 31st Regiment along with some personnel provided by the brigade headquarter were assembled. The regimental commandant's Adjutant, Captain R.S. Garewal volunteered to evict the infiltrators. As the mixed party advanced, the Pakistanis opened up with machine guns. Then the Indian party hesitated to advance. Garewal then abused them in Punjabi for their 'cowardice'. As a result, the men started moving forward firing from their hips. This time the infiltrators ran away.[41]

Complete trust between the higher commanders and the privates is necessary for smooth functioning of the command apparatus in the field of fire. How do the commanders acquire such trust besides showing gallantry and appealing to the soldiers' honour? The British military historian David French suggests that soldiers derive a sense of security from knowing their officers.[42] For this, the commissioned officers have to identify themselves with the men under their command. The officers have to live like common soldiers during the campaign and march on foot with the men and talk to them in the evening around the campfire.[43] In the nineteenth-century US Army, the efficient colonels had to be fair and caring towards the men under them and required oratorial skills for inspiring the soldiers. Generally unpretentious commanders acquired popularity with their men. To identify themselves with the soldiers, the officers were instructed not to go for showy dress while campaigning.[44] These egalitarian actions generated a bond of loyalty between the officers and the men. In the IDF, all the soldiers and the officers serve under similar general conditions: they have the same beds and barracks, similar food and clothes and the same kind of furlough (paid leave). This is noteworthy because an IDF battalion is generally composed of socially heterogeneous elements like students, garage mechanics, managers, store owners, farmers, government clerks, salesmen, electricians, house painters, etc.[45] In contrast, the officers of the Italian Army during the Second World War had better food and alcohol and more leave compared to the private soldiers. This prevented unit cohesion resulting in low combat effectiveness.[46]

The Soviet armed forces' disciplinary structure suffered due to bullying of the junior recruits by the NCOs.[47] However, even the commanders of the armies of the totalitarian states did not depend merely on brute force. The Imperial Japanese Army believed that military discipline could only be sustained if a homelike warm and fraternal atmosphere operated within the units. In this army, the company commanders behaved as fathers and the NCOs as older brothers towards the rank and file.[48] In the Sepoy Army, the commandant of a regiment was a colonel who was a father figure to the men under him.[49]

As regards the divisional commanders, Lieutenant-General P.S. Bhagat wrote in 1971:

> Command Mastery is the state achieved by the Commander wherein his personality and acumen permeates through each sinew of his command and reflects in its make up. . . . Personal Command is the first ingredient

of Command Mastery. The next ingredient of Command Mastery is Command relationship. In the main, it is the superior Commander winning the confidence, respect, loyalty and faith of the subordinates. The esteem of the subordinates for their Commander need not be idolatory, but greater the degree of admiration and affection there is, the more close knit will be the Command and the more ready it will be to go through 'hell and fire' to execute the orders of the Commander. Command relationship is a very intangible quality.[50]

According to Brigadier H.S. Sodhi, the Indian soldiers did not fight so much for the abstract notion of patriotism but for personal loyalty to their commanders. Sodhi was a brigade commander in Bangladesh during the 1971 War. On the night of 4–5 December, when the General Officer Commanding of the two brigades came forward to meet the men, there was jubilation among the latter.[51]

Lieutenant-General S.K. Sinha points out about junior officers:

> As a junior commander one can by sympathetic understanding and genuine endeavour, offset some of the handicaps of man management on account of hardships, isolation and lack of recreation. A few measures which may gain good results are as follows: (a) Officers at company level must always live on picquet with their men and share the hardships of picquet life. A company may be split into platoon picquets but it is better for the company commander to stay with his platoons by turn. . . . Officers must frequently visit picquets.[52]

The moral bond which binds the officers and the soldiers and the affection of the commanders necessary for generating such a bond have been emphasized by several Indian officers. Major K. Brahma Singh had written in 1969:

> Sincerity and love which form the moral bond between the leader and the led is gauged by the amount of sacrifice a leader can make for his men. Considering what a leader expects his men to perform for him, no amount of sacrifice on the part of the leader can be considered too great. Apart from the fact that they trustingly hand over their lives into the hands of their leader, military discipline binds them to such an extent that they literally eat, talk, laugh, sit, stand, and rest only when their leader permits them to do so. This in itself should be sufficient to arouse motherly affection for the men in the heart of the leader. A leader's sacrifice for the men should, therefore, be of a motherly nature. What would a mother not do for the sake of her children? While living with his men he must deny to himself what is being denied to his men. The effect of this self-denial is quite out of proportion to the sacrifices made.[53]

It is necessary that the officers remain in the battalions for long periods to gain personal relationships with the privates. To give an example from the Indian Army, F.L. Freemantle joined the 3rd Battalion of the 5th Gurkha Rifles in 1947 and became the Lieutenant-Colonel of this unit in 1960.[54] Besides the infantry regiment, this tradition also exists in the armoured units. In 1969, Second-Lieutenant Mahesh Chander joined the 69th Armoured Regiment straight from the Indian Military Academy and twenty years later, he commanded this unit.[55] But, during the US Army's war in Vietnam, an officer remained in a unit only for a few months because of the rotation system. The absence of personalized relationship between the officers and the privates resulted in command failure. Absence of cohesive bond resulted in poor training and morale. As a result, about 20 per cent of the American casualties were due to friendly fire and accidents.[56] Even before Vietnam, in Korea between 1950–3, combat effectiveness of the US Army was quite poor. This was because the units were skeletal structures brought to strength by individuals posted from all over the US Army and replaced as death and wounds demanded. Thus, the men in the firing line were strangers to one another.[57]

The battlefield commander should not be a distant figure for his men. During the 1947–8 India–Pakistan War, Cariappa constantly toured the forward areas in order to stamp his personality on the frontline troops and to convince the men in the Forward Edge of the Battle Area (FEBA), that the high command had not forgotten them.[58] K.N. Parik observes that if the men are kept well informed by their commanders then fear can be obliterated from their mind.[59] General K.V. Krishna Rao became the COAS (1981–3) and then the Chairman of the Chiefs of Staff in 1983. He was able to exercise personalized command with a human touch.[60] In 1971, while functioning as a divisional commander, Krishna Rao practised what could be categorized as 'humanistic approach to command'. Krishna Rao deliberately fostered familiarity with the combatants in order to project himself as one of them, to understand their emotion and to earn their loyalty. He narrates in his memoirs: 'Over a cup of tea with the men, I was very much encouraged to find that their morale was on top of the world and that they were looking forward to battle. I made this a practice with the other formations also and addressed every unit before it went into battle'.[61]

The man management technique in the Pakistan Army was not much different from that of the Indian Army. A.A.K. Niazi (later Lieutenant-General) while conducting the Battle for Zafarwal during the 1965 India–Pakistan War, writes about his leadership techniques in the following words: 'One company of 3rd Battalion of the 14th Punjab Regiment was with me and was told to move up and join 4th Frontier Force at Zafarwal. I accompanied them. The officers and the *jawans* were in high spirits. After having a cup of tea with them, I left for my brigade headquarters.'[62]

Men fight better if they feel that their commanding officers really care for them. The commanders have to be in touch with the day-to-day problems of the rank and the file in order to display 'humane personalized command'. It is not possible to treat the men especially in a democratic polity as cannon fodder. Those generals who plan the battles aiming to minimize casualties and hardships of the combatants are able to earn the loyalty of the common soldiers.[63] The Sepoy Army emphasized medical management during campaigns. During the re-conquest of Burma in 1944–5, mobile surgical units and neuro-surgical units were introduced.[64] And the Indian officers who served in the Sepoy Army introduced this trait in independent India's Army. The successful Indian officers displayed genuine concern for the men's welfare. Harbakhsh Singh Commander of the 1st Battalion of the Sikh Regiment during the 1947–8 India–Pakistan War ordered all his company commanders that in every action no matter, however, small, a medical officer with bottles of plasma should accompany the troops. Thus, many lives were saved when soldiers were wounded in action at Tithwal.[65] For raising the morale of the frontline soldiers and to retain their loyalty, Krishna Rao during the 1971 India–Pakistan War used to visit the Advanced Dressing Stations in East Pakistan where the doctors treated casualties fresh from the battle. Rao met a soldier whose stomach was ripped open by shrapnel, but who was still conscious. Rao tried to comfort him by saying that he would personally see to his medical treatment.[66]

The subordinates are not passive onlookers. How they view

their commanders can be gleaned from the nicknames which they give to their leaders. Within an IDF unit, the unit commander is often called by a special name.[67] In the Indian Army, L.P. Sen was known for his indifferent attitude towards the men under his command. His nickname was Bogey Sen, a diminutive of Bogus Sen according to Harbaksh Singh. In contrast, the bearded Wing Commander Mehar Singh who cooperated with the ground troops during the 1947–8 India–Pakistan War earned the nickname of 'Mehar Baba'.[68]

When the senior commanders remained too far away from the field of carnage, disaster caught up with the Indian Army. Just before the Chinese attack on India, General P.N. Thapar was the COAS and Lieutenant-General B.M. Kaul was the Chief of General Staff (CGS). The Eastern Command's Headquarter was at Lucknow and northern Uttar Pradesh and north-east India were under operational command of the Eastern Command. Lieutenant-General L.P. Sen was the General Officer in-charge of the Eastern Command.[69] During the 1962 India–China War, Lieutenant-General B.M. Kaul was given command of the IV Corps in NEFA and he commanded the corps from his sick bed in Delhi. This senior officer was too far away from the front to be of any use. Complete breakdown of communications within the command hierarchy occurred. Major-General Niranjan Prasad, Commander of the 4th Division (which was under the IV Corps) was not even aware of Kaul's whereabouts when the latter went to Delhi. The scenario reminds one of Field Marshal Douglas Haig's *Chateau* Command in France during the First World War. Haig remained a distant figure to his corps commanders who actually ran the trench battles.[70]

Major-General D.K. Palit, the then Director of Military Operations vividly describes the 'command vacuum' in the following words:

First report of the Chinese attack began to arrive . . . early on the morning of 20 October. Niranjan Prasad decided to remain at Zimithang that night. . . . On the morning of 21 October Niranjan Prasad at last decided to move back to Towang. . . . With Niranjan was Brigadier Kalyan Singh, Commander of the Artillery Brigade of 4th Division. . . . Since the laborious plod from Zimithang to Towang would take the best part of two days, this meant the acting Corps Commander,

> the GOC 4 Infantry Division and the local Brigade Commander would all be temporarily *hors de combat* and out of communication with everyone till they reached Towang. There would be no one available to co-ordinate the defence of Kameng during the next two crucial days . . . for the time being there was no one between Lucknow and the battlefields of NEFA to take charge of operations against the enemy.[71]

18 November 1962 was a black day for the Indian Army. Both the General Officer Commanding 4th Infantry Division and the Commander of the 62nd Infantry Brigade left quite early without making any tactical plan of withdrawal from Dirang Dzong. The units were left to fend for themselves resulting in the disintegration of the infantry brigade.[72]

The scene was similar to the command vacuum of Army Group OB West on 22 June 1944. With Erwin Rommel in Germany, *Generalfeldmarshall* Von Rundstedt in Paris and the local commanders engaged in playing a war game at Rennes, there were virtually no senior officers around to control operations on the German side in Normandy.[73] But, Normandy was an exception. In general, the *Wehrmacht* emphasized frontline command, i.e. leading from the front even for the higher commanders. The best example is Colonel-General Heinz Guderian commanding his *panzer* group from his half truck during Operation Barbarossa in the summer of 1941.[74]

An aggressive command culture was displayed by Sam Manekshaw in order to inspire the subordinates and to raise the morale of the command organization in midst of a crisis. Facing disaster, the Indian high command decided to replace the 'Sick General' Kaul with a 'Fighting General'. On 28 November 1962, Lieutenant-General Manekshaw took over the command of IV Corps. Manekshaw entered the conference room with his usual jaunty steps and observed that all his subordinates were looking at him. Manekshaw said, 'Gentlemen, I have arrived! There will be no more withdrawals in IV Corps, thank you.' Then he walked out.[75] This sort of stunt was similar to that practised by *Generalfeldmarshall* Walter Model. To give one instance of Model's leadership style, during the Battle for Arnhem Bridge in late 1944, when Lieutenant-General Wilhelm Bittrich commanding II SS *Panzer* Corps wanted to retreat after destroying the bridge, Model coldly said to him 'I want that bridge'.[76]

Those Indian officers who reached the starry ranks took care to talk with the Junior Commissioned Officers (JCOs) in order to get a 'feel of the battlefield'. The JCOs and the NCOs are very important because small unit level actions depend on them.[77] The British officers of the Sepoy Army realized the importance of the Viceroy's Commissioned Officers (VCOs)[78] and the NCOs. The VCOs in the post-1947 era became the JCOs. Just one year after India's Independence, Colonel C.W. Morton emphasized the importance of these officers in man management[79] and the Indian Army has taken care to raise the status of the JCOs. The term jemadar in civilian society is still in use to designate a sweeper who usually is an untouchable. So, on 1 September 1965, the designation of jemadar was changed to naib-subedar.[80]

One group within the Indian Army was for abolition of the JCOs. They argued that their original function was to act as a link between the British officers and the Indian privates. This was necessary because the white officers' control over the vernaculars was shaky. So, the British officers' orders were interpreted by the VCOs and then passed down to the privates in their vernaculars.[81] In post-colonial India Hindi is the required language for all military officers and this has resulted in greater interaction between the officers and the men.[82] And in the changed circumstances, the JCOs were probably no longer that important. In 1972, Major A.S. Ahluwalia argued that JCOs because of their advancing age and lack of education were incapable of the stress and strain of warfare. Ahluwalia was for replacing the JCOs with the NCOs and giving the latter better pay and service conditions. Ahluwalia considered this necessary because the Indian infantry is generally deployed in jungle and mountainous terrain where the units have to fight in small groups. This required decentralized command and much depended upon the NCOs, the commanders of small unit teams.[83] But, Manekshaw, when he was the COAS (8 June 1969–14 January 1973) resisted this proposal arguing rightly that the post of the JCOs were necessary as these offered some promotion facilities to the privates.[84]

When commissioned officers become casualties, the JCOs had to take on the mantle of leadership. These small unit leaders had to display courage in order to motivate the privates to maintain pressure against the enemy in the FEBA. The colonial military system trained the VCOs to take command of the companies when white officers became casualties and the Indian Army carried on this tradition.[85] In the Sepoy Army, the havildars commanded platoons. In the Indian Army the JCOs are the leaders of the platoons and sections. A section is composed of eleven *jawans* and a platoon is composed of three sections.[86] When captains commanding companies of the Indian Army die or suffer grievous wounds, the subedars are trained to take over command.[87]

During the 1962 War, the Indian politico-military high command bungled badly and the higher command apparatus of the Indian Army just melted away in the high Himalayas. But thanks to the JCOs, the soldiers fought well at the unit and sub-unit level, though disaster occurred at the strategic-operational level. To give an instance, in the Battle of Chusul at Ladakh, a company of the 13th Kumaon Battalion was defending Rezang La. When the Brigade-Major Jagjit Singh visited them, Naib-Subedar Surja told him: '*BM Sahib, ham paltan ki izzat ke khatir, akhri dam tak ladte rahenge*' (Brigade-Major Sahib, for the prestige of the regiment we will fight till death). Two days later the Chinese attacked. Surja died keeping his pledge. Out of a total of 118 men at that post, 106 died.[88]

The JCOs taking over the leadership during crisis was evident in the 1965 War. On 6 September 1965, the 3rd Jat was advancing towards Ichogill Canal. Before the town of Batapur, the C and D companies of the battalion were caught in the open by Pakistani artillery and mortar fire. For a moment, the Indian soldiers were numb with fear and even commissioned officers were confused. At that critical juncture, Subedar Sarup Singh, a short man shouted: '*Kuch nahin hua, Aage bado, tez kadam se*' (Nothing has happened, move ahead quickly). The 3rd Jats re-formed, then set up 2-inch mortars and finally overpowered the Pakistani defenders. Then the Indian unit advanced towards Batapur and the subedar-major performed the task of rounding up the dead and wounded and sending them to the rear.[89]

Episodes from the 1971 War illustrate how heroism of the JCOs galvanized their *saathis* into a battle-hardened team. The Naga Regiment was part of the 4th Mountain Division of the II Corps. In November 1971 they were moved to Shakarpur in Nadia district. On 12 November, two Pakistani companies made an incursion across the border and the Naga Regiment ambushed them. The Pakistanis made a desperate attempt to break the ambush. Havildar Sangram Singh, a Medium Machine Gun Section Commander engaged the Pakistanis at close range. When his machine gun was shot up, he ordered his crew to withdraw while he fought the enemy with hand grenades. In the hand to hand fight he was seriously wounded but finally was able to rejoin his party. When the ambush party was withdrawing, one of the sections got isolated and the section leader Naik Moti Ram laid down his life fighting. The naik set an example of fighting to the last bullet. Encouraged by his example, the privates in the section also held out as long as their ammunition lasted. On 12 December 1971, the 9th Kumaon's A Company was engaged by the Pakistanis at the Lamai sector. Both Major Jesudas and his second-in-command Captain J. Kataria were wounded. Then, the command of the company fell on Subedar Gambhir Chand and he did a fine job.[90]

MANAGERIAL LEADERSHIP IN MOBILE WARFARE

Command: The Technical Angle

How far is the Indian Army's leadership prepared to wage a technology oriented mobile war in an age where technology is becoming increasingly important. Let us assess its performance in the past before making any deductions. One of the basic weaknesses of the French military organization during the Second World War was the overemphasis on classical education and neglect of engineering and technological knowledge. Only after 1960, the French Army officer corps moved towards a more technocratic concept of warfare.[91]

During the Spanish–American War (1895–8), the US Army's lieutenants used couriers for communicating with the senior officers while coordinating their units for advancing against the enemy.[92] Towards the last decades of the nineteenth century, field telegraph replaced the semaphore and the couriers as communication links in the battlefields.[93] During the Second World War, battlefield commanders maintained command through radio and telephone.[94] By the end of World War II, company commanders in the Sepoy Army communicated with battalion commanders through wireless sets manned by the

Madrasi signallers. When all land communications were cut, wireless allowed the higher commanders to know about the condition of their companies, their positions, the reinforcements and supplies they required, etc. The brigade headquarter used Walkie Talkies (R/T sets) for communicating with its subordinate formations.[95] These apparatus for controlling the troops exist even now. As early as 1948, one officer of the Indian Army pointed out the necessity of possessing advanced communications equipment which would enable the commandants at the regimental centres to push reinforcements quickly to the commanders leading troops in the battlefields.[96]

The PLA between 1950 and 1953 in Korea used telephones for communication below divisional level.[97] Let us have a look at the communication system of the Indian Army during the 1962 China War. The communication between the General Officer Commanding 4th Infantry Division at Dirang Dzong and the corps commander occurred through telephone. However, the network of communication was not all that good. On 18 November 1962, when the brigade column composed of the 2nd Sikh Light Infantry Regiment leading, the 4th Garhwal Rifles in the centre and the 13th Dogra Regiment at the rear was retreating from Se La, they were ambushed by the Chinese armed with medium machine guns. The 13th Dogra was attacked from the rear. All coordination between the three regiments was lost and they could not cooperate with each other because of the complete absence of wireless communication between brigade headquarter and the units.[98] In the 1960s the IDF in the Sinai Peninsula and the US Army in Vietnam used helicopters as command vehicles.[99] Helicopters have been integrated in the command apparatus of the Indian Army also. In the Siachen glacier, the artillery observers use helicopters for communicating with the infantry commanders while targeting the Pakistani concentrations.[100]

Computer aids in the acquisition and organization of detailed information by a commander thus reducing the reaction time of the command to an emerging threat. Automation started quite early in the US armed forces. In 1964, the US Department of Defence spent $6 billion on the acquisition, operation and maintenance of computers for the armed forces.[101] In the 1990s, computerized fire control and ammunition management systems were introduced in the US Army to speed up the delivery of fire by the weapons platforms and to reduce wastage of ammunition.[102] Under the leadership of the American armed forces, the Western militaries are currently undergoing the process of digitization—a process by which all elements in the battlefield are bound together by an electronic web that permits rapid transfer of data to commanders, so that they can take decisions and issue orders quickly.[103] This is necessary since highly precise weapons require extensive and accurate intelligence data to be effective. And near real time intelligence requires exceptionally reliable communications.[104] The latest tanks possess infrared or thermal imaging night sights, laser range finders and ballistic computers for fire control, enabling the tank commanders to open accurate fire on a target within seconds of sighting it. A new generation of battlefield surveillance devices has extended the commander's reach. Foliage penetrating radars are now capable of detecting troops hidden among trees and hedges.[105] And all these systems are computer operated.

Till the 1980s, the computerization programme in the Indian Army was *ad hoc* with no systematic long-term plans. Moreover, there was too much centralization. Computerization was concentrated at the Army Headquarter and the lower commands were kept in the dark. There is still no General Staff publications regarding the history of computers and the fields in which it is going to be used. Even small formations need computers. But, the JCOs with only secondary school level education are not equipped to function as operators and specialized training courses for them do not exist.[106]

After the US–NATO air war over Kosovo, the Indian military realized the importance of the Unmanned Aerial Vehicles (UAVs). Non-lethal UAVs are intended to provide blanket imagery collection for tactical and operational commanders in conjunction with other manned and satellite intelligence collection systems. The UAVs are used for surveillance of the terrain before the dispatch of attack helicopters into very high-risk environment.[107] The Indian military is thinking of inducting UAVs in the near future for aerial surveillance to aid the ground-troop leaders.[108]

In order to outfight the enemy it is necessary to think faster than one's opponent and to misinform the enemy. The aim is to implant a picture of defeat in his opponent's mind by manipulating information.[109] Advanced technologies enable the Americans to exploit the information differential, i.e. superior access to and ability to effectively employ information at the strategic and tactical planes. The US Army's Military Satellite Communication Programme provides the commanders with secure strategic communication anywhere in the world. Behind enemy lines, Man Portable Rapid Burst Satellites are used by the Special Forces, like Britain's Special Air Service Commandos, to protect the latter from the enemy's direction-finding equipment.[110] The US Space Command by contributing in the spheres of mapping, charting, geodesy, oceanography and terrain analysis assist the ground force commanders in information processing, navigation and procurement of intelligence.[111] The United States Air Force (USAF) is moving ahead in establishing its Integrated Space Command and Control. This will result in integration of 40 per cent of air, space and missile defence command and control systems into a single communication architecture to provide a common global picture with real time information. This mega system will aid the USAF for integrated air and space contingency planning and for inserting space forces into the Joint Forces Air Component Commander's campaign planning and execution.[112]

Technology has helped in bringing the senior and subordinate commanders closer to each other. Because of digital technology, television has made a giant technological leap since the days of the Vietnam War. Television cameras no longer have to shoot films which are required to be edited before being played on national broadcasts. Now the feed can be shown alive. So, the decision maker at the centre can see the battlefield as well as the theatre commander. The Chairman of the Joint Chiefs of Staff General Colin Powell in 1989, when the US forces invaded Panama, could keep a close watch on the Theatre Commander General Max Thurman through television cameras.[113]

Technology has solved some problems but at the same time has given birth to new obstacles. The rapid advance in information technology has made more information available to the commander leaving him with less time to shift through volumes of data in a quickly changing battlefield.[114] The commanders' ability to manipulate the mass of information within a short span of time and at the same time avoiding too much interference from above depends on the conceptual aspects of command.

Command Concepts for Manoeuvre Warfare

Manoeuvre warfare in recent times, writes Lieutenant-Colonel Clayton R. Newall of the US Army, will aim at surgical strikes with high technology precision weapons in order to make the conflict short. Nevertheless, chaos and friction will characterize such operations. Hence, the outcome of such wars will be unpredictable and uncertain.[115] In manoeuvre warfare, the emphasis is to dislocate the enemy by aiming at its command and control mechanisms.[116] In such combat, asserts the American military theorist William S. Lind, the officers will be frequently surprised by unexpected enemy moves. The officers must learn to handle surprise by using fluid tactics that will not conform to any existing rules and regulations. Battlefield fluidity is to be achieved by frequent shifting of the principal effort during the course of an action.[117] Great military leaders have always harped on the importance of fluidity and friction in warfare. Napoleon Bonaparte emphasized the necessity of uncertainty in mobile warfare. He said that the generals must be ready to change their plans any number of times in rapidly changing contexts. The commander should think about the endless possibilities open to the enemy and prepare himself for responding to any probable situation.[118] In other words, Napoleon meant that prepared plans are not going to work amidst the heat, dust and the resultant fog of mobile warfare.

Interpreting Napoleonic Warfare, Clausewitz has written that the battlefield will be filled with blood, fear and terror. And the ensuing psychological fog would engulf all the participants. The messiness of the battlefield results in the breakdown of the chain of communications and ensures confusion in command. In such circumstances, argue the German interpreters of Clausewitz, the individual commanders must not look behind their shoulders to look for orders from those higher up. Every field commander must display individuality and flexibility. The German command method emphasizes the importance of individual initiative even at the lowest level, a command philosophy known as *Auftragstaktik* (mission oriented command or directive control). This concept enabled the German Army to perform in an excellent manner during the two World Wars.[119] During World War II, in the Eastern Front, the mission-oriented command was one of the principal factors that allowed the *Wehrmacht* to inflict thrice the number of casualties on the numerically stronger Red Army.[120]

By 1988, the Soviet military theorists accepted that due to the fragmented nature of battle, and that there would be overlapping of the units and the sub-units within the zone of combat. The units have to operate independently for a long time[121] and for this to be successful, the local commanders on the spot must possess freedom to determine the course of action without referring to the higher commands. In order to out manoeuvre the enemy, a high tempo which means the rate of rhythm of activity relative to the enemy, needs to be maintained.[122] High tempo refers to the ability to act faster than the enemy can react.[123] And this would be possible if only a decentralized command system is operating. In the IDF, the commanders are supposed to think beyond their organizational slot and comprehend the general picture within which the unit is operating.[124] The junior officers of the IDF like the *Wehrmacht's* officers are trained not to look for continuous guidance from the seniors. Harking back to the superiors in the rear for guidance is not only time consuming but communication lines might be cut due to enemy activities.[125]

For reacting faster than the enemy in a coherent manner, possessing of sophisticated tools alone are not adequate. It demands the capability to observe, orient, decide and act (OODA) faster than the enemy to disrupt the latter's thinking system. It means taking decisions faster than the enemy by getting inside his OODA loop or decision-making cycle. Hence, the necessity of *Auftragstaktik*, i.e. mission-oriented command or directive control. *Auftragstaktik* is the opposite of *Befehlstaktik*. *Befehlstaktik* stands for control by the commanders through detailed orders and regulations and this suits set-piece attritional battles.[126] But, in manoeuvre war, the scenario during combat becomes so unpredictable that within a quick changing battle space detailed orders for disposition and use of military assets would not work.[127] Fast moving manoeuvre campaigns require that senior officers tell their subordinates what to do and not how to do. This in turn allows the commanders of the subordinate formations to create multiple options within the fluid battle scenario for creating uncertainty in the enemy's mind. The net result will be that the enemy will panic first, then become passive and finally disintegrate at the lowest cost in friendly casualties.[128]

Mission-oriented command demands that the subordinate commanders should be allowed to select the objectives within the broad scope of the mission assignment.[129] Directive control depends on loose control by the leaders and requires complete trust and confidence between the lower and the higher level commanders. Under this model of command, the local commanders are allowed to modify their tasks in response to the rapidly changing local conditions provided it result in optimum use of the resources in furthering the superior's general intention.[130] John Keegan hints that the superior authority by allowing space to the juniors to disagree must generate creative tension within the command hierarchy.[131] Current thinking in the British Army argues that a senior commander at the brigadier level should function as a coordinator and intervene in affairs of the battalions only in case of dire emergencies.[132] Fighting manoeuvre warfare requires reorganization of the command hierarchy. This can be achieved by decentralizing the hierarchical command apparatus. The American military theorist Douglas A. Macgregor writes that due to the radical development of surveillance capacity, precision guided munitions and rising tempo of battle, the US Army's pyramidal command system requires to be flattened by abolishing certain intermediate levels of command. This would enable quick to and fro flow of information between the men on the ground and those higher up.[133]

A characteristic of post-Cold War manoeuvre war is jointness in military operations. The British military doctrine of 1995 says that jointness means combining all the combat resources of the air, land and sea in the planning and conducting of an operation.[134] It means that corporate autism of the various branches of the armed service that leads to conceptual dissonance and divergences of military policy must cease. Jointness aims to negate disunity, acrimony and fragmentation.[135]

Jointness is a new term for inter-service cooperation leading to combined arms operations. Combined arms cooperation is an essential ingredient of operational success. In contrast to the

German system, the Allied military organizations before 1945 were not geared towards combined arms approach. During World War II, the principal weakness of the US armed forces was lack of integration between the armour and infantry officers.[136] The backbone of the British system was regimental soldiering. And the regimental system fostered parochialism.[137] However, the scenario is changing. At present each US Division has four brigade headquarters which command combined arms groupings. Each group possesses three to five battalions and one brigade headquarter belongs to the combat aviation brigade for moving firepower rapidly to the required point in the battlefield.[138]

One feature of jointness as conceptualized by the Americans is cross talks among the staff officers of different services. The aim is to encourage cross-fertilization of ideas, for better and faster solutions to combat problems. The joint-force commander aims to obtain leverage from the continuous interaction among the different forces under his command. This requires positive interaction between the joint-force commander and the various component commanders in order to integrate and harmonize their functions. Jointness is teamwork among the various services, to be achieved through continuous interaction of air, land, sea, special operations and space capabilities. It offers the joint commanders an array of command, control and communication measures geared to the applying varying amounts of force from different directions. The objective is to create a shock effect on the opponent's command that would result in the operational paralysis of the enemy force. The most dramatic advances as regards jointness have occurred in the fields of reconnaissance and intelligence. Operation Desert Storm involved fruitful interaction between US Space Command, the US Marines, USAF, United States Navy and the US Army. For example, the US Space Command provided real time intelligence to the US Central Command for launching accurate air strikes against Iraqi ground targets.[139] The American armed forces' jointness requires common radio frequencies for all the disparate military formations (Marines, Army and Air Force units) participating in combat.[140] Jointness is required because deployment in a varying range of environments necessitates inter-service linkages as regards logistical undertakings.[141]

Back in India, close cooperation between the infantry and the artillery is a tradition that developed under British auspices especially when the Sepoy Army conducted 'hit and run' expeditions against the North-West Frontier tribes.[142] This tradition continued in the post-colonial period. In May 1948, at Chowbikal, the gunners gave excellent support to the 1st Sikh Infantry Battalion. The gunners in a show of fellowship managed to move the heavy 25-pounders through narrow, circuitous and slippery roads.[143]

Nevertheless, coordination and cooperation between the IAF and the army in the heat of battle have never been a strong point for the Indian military. The Indian Army commanders are aware of the importance of air–land battle. Lieutenant-General K.P. Candeth, Commander of the Western Front during the 1971 War accept that: 'All wars are inter-service affairs and for success there must be very close inter-service cooperation, and co-ordination of air and ground effort.'[144] But, there was a huge gap between theory and praxis.

Brigadier L.P. Sen commanding the 161st Brigade during the 1947–8 War with Pakistan conveys his feelings about the state of air-ground cooperation on 3 November 1947 at Kashmir in the following words: 'A Liaison Officer with the IAF was with me at the time, and he was ordered to rush back to the airstrip and to inform the Air Force of what had happened and ask for aircraft to strafe the raiders in the Badgam area. The Liaison Officer said that he did not have a 1 inch to 1 mile map, nor did the Air Force.'[145]

During the 1965 War with Pakistan, the Indian Army and the IAF fought their separate battles. The IAF was interested in striking the Pakistani air bases at Sargodha and Peshawar while the army demanded Close Air Support (CAS). The army requested the IAF's intervention against the Pakistani ground forces at the Haji Pir, Sialkot and Lahore sectors[146] but was provided with inadequate CAS preventing a rapid pursuit of the enemy forces. While the Pakistan Air Force flew 27 per cent of its sorties to support the Pakistan Army, only 18 per cent of the IAF's sorties were geared to CAS.[147] On 22 September at about 4 a.m., 41st Brigade ordered the 15th Kumaon Battalion to attack Khem Karan from the left of the 1st Battalion of the 8th Gurkha Regiment. The attack started at 5 a.m. There was no instant response from the IAF which came only at 6.30 a.m. Again due to inadequate liaison with the army, the IAF strafed the Indian troops, breaking the cohesion among the Kumaonis. Worse was to follow. Cooperation among the various services within the Indian Army was also lacking. The Kumaonis failed to contact their supporting armour and the attack stopped at 7.15 a.m.[148] The IAF–Indian Army coordination in the Eastern Theatre was also a fiasco. Instead of fighting the Pakistanis, the generals and air marshals were busy combating each other. The IAF Chief demanded a preemptive strike against East Pakistan in order to destroy the Pakistani aircraft which he was afraid might be transferred to West Pakistan. Sam Manekshaw, the ground commander was against the preemptive strike because he considered his forces unready. Nevertheless, the IAF carried out a dawn raid without informing the army commander in the Eastern Theatre. Manekshaw was furious and wanted to resign. But, the COAS restrained him.[149] Wing Commander K. Advani of the IAF pointed out the defects regarding army–air cooperation in the following words: 'Lack of get-togetherness in the matter of mutual planning between army and air commanders. At present both services seem to site and locate their formations and units without mutual consultations. We can therefore see operational commands of either service located far from each other, or flying units at airfields hundreds of miles from army formations that obviously will require air support.'[150]

Luckily for the Indians, things were not better on the 'other side of the hill'. As regards the high command in Pakistan, Air Marshal of the PAF, M. Asghar Khan writes:

> He [Ayub Khan] understood little of air operations, except that a 'favourable air situation' was a desirable condition for the waging of war on land. In our eight years' association, I had, however, failed to convince him that in our situation, successful land operations would be rendered impossible if the Air Force did not succeed in neutralizing an enemy's Air Force. He preferred to see the Air Force as an arm of the Army, an airborne form of Artillery whose role should be to clear the way for the Infantry and Armour. I had maintained that this concept was fallacious, and that such support could only be provided after the Air Force had removed the threat from an enemy's air power. . . . Ayub Khan's understanding of the country's maritime affairs was even less. . . . He considered any expenditure on the Navy wasteful and did not appreciate that in our peculiar circumstances wherein the two wings

of the country were physically separated by over a thousand miles of territory, the sea provided the only reliable link between them.[151]

Things did not improve much for the Indian military in the next war. During the 1971 War, four Tactical Air Centres (TAC) were attached with four army corps deployed in the Western sector. Each of the TAC had communications link with the forward airfields where the fighter squadrons were located to support the ground forces. The reaction time of the IAF to the demands of the Indian corps remained two hours initially and as the war continued, it was reduced to an hour and a half. By modern standards it was too slow. This was because the staff officers were not conversant with the procedures of calling for air support. Lack of combined training and joint exercises were the real culprits. The IAF which had double the number of fighters and three times the number of bombers than the PAF in the Western sector,[152] was able to dominate the skies through sheer numerical superiority, and this was a characteristic of attritional warfare.

The IAF and the Indian Army blamed each other in midst of the war. The tactical air force was supposed to provide reconnaissance 50 miles ahead of the ground forces. This did not occur due to lack of cooperation between IAF and the army. The IAF claimed that the Indian Army had failed to use effectively the air force assets that were earmarked for the ground forces. The army argued that the IAF had bungled the CAS mission and demanded creation of an integral air arm that would be devoted completely to the demands of ground war, but, this duplication would be uneconomical.[153]

Nevertheless, minor improvement was visible. In 1965, India's conduct of air–land battle involving air and ground forces in the Western theatre was unimpressive because of lack of cooperation and coordination between the IAF, infantry and armour.[154] Later, integration of helicopters with infantry enabled the Indian armed forces to raise the 'speed of battle'. During the 1971 War, the Indian armed forces displayed their heli-lift capacity. On 7 December, a battalion was lifted by helicopters to south-east Sylhet across the Surma River. Again on 9 and 10 December, the helicopters transported troops from Brahamanbaria to Raipura. The heli-lift of infantry also aided the Meghna crossing. Between

11 and 13 December, 2,500 troops were ferried over the Meghna River and landed at Narsingdi.[155]

In 1972, Major Dipes Lahiri in an article written in the *Journal of the United Service Institution of India* argues that armed helicopters are necessary for reconnaissance and fire support to the ground forces. For easier administration and better understanding of the mission, says Lahiri, the attack helicopters should be manned and controlled by the army.[156] The point to be noted is that during the World War II, the German Army did not have any integral air support units. Nevertheless, cooperation between the German Army and the *Luftwaffe* was flawless.[157] The Indian Army won its turf battle and an Army Aviation Corps came into existence.

Inter-service integration requires that all the three Services function under one single commander with a jointness of purpose among the personnel of the different branches of the armed Services. To that effect, in 1955, a National Defence Academy was set up at Khadakvasla near Pune. This institution aims to provide three years, common training to the cadets in order to ensure fellowship among them before they go to specialized institutions which prepare them for joining the different branches of the armed forces. The cadets join at the age of fifteen and are given a general course of education under military discipline. The reason for setting up such an institution was Claude Auchinleck's recommendation for establishing such an academy. During World War II, he found that cooperation between the infantry, navy, air force and armoured officers was necessary.[158] But, a mere three years, training under a single roof to the cadets before they join their service specific institutions remains inadequate for ensuring jointness. The only silver lining is that India's *bete noire*—Pakistan also lacks a joint command system for conducting fast mobile warfare. Both, during 1965 and 1971 wars, there was no integrated planning and cooperation between the three services of the Pakistani armed forces. An overall strategic plan was non-existent and tactical cooperation was limited.[159] The situation remains the same for the Pakistani armed forces even now.

Internecine rivalries harm intelligence assessment. The Research and Analysis Wing (RAW) deliberately enfeebles the

Joint Intelligence Committee of the Indian armed forces. The RAW is never willing to share information with the intelligence wing of the Indian Army. When the Indian Peace Keeping Force operated in Sri Lanka, one divisional commander Lieutenant-General S.C. Sardeshpande claimed that the RAW did not cooperate with the military commanders. No mechanism for collating information from RAW, Intelligence Bureau and the Foreign Ministry exists even now. And all these put a constraint on the operational assessment of the enemy.[160]

Further, India lacks an institution for formulating a joint military strategy which should be a component of a clearly defined national security policy. In the present system, the Chiefs of Staff Committee cannot even enforce a military decision like the use of air power in a specific campaign at a particular time, if it is objected to by one of the Service Chiefs.[161] If the post of Chief of Defence Staff is introduced then he could be made responsible for the inter-service joint training establishments. In addition, he could also command the Defence Intelligence Agency, Defence Planning Staff and Command Control Communications Computers Intelligence and Interoperability systems necessary for conducting joint warfare.[162] In the mid-1980s, when Arun Singh was the Minister of State for Defence, the armed forces set up a Joint Defence Planning Staff. However, the three Service Chiefs refused to transfer any power to this institution which then became defunct.[163] Again, the Indian armed forces need to set up a Defence Logistics Agency (DLA), for unifying and coordinating the logistics efforts of the three Services. This agency will harmonize the logistics system and procedures of the three Services and will also evolve a common logistical doctrine. The DLA should be under the Chief of Defence Logistics who would report to the CDS.[164]

Besides inculcating jointness, setting up of a decentralized command is another requirement for conducting fast mobile warfare. The doctrine of mission command necessary for maintaining a high tempo of operations requires that autonomy should be allowed to the subordinate commanders.[165] Tight command accelerated the decomposition of the Indian Army units in NEFA during the 1962 war. Niranjan Prasad, commanding 4th Infantry Division under the XXXIII Corps, asked for the permission to occupy Thag La Ridge. But Lieutenant-General L.P. Sen, Chief of the Eastern Command, overruled him. In fact,

Sen though in charge of more than 2,50,000 men, started interfering with the deployment of 7th Brigade which was under Prasad. Sen got rid of Lieutenant-General Umrao Singh, Commander of the XXXIII Corps under whom Prasad came. On 4 October 1962, Prasad was threatened by Sen that unless he conform totally with the latter's decisions, Prasad would be removed. Had Prasad been allowed to occupy Thag La Ridge, then the Chinese would have found it difficult to advance to Dhola.[166] Tradition dies hard. In 1970, Freemantle took over the 7th Infantry Division located at Ferozepur. He came under the corps commander Lieutenant-General N.C. Rawley. Freemantle wrote in his autobiography that he was never comfortable with Rawley because of the latter's belief that fear generated by the superior commander was the principal technique that held together the command apparatus.[167]

In 1972, Lieutenant-Colonel Y.A. Mande asserted that higher the level of the command, greater is the freedom necessary for operating. At section (5–11 men) and platoon level (30 men), 'follow me' sort of leadership is necessary. At the company level (100 men), 'I am with you' type of leadership is required. At the battalion level (1,000 men), the leadership required is 'I am watching you, and I will be with you when you need me'. However, at brigade level (5,000 men), leadership becomes 'Well here is a task, think it over and discuss it with me. Let me know what assistance you need from my side'. According to Mande, at this level, *Auftragstaktik* become all the more necessary. Mande asserts that senior commanders ought to be more democratic. Unless the superiors give junior commanders space for manoeuvre, it stifles the latter's initiative. Mande rightly claims that centralization leads to inefficiency, red tape and delay; while decentralized command not only allows the development of personality of the subordinates but also results in their training for higher jobs.[168] Mande writes on decentralized command: 'It is true that delegation of authority does not absolve the commanding officer of responsibility. At the same time, subordinates make mistakes. Those who work make mistakes. Shirkers are generally experts in the game of play safe. The art of management requires balancing of ultimate responsibility and delegation. It can be done by controlling critical points, leaving the rest to the subordinates.'[169]

Individual officers in the Pakistan Army occasionally followed the decentralized command culture. During the 1965 War, A.A.K. Niazi was commanding the 5th Punjabis in Muzaffarabad. His superior officer was Major-General Akhtar Malik commanding the 12th Division. In Niazi's own words, 'He [Malik] asked me to take quick decisions where required and gave me a *carte blanche* to take decisions keeping in view the changing panoramic situation.'[170]

From the 1980s, like their British and American counterparts, the Indian military officers have also been hankering for a decentralized command system for waging short, swift war in the Information Age. Squadron-Leader P.M. Sinha argues that the personnel manning the command apparatus should be free to respond effectively and efficiently to the changing local conditions. This freedom would make the command set up more agile and adaptive to the rapidly changing scenarios. P.M. Sinha accepts the tenor of Douglas A. Macgregor's recommendations that increase in the lethality, accuracy and reach of the weapon systems will force the battle formations to spread out over a wide area. Commanding the widely dispersed units and rapid response to wrest the initiative away from the enemy requires *Auftragstaktik*.[171]

One of the factors behind the IDF's stunning battlefield success is the fact that even section leaders are trained to size up the tactical situations, take decisions and implement them on their own authority.[172] In the Indian Army, not to speak about the JCOs and the NCOs, even the commissioned officers are not trained to take decisions. In order to train the Indian officers in the *Auftragstaktik* command culture, the officers require to have an intelligent anticipation of events by thinking ahead and then acting without specific orders in order to achieve their assigned tasks. The officers should not hang around waiting for orders, Nor should they say, 'Nobody ordered me'. The concept of *Auftragstaktik* cannot be properly implemented in the Indian Army because of the low intellectual capacities of the officer cadre. For the affluent university educated urban youth, joining the army is a low priority, corporate sectors and the bureaucracy being their first choice.[173]

As COAS, Manekshaw realized that the low level of education among the officers was hampering their training in conducting mobile warfare. For lifting the intellectual level of the officers, he requested the government that the defence services' educational institutions be affiliated with the Jawaharlal Nehru University, to equate examinations the servicemen passed with university equivalents. The proposal was to equate graduation at the National Defence Academy with a Bachelor's degree; graduation at the Defence Services Staff College with a Master's degree; and successful completion of the National Defence College with a Ph.D. Manekshaw was also eager to provide more study leaves to the officers.[174] In 1977, a seminar was held at the United Service Institution at New Delhi. Several high ranking officers like Major-General Bhupinder Singh, Deputy Adjutant-General, Army Headquarters, Lieutenant-General K.P. Candeth, Lieutenant-General K.S. Katoch, former Vice Chief of the Army Staff and others participated. A report on the seminar published under the Chairman retired Major-General D. Som Dutt points out the deficiency of the officer corps in the following words:

> Because of the staggering unemployment in our country, there is no shortage whatsoever of volunteers for the forces. The pertinent point, however, is that only those young men apply to join the forces who are ineligible or unlikely to make the grade in commercial life or the civil services. . . . It is quite incorrect to interpret the large numbers of applications for commissions as being indicative of enthusiasm for life in the forces. It is economic necessity which drives them into the Services.[175]

In 1982, Lieutenant-General Jagjit Singh Aurora commented:

> Today very few join the Army because they actually like the profession, they do so purely for economic reasons. . . . If steps are not taken to attract more suitable material we will have mediocre talent. . . . No amount of advanced and sophisticated equipment can make up for weak leadership.[176]

In 1974–5, at the Indian Military Academy, only 70 per cent of the vacancies were filled up and in the Technical Arms about 50 per cent. Even then as regards the quality of intakes, 88 per cent of the candidates concerned were in the lowest acceptable grade. At the National Defence Academy, 30 per cent of the intake were the sons of the JCOs and NCOs; 20 per cent were the sons of civilians who had sent their offspring to the *Sainik* Schools; and 40 per cent were those who failed to make it to the universities.[177] One could contrast the scenario with the IDF where in 1994 about 90 per cent of the battalion commanders were university graduates.[178] The Indian infantry unlike the technical services is the most hard hit in procuring suitable officer material. The Indian Navy is comparatively better off because the naval officers when not serving in the sea are posted near the big cities and have better accommodation facilities than their infantry counterparts. Further, the naval officers after retirement have a lucrative outlet in the merchant navy. In contrast, the army officers have to wait for a minimum a two years before getting accommodation facilities. Further long separation from the families and uncongenial postings create problems in the marital lives and harm children's education.[179] All these factors discourage the cream of the society from entering the army. In 1994, the Indian Army faced a shortfall of 12,972 officers.[180] From the 1960s, the Pakistan Army failed to attract the upper-middle class in its officer cadre. The bulk of the officer recruits came from rural lower-middle class families.[181] And they are no better than their Indian counterparts in grasping the concepts of manoeuvre warfare.

Not only the Indian Army lacks officers of requisite calibre and number but the mode of training is also defective. Harbakhsh Singh asserts in his memoirs: 'We should aim at cultivating in our commanders a spirit and mental attitude which would welcome chances of bold action instead of indecisive wavering when faced with a situation.'[182] However, continues Singh, the training curriculum does not encourage variations from the accepted norms, and the exercises do not include situations that involve taking calculated risks and deviating from the normal procedures and established principles of warfare.[183] This is indeed a serious limitation of the Indian Army because the battlefield will be full of Clausewitzian 'fog'. So, those soldiers and officers who are prepared to think for themselves rather than wait for commands will be better prepared amidst the chaos of battlefields.[184] In 1972, Major-General O.S. Kalkat warned that at the battalion level, the commanders should make simple plans. If the plan is too complicated, though it might appear brilliant on paper, it would fail under real conditions.[185] In other words, Kalkat is saying that the Indian infantry is unprepared for implementing a sophisticated plan involving fire and movement.

However, decentralized command has been operating in the Indian Navy. This is because the ocean is wide and each vessel under its captain operates separately. Admiral S.M. Nanda's (Chief of Naval Staff from March 1970 to February 1973) views about the attitudes of midshipmen and sub-lieutenants fresh from the training ships and schools in 1966 are as follows:

> They were very observant and innocent young officers who did not hesitate to share their views with their fleet commander. The freedom of expression granted to, and frank airing of views by, junior officers represent a very healthy and unique tradition of our Navy. Such a tradition encourages interaction and dialogue amongst officers. It emboldens juniors to ask questions in their quest for knowledge and it infuses a sense of satisfaction in the seniors as they realize that their young 'snottie' (midshipman) is beginning to develop a fertile and inquisitive mind.[186]

In 1969, as Admiral Nanda was preparing the navy for war, he delegated the day-to-day administration to the staff officers he trusted. Nanda only kept a loose watch over them.[187]

John Keegan writes that decentralized command is characterized by 'creative disobedience' by the subordinates.[188] Vice-Admiral S.H. Sarma commanding the Eastern Fleet during the 1971 War displayed 'creative disobedience' which involved taking moderate amount of risk for aggressive actions even at the cost of bending the rule book. Thus, Sarma was able to cause enormous damage to the Pakistani military assets in erstwhile East Pakistan. In Sarma's own words:

> The Pakistanis were most likely to concentrate the 16 Sabre jets they had at Dacca, and order an attack on Vikrant, even if it was a one-way sortie, as the Vikrant symbolized the might of the Indian Navy (and was the fount of all seaward attacks in the Bay). So, as far as possible, I was to keep the aircraft carrier more than a hundred miles from their coast. I discussed this with Captain Prakash [commanding the Indian aircraft carrier Vikrant]. With the Hawks' [planes on board of Vikrant] operational radius of just about a hundred miles, if we carried out these instructions, it would be like two boxers staying in their own corners of the ring and hitting out at nothing. So I decided to operate from about 55 to 60 miles from the coast . . . . This decision paid off, and we were able to carry out over 90 sorties instead of the five initially planned. Had the Sabre jets hit the Vikrant, I would have had to face the consequences.[189]

CONCLUSION

Machines will gradually take up many functions hitherto exercised by the commanders. Whether it will end up with a gigantic super computer exercising theatre-wide command capacity as envisaged by sci-fi writers is yet to be seen. Despite the rising importance of machines, personality and physical presence of the commanders remain crucial ingredients of command at the sharp end of the battle. Scattered small unit-level engagements will remain the order of the day. And command at this level will depend upon the heroic leadership by the JCOs, NCOs and commissioned officers upto the rank of colonels. Display of battlefield bravery by the JCOs, NCOs and the commissioned officers is a tradition fostered by the British and nurtured by independent India's Army. In contrast, the junior and mid-level American Army officers failed to display heroic command in the paddy fields of Vietnam during the 1960s and the early 1970s. This was because the US Army lacked strong regimental traditions. The net result was fragging and combat avoidance on part of the American privates.[190] Regimental *esprit de corps* of the Sepoy Army from which the Indian Army derives its *paltan ki izzat* encourage the officers (both commissioned and non-commissioned) to display heroics in the field of fire.[191] And this aids cohesion in the Indian units. With the passage of time, the Indian Army's officer cadre matured and gradually evolved new techniques for maintaining human relations in the field of fire. However, lack of command coordination in a high-technology environment remains the Achilles heel of the Indian armed forces. This is partly due to resource crunch and also because of the innate conservatism inherent in the military organization. The Indian Army has long way to go before it can integrate the techniques for conducting manoeuvre warfare with high technology weapons.

NOTES

1. Elliot P. Chodoff, 'Ideology and Primary Groups', *Armed Forces and Society*, vol. 9, no. 4 (1983), p. 588. Richard H. Kohn writes that the American soldiers rather than being motivated by patriotism or political ideology, sustained themselves in the stress of battle from the network

of esteem and interdependence that creates solidarity and cohesion within a unit for survival. Kohn, 'The Social History of the American Soldier: A Review and Prospectus for Research', *American Historical Review*, vol. 86, no. 3 (1981), p. 559. Kohn's observations could be applied in the case of army of another democratic polity—India.

2. Lieutenant-General John H. Cushman, 'Challenge and Response at the Operational and Tactical Levels: 1914–45', in Allan R. Millett and Williamson Murray, eds., *Military Effectiveness: The Second World War*, vol. 3, Boston: Allen & Unwin, 1988, p. 326.
3. Sam C. Sarkesian, 'Combat Effectiveness', in Idem, ed., *Combat Effectiveness: Cohesion, Stress, and the Volunteer Military*, Beverly Hills/London: Sage, 1980, p. 11.
4. Eyal Ben-Ari, *Mastering Soldiers: Conflict, Emotions, and the Enemy in an Israeli Military Unit*, New York/Oxford: Berghahn Books, 1998, p. 29.
5. G.D. Sheffield, 'Introduction: Command, Leadership and the Anglo-American Experience', in idem, ed., *Leadership and Command: The Anglo-American Military Experience since 1861*, London/Washington: Brassey's, 1997, p. 1.
6. Stephen Peter Rosen, *Societies and Military Power: India and its Armies*, New Delhi: Oxford University Press, 1996, pp. 197–256.
7. Harold W. Nelson, 'Space and Time in *On War*', *Journal of Strategic Studies* (henceforth *JSS*), vol. 9, nos. 2–3 (1986), p. 134.
8. Marshal G. Zhukov always launched frontal assaults against heavily defended enemy entrenchments regardless of his own losses. But, this was not possible for Dwight Eisenhower or Bernard Montgomery. Especially in the post-World War II era due to the spread of audio-visual media, intense manpower losses are not acceptable to the democracies. For Zhukov see Brigadier Peter Young, ed., *Strategy and Tactics of the Great Generals and their Battles*, 1984, rpt., London: Bison Books, 1986, p. 237.
9. Michael Howard, 'Leadership in the British Army in the Second World War: Some Personal Observations', in Sheffield, ed., *Leadership and Command*, pp. 120–1.
10. Lieutenant-General Harbakhsh Singh, *In the Line of Duty: A Soldier Remembers*, New Delhi: Lancer, 2000, p. 422.
11. Squadron-Leader K.N. Parik, 'Fear and its Effect on Combat Efficiency', *Journal of the United Service Institution of India* (hereafter *JUSII*), vol. CII, no. 427 (1972), p. 171.
12. Fabrizio Battistelli, 'The Postmodern Military: Conscription or Professionalism?', in Stuart A. Cohen, ed., *Democratic Societies and their Armed Forces: Israel in Comparative Context*, London: Frank Cass, 2000, p. 62.
13. S.T. Das, *Indian Military: Its History and Development*, New Delhi: Sagar Publications, 1969, p. 168.

14. General K.V. Krishna Rao, *In the Service of the Nation: Reminiscences*, New Delhi: Viking, 2001, p. 28.
15. K.M. Panikkar, *Problems of Indian Defence*, Bombay: Asia Publishing House, 1960 , pp. 34, 37.
16. Lieutenant-General S.K. Sinha, 'Indian Army before and after Independence: Its Role in Nation Making', in *Field Marshal K.M. Cariappa Memorial Lectures*, New Delhi: Directorate General of Infantry in association with Lancer, 2001, p. 52.
17. 'Field Marshal Cariappa', in *Cariappa Memorial Lectures*, p. 18.
18. Krishna Rao, *Reminiscences*, p. 27.
19. Ian F.W. Beckett, 'Command in the Late Victorian Army', in Sheffield, ed., *Leadership and Command*, p. 49.
20. Brian Holden Reid, 'The Commander and his Chief of Staff: Ulysses S. Grant and John A. Rawlins, 1861–65', in Sheffield, ed., *Leadership and Command*, p. 17.
21. Major-General F.L. Freemantle, *Fred's Foibles*, New Delhi: Lancer, 2000, p. 98.
22. Richard E. Simpkin, *Race to the Swift: Thoughts on Twenty-First Century Warfare*, London: Brassey's, 1985, p. 271.
23. John Keegan, *The Mask of Command*, 1987, rpt., New York: Penguin, 1988, pp. 314–15.
24. John Keegan, *Intelligence in War: Knowledge of the Enemy from Napoleon to Al-Qaeda*, New York: Alfred A. Knopf, 2003, pp. 64–5.
25. Brigadier E.V.R. Bellers, *The History of the 1st King George V's own Gurkha Rifles, The Malaun Regiment, 1920–47*, vol. 2, Aldershot: Gale & Polden, 1956, p. 179.
26. *9 Gurkha Rifles: A Regimental History, 1817–1947*, New Delhi: Vision Books, 1984, p. 134.
27. Jurgen E. Forster, 'The Dynamics of *Volksgemeinschaft*: The Effectiveness of the German Military Establishment in the Second World War', in Millett and Murray, eds., *Military Effectiveness*, pp. 206, 209.
28. Bellers, *Malaun Regiment*, p. 192.
29. 'Matters of Moment', *JUSII*, vol. LXXVII, no. 326 (1947), p. 3.
30. Field Marshal S.H.F.J. Manekshaw, 'Leadership in the 21st Century', in *Cariappa Memorial Lectures*, pp. 30–1.
31. Shankar Roychowdhury, 'Management of Power in the Defence Services: Values Dimensions', in S.K. Chakraborty and Pradip Bhattacharya, eds., *Leadership and Power: Ethical Explorations*, New Delhi: Oxford University Press, 2001, p. 317.
32. Lieutenant-General S.K. Sinha, *Operation Rescue: Military Operations in Jammu and Kashmir 1947–49*, 1977, rpt., New Delhi: Vision Books, 2002, p. 16.
33. Lieutenant-Colonel S.C. Bhattacharya, 'Motivating Forces for Combat', *JUSII*, vol. LXXXXIII, no. 392 (1963), p. 265.

34. Group-Captain Ranbir Singh, *Marshal Arjan Singh: Life and Times*, New Delhi: Ocean Books Pvt. Ltd., 2002, pp. 49, 51.
35. Lieutenant-General M.L. Chibber, *Leadership in the Indian Army during Eighties and Nineties*, United Service Institution Papers, no. 8, New Delhi: United Service Institution of India, p. 38.
36. Lieutenant-General P.S. Bhagat, *The Shield and the Sword* , 1966, rpt., Delhi: Vikas, 1974, p. 42.
37. John A. English, *On Infantry*, 1981, rpt., New York: Praeger, 1984, pp. 192–3.
38. Brigadier Ashok Malhotra, *Trishul: Ladakh and Kargil, 1947–93*, New Delhi: Lancer, 2003, p. 9.
39. K.C. Praval, *Valour Triumphs: A History of the Kumaon Regiment*, Faridabad: Thomson Press, 1976, pp. 312–13, 319.
40. Harinder Baweja and Ramesh Vinayak, 'The Battle Zone', *India Today*, 14 June 1999, p. 29.
41. Major-General Jagjit Singh, *With Honour and Glory: Wars Fought by India, 1947-99*, New Delhi: Lancer, 2001, pp. 32-3.
42. David French, '"Tommy is no Soldier": The Morale of the Second British Army in Normandy, June–August 1944', *JSS*, vol. 19, no. 4 (1996), p. 168.
43. Bruce Vandervort, *Wars of Imperial Conquest in Africa: 1830-1914*, Bloomington/Indianapolis: Indiana University Press, 1998, p. 66.
44. Joseph G. Dawson III, '"Zealous for Annexation": Volunteer Soldiering, Military Government, and the Service of Colonel Alexander Doniphan in the Mexican-American War', *JSS*, vol. 19, no. 4 (1996), pp. 20, 23.
45. Ben-Ari, *Mastering Soldiers*, p. 29.
46. MacGregor Knox, 'The Italian Armed Forces, 1940-43', in Millett and Murray, eds., *Military Effectiveness*, p. 164.
47. Jane Lennox, 'Morale Implications of *Perestroyka*', in Air-Commodore E.S. Williams, ed., *Soviet Air Power: Prospects for the Future, Perestroyka and the Soviet Air Forces*, London: Tri-Service Press, 1990, p. 96.
48. Alvin D. Coox, 'The Effectiveness of the Japanese Military Establishment in the Second World War', in Millett and Murray, eds., *Military Effectiveness*, p. 36.
49. J.M. Bourne, 'British Generals in the First World War', in Sheffield, ed., *Leadership and Command*, p. 95; Philip Mason, *A Matter of Honour: An Account of the Indian Army, Its Officers and Men*, 1974, rpt., Dehra Dun: Educational Private Limited, 1988, pp. 375–92.
50. Lieutenant-General P.S. Bhagat, 'The Divisional Commander', *JUSII*, vol. CI, no. 425 (1971), pp. 295–6.
51. Brigadier H.S. Sodhi, 'The Higher Echelons of Leadership', *JUSII*, vol. CIV, no. 436 (1974), p. 263.
52. Sinha, *Operation Rescue*, p. 163.
53. Major K. Brahma Singh, 'Military Leadership: An Objective Study', *JUSII*, vol. 99, no. 416 (1969), p. 276.

54. Freemantle, *Fred's Foibles*, p. 78.
55. Major-General Gurcharan Singh Sandhu, *69th Armoured Regiment: 1968-1993*, New Delhi: Lancer, 1996, p. 17.
56. S.P. Mackenzie, *Revolutionary Armies in the Modern Era: A Revisionist Approach*, London/New York: Routledge, 1997, pp. 168-9.
57. Anthony Farrar-Hockley, 'Korea, 1950-53: The Far Side of the World', in Major-General Julian Thompson, ed., *The Imperial War Museum Book of Modern Warfare: British and Commonwealth Forces at War, 1945-2000*, 2002, rpt., London: Pan Macmillan, 2003, p. 111.
58. 'Introduction', in *Field Marshal K.M. Cariappa Memorial Lectures*, p. 15.
59. Parik, 'Fear and its Effect on Combat Efficiency', p. 171.
60. Krishna Rao, *Reminiscences*, p. 14.
61. Ibid., p. 91.
62. Lieutenant-General A.A.K. Niazi, *The Betrayal of East Pakistan*, 1998, rpt., Karachi: Oxford University Press, 2000, p. 28.
63. Sodhi, 'Higher Echelons of Leadership', pp. 269–70.
64. Lieutenant-Colonel B.L. Raina, 'Military Medical Services in War', *JUSII*, vol. LXXVIII, no. 330 (1948), p. 30.
65. Harbakhsh Singh, *In the Line of Duty*, p. 261.
66. Krishna Rao, *Reminiscences*, p. 95.
67. Ben-Ari, *Mastering Soldiers*, p. 29.
68. Harbakhsh Singh, *In the Line of Duty*, pp. 200–1, 207.
69. Neville Maxwell, *India's China War*, 1970, rpt., Bombay: Jaico, 1971, pp. 294–5.
70. Keegan, *Mask of Command*, p. 316; Major-General D.K. Palit, *War in High Himalaya: The Indian Army in Crisis, 1962*, New Delhi: Lancer, 1991, p. 241.
71. Palit, *War in High Himalaya*, pp. 240, 242–3.
72. S.N. Prasad (Chief Editor), P.B. Sinha and Colonel A.A. Athale, *History of the Conflict with China: 1962*, New Delhi: History Division, Ministry of Defence, Government of India, 1992; www.timesofindia.com, p. 199.
73. Samuel W. Mitcham Jr., *The Desert Fox in Normandy: Rommel's Defence of Fortress Europe*, London: Praeger, 1997, pp. 57–92.
74. Colonel-General Heinz Guderian, *Panzer Leader*, tr. from German by Constantine Fitzgibbon, 1952, rpt., London: Arrow Books, 1990, pp. 140–271.
75. Lieutenant-General Depinder Singh, *Field Marshal Sam Manekshaw: Soldiering with Dignity*, Dehra Dun: Natraj, 2002, p. 10.
76. Cornelius Ryan, *A Bridge Too Far*, 1974, rpt., London: Coronet Books, 1975, p. 327.
77. Sergeants and sergeant-majors are the backbone of the regiments in the 100,000 strong Bangladesh Army. P.B. Sinha, *Armed Forces of*

Bangladesh, New Delhi: Institute for Defence Studies and Analyses, 1979, pp. 1, 32.

78. The Viceroys Commissioned Officers were mostly promoted from the ranks on the basis of seniority. They commanded sections and platoons. David Omissi, *The Sepoy and the Raj: The Indian Army, 1860-1940*, Houndmills, Basingstoke: Macmillan, 1994, pp. 153–60.
79. Colonel C.W. Morton, 'Man Management', *JUSII*, vol. LXXVIII, no. 330 (1948), p. 45.
80. Lieutenant-Colonel Gautam Sharma, *Path of Glory: Exploits of 11th Gorkha Rifles*, Ahmedabad: Allied, 1988, p. 55.
81. Omissi, *Sepoy and the Raj*, pp. 154–6.
82. Daljit Singh, 'Military Education in India: Changes from the British Tradition', *JUSII*, vol. CIV, no. 436 (1974), p. 234.
83. Major A.S. Ahluwalia, 'Junior Leadership Training', *JUSII*, vol. CII, no. 427 (1972), pp. 157–9.
84. Depinder Singh, *Field-Marshal Manekshaw*, pp. 68–9.
85. Bellers, *Malaun Regiment*, pp. 312–13.
86. Rajesh Kadian, *India and Its Army*, New Delhi: Vision Books, 1990, p. 164.
87. *9 Gurkha Rifles*, pp. 131, 211, 166.
88. Jagjit Singh, *With Honour and Glory*, pp. 25–6. The quotation is from p. 25.
89. Brigadier Desmond E. Hayde, *The Battle of Dograi*, 1984, rpt., Dehra Dun: Natraj, 1991, pp. 76–8, 84. The quotation is from p. 78.
90. Praval, *Valour Triumphs*, pp. 310–11, 334.
91. Ronald Chalmers Hood III, 'Bitter Victory: French Military Effectiveness During the Second World War', in Millett and Murray, eds., *Military Effectiveness*, p. 250.
92. Joseph Smith, 'The Spanish-American War: Land Battles in Cuba, 1895–98', *JSS*, vol. 19, no. 4 (1996), p. 52.
93. Vandervort, *Wars of Imperial Conquest*, p. 114.
94. Bruce Allen Watson, *When Soldiers Quit: Studies in Military Disintegration*, Westport, Connecticut: Praeger, 1997, p. 98.
95. Bellers, *Malaun Regiment*, pp. 193, 201, 210.
96. Forward, 'Control of Reinforcements', *JUSII*, vol. LXXVIII, no. 330 (1948), pp. 55–7.
97. Farrar-Hockley, 'Korea, 1950–53', in Thompson, ed., *British and Commonwealth Forces at War, 1945-2000*, p. 107.
98. Prasad (Chief Editor), *History of the Conflict with China*, pp. 191, 193, 195.
99. Martin Van Creveld, *Command in War*, Cambridge/Massachusetts: Harvard University Press, 1985, pp. 232–60.
100. Lieutenant-General V.R. Raghavan, *Siachen: Conflict Without End*, New Delhi, Viking, 2002, p. 68.

101. J.R. Isaac, S.C. Gupta, E.M.E. and S. Ramani, 'Computer Aids to Command and Control', *JUSII*, vol. CVI, no. 4 (1976), pp. 135, 145–6.
102. Tim Ripley, *The New Illustrated Guide to Modern US Army*, London: Salamander, 1992, p. 18.
103. Alistair Irwin, 'The Buffalo Thorn: The Nature of the Future Battlefield', *JSS*, vol. 19, no. 4 (1996), p. 234.
104. Chris C. Demchak, 'Numbers or Networks: Social Constructions of Technology and Organizational Dilemmas in IDF Modernization', *Armed Forces & Society*, vol. 23, no. 2 (1996), p. 207.
105. Major Gurmeet Kanwal, 'Weapons and Technology', *IDR*, vol. 2, no. 1 (1987), pp. 128, 133.
106. Colonel B.S. Ramdas, 'The Computer and Its Effect on Management from Army to Regimental Level', *IDR*, vol. 2, no. 1 (1987), pp. 107–8.
107. G. Jacobs, 'Unmanned Aerial Vehicles: Some Programmes', *IDR*, vol. 14, no. 2 (1999), pp. 101–2.
108. C.N. Ghosh, 'Application of Unmanned Combat Aerial Vehicles in Future Battles of the Subcontinent', *Strategic Analysis*, vol. 25, no. 4 (2002), pp. 599–605.
109. Simpkin, *Race to the Swift*, p. 227.
110. Ripley, *US Army*, p. 150; 'Who Gains', *The Economist*, 29 September–5 October 2001, p. 57.
111. *Joint Warfare of the Armed Forces of the United States*, Ch. IV, Joint Pub.-1, 10 January 1995, p. 9.
112. Bryan Bender, 'TRW, Lockheed Move Ahead in $2 Billion Space Deal', *Jane's Defence Weekly*, vol. 33, no. 8, 23 February 2000, p. 3.
113. Eliot A. Cohen, 'Technology and Supreme Command', in Cohen, ed., *Democratic Societies and Their Armed Forces*, pp. 96–7.
114. Creveld, *Command in War*, p. 267.
115. Clayton R. Newall, *The Framework of Operational Warfare*, London: Routledge, 1991, pp. 94–7.
116. J.J.A. Wallace, 'Maneuver Theory in Operations other than War', *JSS*, vol. 19, no. 4 (1996), p. 210.
117. William S. Lind, *Maneuver Warfare Handbook*, Boulder/London: Westview, 1985, p. 109.
118. *The Military Maxims of Napoleon*, tr. from the French by Lieutenant-General George C.D'Aguilar, Introduction and Commentary by David G. Chandler, 1987, rpt., New York: Da Capo, 1995, pp. 55, 58.
119. Williamson Murray, 'Clausewitz: Some Thoughts on What the Germans got Right', *JSS*, vol. 9, nos. 2–3 (1986), pp. 274–7.
120. Niklas Zetterling and Anders Frankson, *Kursk 1943: A Statistical Analysis*, London: Frank Cass, 2000, pp. 135–6.
121. David M. Glantz, *The Soviet Conduct of Tactical Maneuver: Spearhead of the Offensive*, London: Frank Cass, 1991, p. xxiv.
122. Lind, *Maneuver Warfare*, pp. 127–8.

123. John Kiszely, 'The British Army and Approaches to Warfare since 1945', *JSS*, vol. 19, no. 4 (1996), p. 180.
124. Ben-Ari, *Mastering Soldiers*, p. 40.
125. Lind, *Maneuver Warfare*, p. 111.
126. Andrew Gordon, 'Ratcatchers and Regulators at the Battle of Jutland', in Gary Sheffield and Geoffrey Till, eds., *The Challenges of High Command: The British Experience*, Basingstoke: Palgrave Macmillan, 2003, p. 32.
127. Simpkin, *Race to the Swift*, pp. 227–9.
128. Lind, *Maneuver Warfare*, pp. 5–7, 13–14.
129. Ibid., p. 125.
130. Simpkin, *Race to the Swift*, pp. 229, 232.
131. John Keegan, 'Introduction', in idem, ed., *Churchill's Generals*, 1991, rpt., London: Weidenfeld & Nicolson, 1993, pp. 12–13. Keegan is speaking of the relationship between politicians and the generals. But his idea could be transplanted between the senior and junior commanders.
132. Nigel de Lee, '"A Brigadier is only a Coordinator": British Command at Brigade Level in North West Europe, 1944: A Case Study', in Sheffield, ed., *Leadership and Command*, pp. 137–9.
133. Douglas A. Macgregor, *Breaking the Phalanx: A New Design for Landpower in the 21st Century*, Westport: Praeger, 1997, pp. 59–89.
134. Robert Lyman, 'The Art of Maneuver at the Operational Level of War: Lieutenant-General W.J. Slim and Fourteenth Army, 1944–45', in Sheffield and Till, eds., *Challenges of High Command*, p. 91.
135. Nigel de Lee, 'Scandinavian Disaster: Allied Failure in Norway in 1940', in Sheffield and Till, eds., *Challenges of High Command*, pp. 59–60.
136. Allan R. Millett, 'The United States Armed Forces in the Second World War', in Millett and Murray, eds., *Military Effectiveness*, p. 69.
137. Williamson Murray, 'British Military Effectiveness in the Second World War', in Millett and Murray, eds., *Military Effectiveness*, p. 111.
138. Ripley, *US Army*, p. 16.
139. *Joint Warfare of the Armed Forces of the United States*, Ch. IV, pp. i, ix, 4, 10-11, Appendix A.
140. Grant T. Hammond, 'The Perils of Peacekeeping for the US: Relearning Lessons from Beirut for Bosnia', in Edward Moxon-Browne, ed., *A Future for Peacekeeping?* Basingstoke: Macmillan, 1998, pp. 83–4.
141. *Joint Warfare of the Armed Forces of the United States*, pp. v–vi.
142. T.R. Moreman, '"Small Wars" and "Imperial Policing": The British Army and the Theory and Practice of Colonial Warfare in the British Empire, 1919–39', *JSS*, vol. 19, no. 4 (1996), p. 116.
143. Harbakhsh Singh, *In the Line of Duty*, p. 263.
144. Lieutenant-General K.P. Candeth, *The Western Front: The Indo-Pakistan War 1971*, New Delhi: Allied, 1984, p. 157.

145. Lieutenant-General L.P. Sen, *Slender was the Thread: Kashmir Confrontation, 1947–48*, 1969, rpt., New Delhi: Orient Longman, 1994, p. 72.
146. P.K. Vasudeva, 'The Role of the Chief of Defence Staff', *JUSII*, vol. CXXX, no. 542 (2000), p. 566.
147. P.V.S. Jagan Mohan and Samir Chopra, *The India-Pakistan Air War of 1965*, New Delhi: Manohar, 2005, pp. 301–2.
148. Praval, *Valour Triumphs*, p. 284.
149. Freemantle, *Fred's Foibles*, p. 92.
150. Wing-Commander K. Advani, 'Why the Army Should Not Have an Air Corps', *JUSII*, vol. 69, no. 417 (1969), p. 414.
151. Mohammad Asghar Khan, *The First Round: Indo-Pakistan War 1965*, Ghaziabad: Vikas, 1979, pp. 3–4.
152. Candeth, *The Western Front*, pp. 157–8.
153. Fang La, 'What's Wrong with Our Army-Air Cooperation', *JUSII*, vol. CI, no. 425 (1971), pp. 299, 301–2.
154. Shankar Roychowdhury, *Officially at Peace: Reflections on the Army and its Role in Troubled Times*, New Delhi: Viking, pp. 270–1.
155. Singh, *With Honour and Glory*, pp. 56–58; *HODAR 1971–72*, p. 37.
156. Major Dipes Lahiri, 'Mountains and Armed Helicopters', *JUSII*, vol. CII, no. 428 (1972), pp. 273–4.
157. John Killen, *The Luftwaffe: A History*, 1967, rpt., London: Sphere, 1970, pp. 102–29, 177–89.
158. Daljit Singh, 'Military Education in India', pp. 232–3; Panikkar, *Indian Defence*, p. 67.
159. Stephen P. Cohen, *The Pakistan Army*, 1882, rpt., Karachi: Oxford University Press, 1993, p. 125.
160. Rear-Admiral Satyindra Singh, 'The Kargil Review Committee Report: Much to Learn and Implement', *IDR*, vol. 15, no. 1 (2000), p. 27.
161. Vinod Anand, 'India's Military Response to the Kargil Aggression', *Strategic Analysis*, vol. 23, no. 7 (1999), p. 1055.
162. Vinod Anand, *Joint Vision for the Indian Armed Forces*, Delhi Papers, no. 16, New Delhi: The Institute for Defence Studies and Analyses, 2001, pp. 90, 94.
163. Rear Admiral Raja Menon, 'Reflections on India's Nuclear Doctrine and Command and Control', *JUSII*, vol. CXXXIII, no. 552 (2003), p. 204.
164. Vinod Anand, 'Joint and Integrated Logistics System for the Defence Services', *Strategic Analysis*, vol. 25, no. 1 (2001), pp. 102–3.
165. Robert Fox, 'Bosnia and Kosovo, 1992–2000: Balkan Stew', in Thompson, ed., *British and Commonwealth Forces at War, 1945–2000*, p. 335.
166. Lieutenant-General B.M. Kaul, *The Untold Story*, Bombay: Allied, 1967, pp. 354, 374; Freemantle, *Fred's Foibles*, p. 121.

167. Freemantle, *Fred's Foibles*, pp. 107, 109.
168. Lieutenant-Colonel Y.A. Mande, 'Commanding Officer and His Team', *JUSII*, vol. CII, no. 427 (1972), p. 150; idem, 'Our Training Establishments', *JUSII*, vol. CVI, no. 442 (1976), pp. 64–5.
169. Mande, 'Commanding Officer and His Team', p. 150.
170. Niazi, *Betrayal of East Pakistan*, p. 19.
171. P.M. Sinha, 'Air Power in the Information Age', *IDR*, vol. 16, no. 3 (2001), pp. 46–7; Macgregor, *Breaking the Phalanx*, pp. 225–35.
172. English, *Infantry*, p. 191.
173. Chibber, *Leadership*, pp. 35, 38.
174. Depinder Singh, *Field Marshal Sam Manekshaw*, pp. 79–80.
175. *Recruitment into the Officer Corps of the Armed Forces*, New Delhi: USI, 1977, p. 4.
176. Quoted in Chibber, *Leadership*, p. 18.
177. *Recruitment into the Officer Corps of the Armed Forces*, p. 11.
178. Stuart A. Cohen, *Towards a New Portrait of the (New) Israeli Soldier*, Ramat Gan: Bar Ilan University, BESA Centre, 1997, p. 81.
179. *Recruitment into the Officer Corps of the Armed Forces*, pp. 6, 8.
180. Leena Parmar, 'Resettlement of Ex-Servicemen in India: Problems of Army Socialization and Adjustment', in idem, ed., *Military Sociology: Global Perspectives*, Jaipur/New Delhi: Rawat, 1999, p. 448.
181. Cohen, *Pakistan Army*, p. 53.
182. Harbakhsh Singh, *In the Line of Duty*, p. 393.
183. Ibid., p. 393.
184. English, *Infantry*, p. 221.
185. Major-General O.S. Kalkat, 'Approach to Tactics', *JUSII*, vol. CII, no. 427 (1972), p. 137.
186. Admiral S.M. Nanda, *The Man Who Bombed Karachi: A Memoir*, New Delhi: HarperCollins, 2004, p. 149.
187. Ibid., p. 167.
188. Keegan, *Intelligence in War*, p. 89.
189. Vice-Admiral S.H. Sarma, *My Years at Sea*, New Delhi: Lancer, 2001, pp. 177–8.
190. Richard A. Gabriel and Paul L. Savage, *Crisis in Command: Mismanagement in the Army*, 1981, rpt., New Delhi: Himalayan Books, 1986, pp. 12–13, 39–40, 43–7, 66, 202, 230.
191. R. Prasannan, 'Colonial Hangover', in *The Week*, 28 January 2001, pp. 21–2.

FOUR

Indian Army and the Art of War

War fought by the professional armed forces using non-nuclear weapons may be categorized as conventional war. This sort of warfare focuses on battlefield victory.[1] Conventional war could be of two types: limited war and total war. Carl Von Clausewitz conceptualizes conventional war as a duel between two forces equipped with conventional weapons, and the confrontation because of mutual escalation could transform itself into *der totale krieg* (total war). The aim of total or absolute war is complete annexation of enemy territory through the application of the maximum amount of brute force resulting in massive slaughter and bloodletting. Conducting total war requires total mobilization of the belligerant societies for raising and sustaining mass armies.[2] The two World Wars are examples of total war.

Limited wars are characterized by moderate objectives. Unlike total war, the aim is not complete destruction of the enemy's combat potential resulting in annexation of the adversary's homeland. Since the goals are limited, limited wars do not result in complete mobilization of the protagonist countries' manpower and economies. The short duration of such conflicts also prevents total mobilization of the societies engaged in such combat. The area over which such a conflict is conducted along with the means of engagement remain restricted.[3] Till 2003, India had fought only limited wars with China and Pakistan. General K. Sundarji, Chief of the Army Staff of India from February 1986 to April 1988, points out the limited nature of India's conflicts with Pakistan in the following words: 'During the last three wars fought by the two of them against each other, they have displayed enormous restraint and refrained from wilfully targeting civilians, industry or economic infrastructure which is more than many in the West have done.'[4]

Limited conventional war can be attritional as well as manoeuvre oriented. Attritional strategy seeks victory by harnessing the weight of resources. Attritional warfare rests on estimate driven planning rather than on daring skill. Attrition is a gradual and piecemeal process of destroying an enemy's military capabilities. The goal is not complete destruction of the enemy at one go but wearing down the enemy forces in prolonged warfare. Attritional war is characterized by massive set-piece battles based on massed artillery firepower. Commanders start these battles only after careful deliberations and massive concentration of resources. The battle is conducted cautiously with reference to a previously prepared master plan. The aim is to wear out the enemy through firepower and superiority of materials in a step-by-step approach.[5] A recent American theorist Carter Malkasian asserts that attrition warfare is more relevant for the near future. Because of nuclear deterrence, the age of total war aiming at decisive victory is out. Under the nuclear umbrella only limited wars can occur. Attrition warfare due to its slow pace and within the format of limited war, says Malkasian, by avoiding a quick decision contains rapid escalation. And this allows negotiations by the diplomats for deescalating the conflict.[6]

In contrast, manoeuvre war is designed to confuse and disrupt the enemy through a judicious combination of fire and movement.[7] The aim is to focus on the operational level of warfare. Operational art has been defined by Brian Holden Reid, a British military theorist as 'the area of military activity which lies between tactics and grand strategy and seeks to relate the tactical means to strategic ends.'[8] From the very beginning, the Indian Army followed attritional warfare. Despite some sporadic attempts, the Indian Army has proved incapable of conducting manoeuvre operations. Pradeep P. Barua claims that because of the British influence, the Indian Army always emphasized carefully planned defensive-offensive operations rather than high risk mobile warfare.[9] Let us take a look at the evolution of Indian Army's military doctrine[10] and combat effectiveness in the battlefield.

CRESCENT AGAINST TRISHUL: INDIA'S LIMITED WARS WITH PAKISTAN

When India became independent, the military officers had high hopes regarding the future structure of the army in the newly independent country. One such officer, Lieutenant S.G. Chaphekar was heavily influenced by the Allies' Operation Market Garden as well as by the exploits of General Student's German paratroopers[11] and the German development of V 1 and V 2 flying bombs.[12] Chaphekar visualized independent India's army in the following words:

> Modern wars are very mobile, scientific and mechanical. Distances are not insurmountable obstacles. They can be covered by transport planes, gliders and parachute troops. Defence lines on the model of the Maginot Line will be useless in future wars. During the late War parachute troops were landed several hundred miles behind enemy lines; gliders and transport planes carried German troops to Crete, and British troops to Holland. If, therefore, such troops were carried 200 or 300 miles in this late War, it may well be that in the next they will be able to be send 1,000 or maybe 2,000 miles. Moreover, by that time radio-directed planes and rockets will be perfected, and distant objectives may be bombed without the risk of loss of pilots.[13]

In a similar vein, Lieutenant-Colonel C.B. Ponnappa dreamed about a mechanized army. In 1947, Ponnappa wrote: 'The army must be strong and mobile, containing the nucleus of all arms; it must be strong not only numerically, but also in firepower and mobility. Its units must be stationed in strategic places; its motor and air transport sections will ensure rapid movement to any specified locality.'[14]

However, reality was different. Just after Independence, the Indian Army emerged as an unbalanced force. It was an infantry-oriented army lacking essential support services and administrative infrastructure. The government decided to maintain an army of 3,00,000 men and the divisional organization was a copy of the British Army. Each infantry division had three infantry brigades plus one artillery brigade and the army's equipment was obsolete. The infantry had .303 Lee Enfield rifles and the armour consisted of outdated Churchill tanks. The Churchill Armoured Fighting Vehicle (AFV) designed as an infantry support weapon was a forty ton tank with a six pounder gun which came into service in 1941. However, the Churchills were out-gunned and out-manoeuvred by the *panzers* in the North African desert. The condition of the Pakistan Army was not much better either. Pakistan inherited six infantry divisions and one armoured brigade. And all these formations were under-equipped.[15]

In October 1947, Pakistan armed 10,000 tribesmen of the North-West Frontier Province with rifles, machine-guns and mortars. On 24 October they descended on Jammu, Poonch and Kashmir Valley. Sending Indian troops by the land route was time consuming and would have allowed the raiders to occupy Srinagar. The key to the campaign was the airfield at Srinagar. It was necessary to take over the airfield at Srinagar before the raiders moved in to capture it. The Indian Army at this stage lacked adequate number of trained parachutists for carrying out an airborne operation. India responded by airlifting battalions into Srinagar with Sardar Patel requisitioning all available civil aircraft. This campaign showed that Chaphekar and Ponnappa's view of making the army units airmobile was all the more necessary. On 7 November 1947, the Indian Army units managed to launch a limited counter-attack using armoured cars and infantry.[16]

By 1948, when Pakistan inserted regular soldiers into Kashmir, this low-intensity conflict was transformed into a low-intensity conventional war. India responded by deploying two divisions. In 1947–8, the Indian Army's tactics in the Jammu and Kashmir conformed to the Sepoy Army's mountain warfare pattern: occupying the heights dominating the sensitive axes to enable the movement of supply and administrative columns along them. In early 1948, Britain reduced India's oil supplies to prevent India from attempting to launch an all out invasion against Pakistan thereby forcing Nehru's government to the negotiating table. The campaign which lasted for fourteen months resulted in a stalemate. India failed to defend the whole of the Jammu and Kashmir province, but, Pakistan too, had failed to conquer whole of the mountain kingdom. Pakistan was able to capture only one-third of the erstwhile Jammu and Kashmir kingdom. The principal problems faced by the Indian troops were inadequate supplies and insufficient availability of high-altitude warfare equipment. Many of the troops airlifted to the high-altitude area were from the plains and they lacked training to cope with the high altitude and the low temperature prevailing in Kashmir.[17] Nevertheless, Pakistan suffered heavier losses. Between 24 October 1947 and 1 January 1949, Pakistan's casualties (killed, wounded and prisoners) was 20,000, and India's tally was 6,000.[18]

The stalemate of the 1947–8 War encouraged some elements within the Indian Army to come up with a more ambitious operational plan. The spectre of a two front war both with East and West Pakistan haunted the Indian military leaders. Influenced by Napoleon Bonaparte's emphasis on concentration and economy of forces,[19] General J.N. Chaudhuri in 1951 emphasized the necessity of these twin principles to overcome India's strategic predicament in the following words:

> Concentration and economy of force being complementary are best taken together. . . . The modern concept does not necessarily imply a massing of force in one place but it does envisage having them so disposed as to be able to unite and deliver a decisive attack at the correct time and place. Economy of force is perhaps better expressed as economy of effort and implies the correct grouping of resources to achieve concentration where it is required. The holding of large reserves is generally speaking uneconomical unless there is an opportunity of using them speedily to effect the attainment of the aim. The main difficulty of a country fighting on two or more fronts is to obtain decisive concentration with economy of force when it is required.[20]

A framework for an aggressive military strategy was put forward in 1954 by a Brigadier named Rajendra Singh. He asserted:

> It is, therefore, obvious that the national goal must always be defended. In every country there are certain vital points whose occupation will cause general paralysis of the body such as capital, industrial centres, communication nerves and the sources of energy. . . . It is not possible to ensure the security of these goals by Siegfried Lines because not only are such defensive measures expensive but they are not even foolproof as was proved in the last war.[21]

Taking a cue from history, Rajendra Singh continued, 'the decisive battles for the possession of India were fought well within the frontiers of the land. A battle fought, not against the enemy at the gates, but with the enemy marching swiftly many hundreds of miles within the ramparts is a lost battle indeed.'[22] So, he was for fighting the invaders as near the international border as possible. For Singh, robust military response could be achieved if the Indian Army was organized as a three-tier force. The local units at the border would function as a screening force in delaying the enemy. Then, the second echelon force would rush to the battlefield to counter enemy formations and to allow time for general mobilization of the country. The mass mobilized soldiers constituting the third echelon force would defeat the enemy completely.[23] Three years before Rajendra Singh, J.N. Chaudhuri also argued for two types of forces: a mechanized striking force supported by a semi-mobile field army. Chaudhuri wrote:

> One suggestion, however, is to divide the army into two groups during a given period of development. These groups could perhaps be named the striking force and the field army. The former could be small, compact, highly mechanized, equipped with the latest weapons the country can afford and meant for instant use in a military emergency. The striking force might also be given the pick of the officer material, particularly in the higher ranks, so that apart from the technical competence required, modern military thought suited to the needs of the country could be developed.[24]

Both Rajendra Singh and Chaudhuri's views were somewhat influenced by the Soviet doctrine during World War II. The Soviets believed that the skirmishing forces at the frontier would disrupt the deployment of the enemy allowing time for mobilizing the main force geared for annihilating the invaders. But, such a scheme was never implemented in the Indian military. The Indian strategy was to contain Pakistan's military thrusts and hold the Pakistani forces in Kashmir while the main Indian field army could attack the Lahore–Sialkot sector. The primary aim was to destroy as many Pakistani units as possible in order to compel the Pakistani government to sue for peace.[25] Nevertheless, the possibility of a Sino–Pak alliance forced the uniformed personnel to think about reorganizing the Indian military.

In 1965, just before the Pakistani attack materialized, Major Harish Bhatia pushed the concept of jointness. He wrote:

> With the present resources of our country and even taking into account further improvement in these resources and potential available, we cannot expect to have such a large force which can take on any of the enemies on different parts of our borders at all times. Thus we will have to depend on smaller forces, well integrated and mobile, which can be moved swiftly into battle on any front. Such swiftness also requires a

completely integrated command at the highest level. In addition, it calls for integration of logistic support capable of maintaining an integrated force anywhere. Separate headquarters with their own administrative services cannot meet the situation likely to be created on a sudden fast moving attack. Here again, separate administrative services will create confusion and disorganization.[26]

As a prelude to the full-scale offensive, Pakistan in August 1965 sent 10,000 well-armed and well-trained guerrillas into Jammu and Kashmir under the command of Major-General Akhtar Malik. The operation was code named Gibraltar and the objective was to damage communications in order to tie up large number of Indian infantry.[27] On 1 September 1965, Field Marshal Ayub Khan, the dictator of Pakistan initiated Operation Grand Slam.

During the 1965 War, Pakistan enjoyed both quantitative and qualitative superiority in armour as well as in field artillery. Ayub Khan tried his best to raise the quantity of motorized infantry and armour within the armoured division. His idea was to raise the volume of firepower at the cost of manpower both for attack as well as defence. Against India's 608 tanks, Pakistan deployed 756 tanks. India had 90 AMX light tanks against Pakistan's 96 Chaffee light tanks. The French built AMX tanks were designed to function as light cavalry of the modern age. These tanks weighing only fifteen tons each were good for reconnaissance. The Chaffee was a four man light American tank with a 75-mm. gun. Both Pakistan and India had 308 and 332 Sherman tanks respectively. The Shermans were of Second World War vintage. The Shermans were American tanks adopted by the British Army towards the end of World War II. The Shermans had either a 17-pounder gun or a 76 mm. cannon. The core of Pakistan's armoured punch was provided by the 352 Patton tanks. These had 90-mm as well as 105-mm guns. Further, the Pakistanis had improved the Pattons by installing diesel engines. Against them India mobilized 186 Centurions (which were leftovers from World War II) each equipped with a 17-pounder gun. It was much heavier and possessed more firepower punch compared to the Shermans.[28]

Only in the infantry branch, India had a quantitative edge. Against Pakistan's 552 battalions, India pitted 628 battalions. Nevertheless, each Pakistani infantry battalion vis-à-vis an Indian infantry battalion possessed twice the amount of anti-tank and automatic firepower. For instance, each infantry section of the Pakistan Army had two light machine-guns compared to one in the Indian infantry section. Under the new concept of using mobility and firepower, each Pakistani infantry company was trained to hold about 4,000 yards of ground. Ayub Khan took care to equip his forces with 106-mm anti-tank recoilless rifles. Both the Pakistani and the Indian armies used mobile teams of jeep—mounted recoilless guns which provided firepower support to the infantry.[29] These mobile guns also functioned as tank hunter and destroyer units.

At the strategic level, India's objective was to conduct a defensive operation. But, at the operational level, offensives with corps size formations were planned for destroying Pakistan's reserves through attrition. The Indian Army decided to threaten both Sialkot and Lahore. The COAS General J.N. Chaudhuri's pet operational idea was to launch multi-pronged thrusts along several routes into enemy territory. The 1965 operations show that the Indian Army had adopted the Soviet formula which emphasized that operational success could only be achieved by conducting successive attacks along multiple axes.[30]

Lieutenant-General Harbakhsh Singh was in-charge of the crucial Western Command that faced the brunt of Pakistani challenge in September 1965. He elaborates the operational aim of his command in the following words:

> My strategic concept for the offensive had the dual aim of destroying the PAK forces along with the acquisition of a large chunk of enemy territory which, I thought, would give us an invaluable bargaining lever in the political parleys inevitable on the conclusion of hostilities. I had decided upon the capture of ICHOGIL Canal. . . . The successful conclusion of this offensive would, therefore, gain for us a vast expanse of enemy territory simultaneously with the whittling down of her military potential. Moreover, our forces would be well established on a contiguous massive water obstacle, readily poised to pose a threat either towards LAHORE or the strategically vital road link, LAHORE-WAZIRABAD, should we intend to pursue the offensive further.[31]

In the Chamb region existed the Akhnur 'dagger' (a piece of Pakistan's territory) pointing towards the one and only bridge across the river Chenab. This bridge connected Poonch, Rajouri and Naushera with Jammu and the rest of India. If Pakistan could capture this bridge then the western part of Jammu would be at its mercy and the town of Jammu could also be threatened. Further, the flat ground was highly suitable for Pakistan to deploy the much vaunted Pattons. At 4 a.m. just before dawn, one Pakistani infantry division supported by two armoured regiments attacked Chamb. The Pakistanis seemed to be repeating the classic *Wehrmacht* formula of launching an armoured attack at dawn to surprise the enemy defence. The aim was to capture the strategically crucial Akhnur bridge so that the Indian troops stationed in Naushera, Rajouri and Poonch would be isolated. Within five days, the Pakistanis had reached Jaurian not far from Akhnur. At this juncture, the Indian Army launched twin thrusts towards Lahore and Sialkot, as a result of which the front stabilized. The capture of Sialkot by India would have enabled the Indian armoured elements to race south along the Grand Trunk Road and then Lahore could be isolated from the west.[32] So Pakistan was forced to re-deploy troops from Chamb to check the Indian thrusts in the Lahore–Sialkot sector.

The elite Pakistani 1st Armoured Division moved into Khem Karan sector and pushed back the 4th Indian Infantry Division towards the village of Asal Uttar. However, the Pattons got bogged down in the water logged sugar cane fields of Kasur–Khem Karan just as the mighty Tigers of the *Wehrmacht* got bogged down in the marshes of Kursk in the summer of 1943. And this allowed Harbakhsh Singh to turn the whole area into a tank trap. The Pakistanis lost about 100 Patton tanks from their 1st Armoured Division. In 1973, during the Yom–Kippur War, the Israeli armour had attacked the Egyptian positions on the Suez Canal without infantry and mortar and had suffered heavy losses. The Pakistani armour also made the mistake of advancing into Khem Karan without supporting infantry riding in the Armoured Personnel Carriers (APCs).[33] Operation Grand Slam was turning out to be 'Operation Grand Shame' for the Pakistan Army. However, the Indian Army was no better in exploiting the potential victory lurking in the horizon.

The Pakistanis used recoilless guns and Cobra anti-tank guided missiles to stop the armour attack launched by Harbakhsh Singh's command. The Ichogil Canal which runs almost parallel to the India–Pakistan border connects the Ravi with the water systems of the Sutlej. Along the Ichogil Canal, the Pakistani infantry took up a defensive position. Some armed with machine-guns manned pillboxes and others with light machine-guns and grenades supported by artillery and mortars took positions within the buildings of the small town named Dograi. The entrenched Pakistani infantry proved to be a tough nut to crack for the Indian infantry. The Indian armour unlike Guderian's *panzers* at river Meuse in 1940 could not cross a water obstacle defended by a well-entrenched enemy infantry.[34] The Indian Army was not yet proficient in conducting breakthrough battle and a stalemate followed.

India's casualties in the 1965 War were 12,000. On the Lahore front the Indians lost 2,000 killed and 5,000 wounded. And the Pakistani combat casualties including prisoners captured by India numbered 4,500. In terms of manpower losses, the whole campaign was quite small when compared to the war between the North Vietnamese and the French. For instance between March and April 1954, in a single battle at Dien Bien Phu, General Vo Nguyen Giap lost 19,000 men (6,000 killed, 8,000–10,000 seriously wounded, 2,500 captured and the rest missing). Against 126 tanks lost by India, Pakistan lost about 471 tanks including 252 Pattons, 161 Chaffees, 46 Shermans and 12 other tanks.[35] G.D. Bakshi sums up the Indian offensive thrusts during the 1965 campaign as straightforward and disjointed frontal pushes. In addition, these thrusts were not inter-linked with each other to generate a systemic crisis for the enemy.[36] Unlike the IDF, the Indian Army had failed to emerge as a master of mobile warfare.

At least a segment of the Indian officers were thinking seriously about conducting manoeuvre warfare centred on an armoured offensive in the Rajasthan desert. Brigadier Surinder Harsh was one such officer. In an article which he wrote in 1969 in the *Journal of the United Service Institution of India*, he contrasted the actions and writings of Field Marshal A. Wavell and Field Marshal Bernard Montgomery with the dazzling performance of the Israeli armoured units in the 1967 War in the Sinai Peninsula. Harsh argued that the Indian Army's focus on positional warfare which was a legacy of British influence, should be replaced by mobile warfare. Within the tactical format of the Indian Army, armour is used as an infantry support weapon, a military doctrine followed by the British during World War II. Harsh asserted that it should be the other way round mobile and mechanized infantry should cooperate with armour operations. Probably

Harsh was influenced by the *Wehrmacht* and the IDF. It is to be noted that during the 1956 campaign while the Egyptians used tanks as infantry support vehicles, the Israelis like the *Wehrmacht* used tanks for conducting long distance rapier like thrusts. Harsh pointed out that the Indian armour should be capable of conducting night operations. The tactics of the Indian Army, concluded Harsh, ought to emphasize rapid long-range armoured strikes supported by infantry carried in the APCs inside the depth of enemy defence.[37] So, Harsh was a pioneer in the Indian context in theorizing about the necessity of conducting 'deep battle'. However, the next campaign showed that the Indian Army was incapable of translating such sophisticated theoretical concepts into reality.

In the late 1960s, India had twenty infantry divisions against Pakistan's seventeen infantry divisions. After the 1962 debacle against China, India maintained eleven mountain divisions for deterring the Dragon. During a crisis with Islamabad, at least some of these mountain divisions could be re-deployed by India against Pakistan. After being denied victory in the 1965 War, Pakistan started rebuilding its armoured strength. Pakistan imported T-59 tanks from China. The size of the Indian Army remained static at 8,50,000. India settled for the indigenously manufactured Vijayanta tanks. As regards light tanks, the Indian armoured regiments replaced the Stuart (US M-3) tanks with PT-76 tanks acquired from the USSR. The PT-76 was an amphibious AFV with a 76-mm gun. The Indian Army acquired these tanks with the aim of using them in the swampy and riverine tracts of East Pakistan. Just before the 1971 War, the PT-76 tanks were issued with High Explosive Anti-Tank ammunition. This not only raised the anti-tank capability of the AFVs but also strengthened the morale of the crews. Towards the end of the 1960s, the Soviet military influence on the Indian Army continued to increase. Several Indian tank officers were sent to the Soviet Union for training. In September 1969, Major Pawittar Singh belonging to the 69th Armoured Regiment was sent to the Tank Academy at Vistryl to attend a six-month long tactical course. The Indian Army took care to jointly train the infantry and the armoured regiments that were to be deployed against East Pakistan. For cooperation with the rapidly moving tank columns, some infantry regiments like the 20th Maratha Light Infantry were equipped with wheeled Skot APCs. Tactical concepts, standardized battle procedures and drill for infantry-tank joint operations were worked out.[38]

In late 1971 Lieutenant-General Jagjit Singh Aurora commanding the Indian Army contingents against East Pakistan had six infantry divisions and three independent infantry brigades. Of his three armoured regiments, one was equipped with T-54 and the other two with PT-76 tanks. The T-54s were Soviet built medium tanks, which the North Vietnamese Army used with good effect against the American M-41 light tanks. Aurora's task was to smash the Pakistani war machine in three weeks and to liberate the whole of East Pakistan. However, the Indian Army's high command was divided regarding the importance of the enemy's capital. General Sam Manekshaw (the COAS) believed that the capture of the enemy's capital was not the prime objective. Manekshaw was against giving first priority to the capture of Dacca. He argued that the capture of the important ports like Chittagong and Khulna, etc., would end the war. But, Manekshaw came under the pressure of the field commanders who wanted to concentrate on the enemy's capital. The Chief of Staff of the Eastern Army Lieutenant-General J.F.R. Jacob assumed that capture of Dacca would result in the collapse of Pakistani military resistance in East Pakistan. Jacob argued that India should use its superior naval capability to block the above-mentioned ports. Finally, Manekshaw bowed to the pressure of the field commanders.[39] With hindsight, it could be argued that the Jacob plan was better than the Manekshaw plan. Since the Indian Eastern Army followed roughly the Jacob plan, its salient features are worth examining in detail. Jacob in his draft plan submitted in March 1971 pointed out:

> The final objective was to be Dacca, the geopolitical and geostrategic heart of East Pakistan. Thrust lines were to be selected and to isolate and bypass Pakistani forces to reach the final objective. Subsidiary objectives were to be selected with the aim of securing communication centres and the destruction of the enemy's command and control capabilities. Fortified centres of resistance were to be bypassed and dealt with later. Preliminary operations were aimed at drawing out the Pakistani forces to the border, leaving key areas in the interior lightly defended.[40]

The Indian attack against East Pakistan started on 3 December

1971. The monsoon was over and the ground was somewhat suited for mobile operations from November onwards. Three Indian corps invaded East Pakistan—Lieutenant-General T.N. Raina's II Corps advanced into the western part of East Pakistan, simultaneously, XXXIII Corps under Lieutenant-General M.L. Thapan attacked along the north-west portion of East Pakistan. Thapan was a veteran of the 1965 War when he commanded a division. And, Lieutenant-General Sagat Singh's IV Corps operated along the north-east portion of East Pakistan. The Pakistani high command deployed thirty-five infantry battalions, six artillery regiments and one armoured regiment plus two armoured squadrons for defending East Pakistan. In 1969, the North Vietnamese Army used light Soviet PT-76 tanks in the paddy fields of South Vietnam. Similarly, the Indian Army also used PT-76 while fighting in the paddy fields of East Pakistan criss-crossed by numerous rivers and streams. In East Pakistan, the PT-76 and T-54/55 tanks belonging to the Indian Army clashed with the Chaffee tanks of the Pakistan Army.[41]

During the 1971 War, while the Indian Army adopted an offensive posture in East Pakistan, it took the stance of strategic defence against West Pakistan. Lieutenant-General K.P. Candeth was in-charge of the Western Command and he displayed a slow, deliberate and cautious approach as the war unfolded and he had reasons not to take a bold approach. The Chinese maintained one division in western Tibet could reinforce this theatre faster than India. The Indian Army feared that if Beijing decided to aid Pakistan actively, then the People's Liberation Army could advance towards Leh in Ladakh with two divisions.[42]

The Pakistani military strategy for defending East Pakistan, was to launch several heavy offensive thrusts from West Pakistan. The Pakistani army's senior officers were probably influenced by Moshe Dayan's armoured thrust across the Sinai Peninsula. The Pakistani high command nursed the dream of launching a massive surprise attack from West Pakistan backed by armoured forces designed to strike deep inside India's heartland—Punjab.[43] And this, perceived the Pakistani high command, would result in engaging and containment of many units in India's Western Command and prevent their transfer to the crucial Eastern Command facing East Pakistan. The Pakistani generals probably hoped that their attack on India's Western Command would also force the Indian high command to transfer units from East Pakistan towards the West.

The Pakistan Army in the Western Front (the region stretching from Kashmir to Sind) deployed ten divisions. Of them two were newly raised without their full complement of equipment and manpower and therefore their combat effectiveness was not that high. Two armoured divisions and an independent armoured brigade backed the infantry divisions. The 12th Pakistan Division was at Pakistan occupied Kashmir. At the Kotli–Poonch sector was the 23rd Pakistan Division. The II Pakistan Corps was in-charge of the Sialkot Sector. This corps had the 8th, 15th, and 17th Infantry Divisions plus the 6th Armoured Division. The IV Pakistan Corps was in Lahore. It consisted of the 10th and 11th Infantry Divisions backed by the 8th Independent Armoured Brigade. This corps' objective was to threaten Amritsar and Khem Karan. The headquarter of the I Pakistani Corps was Multan. It had 7th and 33rd Infantry Divisions with 25th Infantry Brigade plus the 1st Armoured Division. This corps threatened Rajasthan. Finally, the Pakistani 18th Infantry Division with two armoured regiments were located at Hyderabad in Sind.[44]

At 8.30 p.m. on 3 December 1971, the Pakistan Army launched its principal attack on the Poonch and Chamb sectors. The Indian collapse at the Chamb Sector was due to want of anti-tank defence. Lack of time and inadequate availability of mines meant that the minefields were not prepared properly. And India's one squadron consisting of eleven AMX tanks was destroyed by the two Pakistani tank regiments. Further, the Indian infantry were not trained to stop enemy tank attacks.[45] Table 4.1 gives an indication of the casualties suffered by the Indian Army in the Western Command. However, the Pakistanis were not Israelis.

TABLE 4.1: MANPOWER LOSSES OF THE INDIAN ARMY IN THE WESTERN COMMAND DURING THE 1971 WAR

Ranks	Killed	Wounded	Missing	Prisoners of War
Officers	98	222	10	9
JCOs	70	142	8	16
Other Ranks	1,589	3,996	245	504
Total	1,765	4,360	263	529

Source: Ministry of Defence, Government of India Annual Report: 1971-72, p. 33.

And certainly the Pakistan Army was not like the super professional *Wehrmacht.* The Pakistani attacks penetrated only 20 to 30 miles into India.[46]

The Indian Army's combined casualties both in the Eastern and the Western Commands came to about 3,153 killed and 8,192 wounded.[47] Table 4.4 shows the equipment loss of the Indian Army, negligible compared to the strategic advantages gained by India. Despite romanticization by some, the role of the *Mukti Bahini* in causing the military collapse of Pakistan was minor.[48] 1971 was definitely the Indian Army's 'finest hour'. For the first time in the post-World War II era, a country's armed forces had been able to attack and dismember an enemy country. The Indian Army's campaign in East Pakistan changed the balance of power in South Asia. The enormity of the Pakistani collapse in

TABLE 4.2: PAKISTANI PRISONERS OF WAR IN INDIA AFTER THE 1971 WAR

Types of Prisoners	Numbers
Army	54,753
Navy	1,409
Air Force	836
Total	56,998
Paramilitary	18,287
Grand Total	75,285

Note: Police and civilian prisoners not included.

Source: Ministry of Defence, Government of India Annual Report: 1971–72, p. 38.

TABLE 4.3: LOSS OF EQUIPMENT BY THE PAKISTAN ARMY IN THE 1971 WAR

Equipment	Numbers	
	East Pakistan	Western Front
Tank	72	181
25-pr. Gun	4	–
37-mm. AA Gun	34	–
.30 AA Gun 145-mm.	5	–
RCL Gun 106-mm.	51	8
Gun 37-mm. Super Long Range	1	–
105-mm. Pack Howitzer	43	1
120-mm. Brandt Mortar	44	3
3.7-inch Howitzer	11	–

Source: Ministry of Defence, Government of India Annual Report: 1971–72, pp. 34, 39.

TABLE 4.4: EQUIPMENT LOSS OF THE INDIAN ARMY BOTH IN THE EASTERN AND WESTERN SECTORS DURING THE 1971 WAR

Items	Numbers
120. mm-Brandt Mortar	7
106-mm-RCL Gun	34
40-mm-L60 Gun	3
40-mm-L70 Gun	1
25-pr. Gun	9
5.5-inch Gun	2
130-mm Gun	6
Tank	69*

Note: *Twenty-four tanks lost in East Pakistan and the rest in the Western Sector.

Source: *Ministry of Defence, Government of India Annual Report: 1971–72*, p. 33.

East Pakistan as shown in Tables 4.2 and 4.3 allows one to categorize the campaign centred round Dacca as an Asian Stalingrad. Even the IDF with its sterling battlefield effectiveness has till date failed to construct such a scenario against its Arab opponents. Then can we categorize the Indian Army as the super army of post-World War II period, i.e. the true heir of the *Wehrmacht*?

Both the Indian Army and the IDF fought limited wars. However, the Indian Army's performance during both the 1965 and 1971 wars cannot be compared with the IDF's rapid thrusts during the Sinai Campaign of 1956 and later in 1967 during the Six Day's war. In both the 1965 and 1971 wars, the Indian Army failed to surprise its opponent through rapid mobile operations. The Israeli ground forces were able to disrupt the enemy through shock effect of relentless movement and surprise. The Israeli officers unlike the Indian officer cadre closely studied the *Wehrmacht's* operations. Lieutenant-Colonel Uri Ben Ari after studying the *Blitzkrieg* operations adopted Guderian-style methods. Again, both the Indian and the Pakistan armies used tanks in penny packets as infantry support vehicles. To give an example about the Indian Army, during the 1965 War, when the 3rd Jat advanced, it was supported by three tanks of the 14th Horse. Both the Indian and the Pakistani armies were uncomfortable with rapier like armoured thrusts as conducted by *Generalfeldmarschall* Von Manstein. In contrast to the Pakistani and Indian generals, General Haim Laskov of the IDF accepted that the armoured attack columns should operate independently. Generals Yigael Yadin and Moshe Dayan emphasized battlefield mobility using mechanized infantry-tank combine.[49] During the 1971 War, Major-General Dalbir Singh commanding the 9th Infantry Division under the II Corps of General Raina in East Pakistan attacked the Pakistani strong points frontally. Major-General Sukhwant Singh, the Deputy Director of Military Operations at the Army Headquarter during the 1971 War writes that, in characteristic Indian Army style, Dalbir Singh cleared the Pakistani delaying positions step by step conducting deliberate attacks supported by heavy concentration of artillery. His division took four days to cover 30 miles defended by one Pakistani battalion.[50] In general, the Indian attack was characterized by artillery softening up the chosen enemy strong points for about four hours and then infantry and tanks advancing behind a creeping barrage.[51] All these techniques are characteristics of attritional paradigm of warfare as developed by Bernard Montgomery in the North African desert from late 1942 onwards.[52]

Nevertheless, there were elements of commonality between the battles of Indo–Pak wars and Arab–Israeli wars. The resolute fighting during the 1965 War by the entrenched Pakistani infantry at Ichogil Canal which stopped the onrush of Harbakhsh Singh's tank fleet and infantry could be compared to the *Pakfront* prepared by the Egyptian infantry on the banks of the Suez Canal in October 1973. Both the Pakistani and the Egyptian infantry were able to dent tank charges at close quarter.[53]

The military collapse of the Pakistan Army in East Pakistan was more due to the strategic blunder of the Pakistani high command and tactical ineptitude of Lieutenant-General A.A.K. Niazi rather than to the brilliant generalship of the Indian high command. East Pakistan was totally cut off from West Pakistan; no land route existed for aiding the Pakistani garrison in East Pakistan from West Pakistan. The air route was lengthy and non-operational because the Indian Air Force had gained air superiority. The sea route was also very lengthy and there also the Indian Navy was in a superior position vis-à-vis its Pakistani counterpart. Moreover, Pakistan lacked cargo capacity for large-scale supply to East Pakistan via the sea. To some extent, the position of the Pakistan Army in East Pakistan was somewhat similar to the position of the German Sixth Army in Stalingrad during late 1942. In fact, Niazi's position was worse than that of Von Paulas during the winter of 1942. For Paulas there was a possibility of *der Manstein kommt* (that Manstein might come to his rescue). But, there was no possibile hope for a successful Pakistani ground attack from West Pakistan breaching the Indian units that were encircling Niazi at Dacca. Manstein's ground forces reached within 20 km. of *festung* Stalingrad, whereas the attacking Pakistani ground forces in West Pakistan were more than 1,000 km. away from fortress Dacca (if at all the term could be applied). Paulas unlike Niazi got at least some amount of air supply.[54] The Pakistani high command believed that the Indian Army would attempt to occupy some areas of East Pakistan and then set up a government under *Mukti Bahini*. So, Niazi's deployment was designed to defend every possible inch of East Pakistan's territory to prevent any attempt to set up a satellite Bangladesh government. He ordered towns like Jessore, Bogra, Rangpur, Comilla and Chittagong to be turned into fortresses.[55] In addition, Niazi had to guard the 4,000 km. long border with India. Like the 1956 Egyptian deployment along Sinai, Niazi deployed his troops in isolated positions which were not interconnected to provide mutual aid to each other. This enabled the Indian troops like the Israelis to isolate some strong points and destroy them one by one and bypass several strong points.[56]

The terrain of East Pakistan is suited for infantry-oriented defensive fighting. East Pakistan covers an area of 56,000 sq. miles with a population of 75 millions scattered in 60,000 villages. The region is a flat deltaic plain, about 40 ft. above sea level, interspersed with big rivers—Ganga, Brahmaputra and Meghna running north to south. These rivers give birth to numerous streams as they approach the Bay of Bengal. Most of these streams are more than 1,000 ft. wide at places. The heavy monsoon leaves the entire region covered with lakes and vast stretches of marshes make cross-country movement very difficult. Again, the hills of Sylhet district and Chittagong area are thickly wooded.[57] East Pakistan is somewhat similar to the Pripet Marshes, the region from where the Soviet infantry continuously harried the German columns.[58]

Niazi's faulty tactical deployments and his lack of 'will to war' accelerated the Pakistani collapse. Motorized columns could not reach Dacca without crossing at least one of the two rivers—Brahmaputra and Meghna.[59] Lieutenant-General 'Jack' Jacob who was the Chief of Staff of the Eastern Army under General Aurora says in his memoirs: 'Fortunately for us, the Pakistanis had concentrated their troops in the towns. Had they chosen to defend approaches to the river crossing sites, we would not have been able to cross the river and reach Dacca.'[60] In accordance with the Soviet doctrine which was current in the Egyptian Army of the 1970s, the Pakistan Army could have constructed a series of fortified lines taking advantage of the natural obstacles along the major axis of advances which the Indian formations could not afford to bypass. Behind these lines Niazi could have concentrated mobile troops to launch counter-attacks. The Pakistani infantry could have formed a series of anti-armour urban hedgehogs in Dacca supported by an armoured force geared for counter-attack. Infantry oriented urban struggle in which the Pakistani soldiers could have followed the Soviet Union's 'hugging tactics' would not only have slowed down the Indian advance but might also have caused considerable loss among the Indian infantry.[61] And perhaps a long drawn out campaign would have resulted in intense international pressure forcing Indira Gandhi to suspend military operations.

However, Niazi, a portly Punjabi did not go on fighting till 'five minutes past midnight'. As the Indian troops approached Dacca, he was shaky and in tears. Yahya Khan in order to prolong the Pakistani resistance was telling lies to Niazi that very soon the Americans and the Chinese would come to help him.[62] History seemed to be repeating itself. In late 1942, Hitler tried to revive Paulas' spirit by falsely saying that Goering's air supply and Manstein's armoured counter-attack would save the Sixth Army from the Soviet pincers closing around Stalingrad.[63] When Niazi met his Indian captors, he sobbingly said '*Pindi men baithe hue haramzadon ne marwa diya.*' (Those bastards sitting at Rawalpindi [the Pakistan Army Headquarter] have let me down).[64] If Niazi had asked the world 'how well did I fight?', then history's verdict would have been 'very badly indeed'.

As the war progressed successfully in East Pakistan, there was a ferment of ideas among the military theoreticians of India. Major Yogi Saksena who wrote one year after the 1971 India–Pakistan War concluded that the superpowers would not tolerate radical alteration in the balance of power of any region. So, any protracted conflict between two nations would not be allowed by the big powers. Hence, all future wars are going to be of limited duration. Thus, the current stocks of materials available in hand rather than the actual industrial potentials of nations, says Saksena, are going to shape the nature of the limited war in near future.[65]

One civilian analyst, Ravi Rikhye, asserted that the 1971 War showed the primacy of AFVs and mechanized infantry battalions. He wanted to increase the number of mechanized infantry battalions and armoured regiments. Rikhye rightly conjectured that close cooperation between the mechanized infantry battalions and armoured regiments is the path to success. Rikhye wanted the APCs not only to function as a medium of transport for the mechanized infantry but also to provide extra firepower. So, he wanted the APCs to be fitted with Anti-Tank Guided Missiles. The APCs, continued Rikhye, should have the agility of a light tank and firepower of a medium tank.[66] After the Six Days' War, the IDF discarded the principle of mechanized infantry cooperating with tanks. The IDF's high command was also reluctant to arm the infantry with APCs and anti-tank weaponry. The net result was that the Egyptian Army licked the IDF badly during the 1973 War. It is to be noted that unlike the arrogant IDF's theorists who propounded 'all tanks' doctrine,[67] Rikhye did not neglect infantry–armour cooperation, the cardinal principle of *Wehrmacht's* success during World War II.[68] Next, Rikhye introduced a much more sophisticated concept named Tricap. It stood for integration of tanks with mechanized and airmobile infantry into a composite divisional size formation capable of deep penetration in enemy territories. Rikhye also accommodated Chaphekar's focus on the necessity of making the troops airmobile in order to increase the 'speed of battle'. A Tricap division with 13,000 men included helicopters in an anti-tank role, plus medium tanks and towed artillery supported by infantry transported by aircraft. However, Rikhey's Tricap concept is not totally a piece of original thinking but derived from the US and the NATO's concepts designed to counter the Soviet armoured juggernaut in central Europe.[69]

The cooperation of tank and mechanized infantry was displayed by the Syrians who during the 1973 War detached attack columns composed of tanks supported by infantry riding APCs. It must be admitted that the Syrian tank force was much more advanced compared to the Indian armoured formations. During the 1973 War, the Syrians used night fighting optical and infrared devices.[70] Despite Brigadier Harsh's urgings as early as 1969, the Indian Army was not equipped with all these devices.

Though the Indian armed forces failed to introduce concepts for conducting air–land war, the flow of new ideas ultimately derived from the West, was refreshing indeed. In contrast to the concept of manoeuvre warfare as chalked out by Rikhye, Major-General O.S. Kalkat preached positional attritional warfare. In 1975, the manpower ceiling of the Indian Army remained static at 8,37,800.[71] For defending India's western border against Pakistan, Kalkat came up with a defensive-offence plan to be implemented at the level of battalion-sized units. He commented that defence is going to be a war of attrition. A string of mutually supported localities would inflict heavy casualties on the invaders. Kalkat emphasized fire support to back up the defensive positions. However, he was not for pure defence and also wanted the defenders to launch limited counterattacks to clear the dent made by the invaders along the defensive line.[72] In 1974, a civilian strategic analyst, Ravi Kaul, had written about conducting limited counter-attacks along the border. He says:

> Deterring Conventional War is also essential. We must also be prepared to fight localized limited wars and meet border incursions. While meeting these we must make counter-challenges at points of our own choosing. This is the only effective method of countering nibbling tactics by an enemy, for when he stands to lose at least as much or more than he might gain, the likelihood of incursions by the enemy would be greatly reduced.[73]

From the 1980s, the Western militaries underwent a renaissance in military thought. The focus shifted from attritional war to manoeuvre war. The Soviets believed that by conducting merely a passive defence, the enemy could not be defeated, and only a strong counter-attack could destroy the enemy force. For this purpose the Russians came up with highly mobile reserves capable of rapid manoeuvre towards threatened axes. As part of the new force structure, the Soviet military introduced the concept of Operational Manoeuvre Group (i.e. OMG). With motor rifle and tank divisions (consisting of T-72 tanks), the

OMGs in order to prevent the enemy any breathing space were geared to fight a single gigantic breakthrough battle at the very beginning of the operation. According to the Soviet doctrine, victory is to be gained by forcing breakthroughs against the weakest hostile ground forces and then encircling the strongest of the enemy groupings.[74] This is a sort of Soviet modification of the *Blitzkrieg* doctrine.

During the 1980s, India's operational doctrine against Pakistan emphasized penetration by high-speed mobile armoured columns supported by mechanized infantry riding the APCs. Integration of tanks with infantry riding in the BMPs was a technique adopted from Soviet military thought. From 1980 onwards, the Defence Ministry took steps to equip selected infantry battalions with the APCs to provide the infantry with battlefield mobility, protection and firepower. Towards the end of 1980s, the Indian Army came up with integrated combat formations designed for deep penetration into Pakistan, in case of war. The combat team included T-72 tanks, with mechanized infantry in BMPs supported by self propelled artillery. And above these steel monsters flew helicopters of the Army Aviation Corps.[75] The Indian generals were keeping track of what was going on all over the world. The importance anti-armour attack helicopters became clear during the 1982 Bekaa Valley War, when the French-built Gazelle helicopters of the Syrian armed forces using missiles destroyed several Israeli tanks.[76]

The most famous proponent of conducting limited manoeuvre war was General K. Sundarji. Sundarji while planning his offensive focused on the twin principles of fire and movement[77] which seems he lifted from the theory and practices of the *panzer* knight Colonel-General Heinz Guderian.[78] For testing the integrated armour-infantry teams for conducting mobile warfare, Sundarji held an exercise named Operation Brasstacks in 1986–7. This exercise involved 1,50,000 troops and took place east of the Indira Gandhi Canal in Rajasthan. The Indian military doctrine emphasized that unlike the 1965 and 1971 wars, Punjab would no more be the chief theatre of operations against Pakistan. Deep penetration by fast moving Indian *panzers* anyway would have been impossible because of the ditch-cum-bund defence erected by Pakistan in Pakistani portion of Punjab. The Indian Army decided to assume a defensive posture in Jammu & Kashmir and use its qualitative and quantitative superiority in armour by conducting a *Kesselschlacht* (German-style cauldron battle) in the Rajasthan–Sind theatre.[79] The point to be noted is that deep operations by armoured columns had never been a strongpoint with the Indian Army. In terms of scale of operation (which included frontage and depth) and average rate of advance per day, etc., the Indian Army was no match compared to the Soviet Army's deep offensive thrusts during the Second World War.[80] The Indian armour theoreticians planned that armoured advance of 40–60 km. a day would be maintained. But, both in the 1965 and in the 1971 wars, the Indian armoured formations averaged less than 10 km. a day.[81]

During the last decade of the twentieth century, India's edge in the sphere of conventional land power started eroding slowly.

TABLE 4.5: STRENGTH AND EQUIPMENT OF THE PAKISTAN ARMY IN 1995

Total Troop Strength	5,20,000	
Corps HQ	9	
Armoured Division	2	
Independent Armoured Brigade	7	
Infantry Division	19	
Independent Infantry Brigade	9	
Equipment	*Numbers*	*Types*
Main Battle Tank	1,950	120 M-47 + 280 M-48 A-5 + 50 T-54/55 + 1,200 Chinese Type 59 + 200 Chinese Type 69 + 100 Chinese Type-85
Armoured Personnel Carrier	820	M-113
Towed Artillery	1,572	85 mm. 200 Chinese Type 56 + 105 mm. 300 M-101 + 56 M-56 Pack + 122 mm. 200 Chinese Type 60 + 400 Chinese Type 54 + 130 mm. 200 Chinese Type 59-1 + 155 mm. 30 M-59 + 60 M-114 + 100 M-198 + 203 mm. 26 M-115
Self-Propelled Artillery	240	105 mm. 50 M-7 + 155 mm. 150 M- 109 + 203 mm. 40 M-110
Air Defence Guns	2,000	14.5 mm., 35 mm., 37 mm., 40 mm., M-1, and 57 mm.
Surface to Air Missiles	350	Stinger, Arza, etc.
Surface to Surface Missiles	18	Haft 1 and Haft 2

Source: J. Baranwal, ed., *SP's Military Yearbook: 1995,* New Delhi: Guide Publications, 1995, p. 213.

TABLE 4.6: ORBAT OF THE INDIAN ARMY IN 1998

5	Regional Commands
11	Corps (3 Strike Corps and 8 Holding Corps)
10	Mountain Divisions
18	Infantry Divisions + 7 independent Infantry Brigades
4	Rapids
1	Artillery Division
3	Armoured Divisions + 5 independent Armoured Brigades
	Armour Strength = 3,600 (1,900 T-72 + 700 T-55 + 1,000 Vijayanta. 500 Vijayanta in store)
1	Parachute Brigade

Source: Sanjay Badri-Maharaj, 'The Nuclear Battlefield: India vs Pakistan', *Indian Defence Review*, vol. 14, no. 2 (1999), pp. 75–6.

In 1993, the Indian Army had 1,250,000 troops. These were organized in twelve corps. Of these eight were deployed in the Western Front against Pakistan. In response, Pakistan also deployed eight corps which were better equipped than their Indian counterparts. The total manpower of the Pakistan Army was 5,20,000 and the bulk of them were deployed against India. So, India did not enjoy any significant quantitative or qualitative superiority over Pakistan. The cutting edge of the Indian Army was the 1st, 2nd and 21st Strike Corps. Each such corps was built around a nucleus of a single armoured division and two infantry divisions. The Indian armoured punch was provided by the fifty-eight armoured regiments and thirty-seven of them had T-72 tanks. To counter the threat posed by the Indian armour, in 1996 Pakistan ordered 320 T-80 Main Battle Tanks armed with 120-mm. smoothbore gun from Ukraine. By 1999, these tanks were delivered to the Pakistan Army. In response to Pakistan's acquisition of T-80s, the Indian Army procured T-90s.[82] The strength of the Indian Army remained constant in 1995. Due to the financial crunch, the balance of armour between India and Pakistan has fallen from 1.99: 1 in 1993 to 1.4: 1 in 1997.

From the 1980s, the Indian Army is preparing to fight both China and Pakistan in the mountains. A High Altitude Warfare School at Gulmarg trains military mountaineers. The fight for Siachen Glacier started when the Indian troops occupied the 15,000 ft. high Saltoro Ridge. The Indian and the Pakistani armies continue to fight each other along the Karakoram Glaciers with the help of long-range artillery. Ammunition for the Indian contingent has to be air-dropped near the guns. In contrast, at Siachen, Pakistan possesses a network of roads to move the guns forward and supply them with ammunition. However, artillery skirmish does not prove to be deadly for either party because the munitions explode only after penetrating the soft snow.[83]

Quasi-conventional warfare represents a conflict somewhat less intensive than conventional warfare but certainly more intensive than low-intensity warfare. Quasi-conventional warfare might involve irregular forces somewhat supported by regulars against the army of the enemy state. The action at Kargil during 1999 is an example of quasi-conventional war. In the 1999 Kargil Operation, the Pakistani COAS General Pervez Musharraf's plan was to occupy the dominating heights overlooking the Srinagar–Kargil–Leh road which were temporarily vacated by the Indian soldiers during the winter. In the next phase the aim was to cut off the Indian Army's line of communications in the Ladakh Sector thereby undermining the Indian position at Siachen. Pakistan sent 3rd, 4th, 5th and 7th Northern Light Infantry (NLI) battalions into the Kargil Sector and the 6th, 11th and 12th NLI battalions were deployed in the Drass Sector. The 3rd and 8th Indian divisions were deployed against the intruders. The Indian XV Corps based in Srinagar controlled the operation.[84]

The fight was as much against nature as against the enemy. The battleground was full of knife-edge ridges. In winter avalanches, of about 40 to 60 ft. of snow are quite common. To silence the enemy machine gun nests at the top of the peaks, the *jawans* weathering sub-zero temperature, tried to climb the peaks with pitons and ropes carrying survival rations of *gur* and *chana*.[85] Nature was equally cruel to both the invaders as well the defenders. The diary of Captain Husain Ahmad belonging to the 12th NLI portrays the inhuman face of war in the high Himalayas. The jottings of 12 April ran as follows: 'The weather cruellest, harshest and most nasty. Disappointed to the lowest ebb of hope and courage. No mercy from Allah Almighty received yet despite five days rigorous treatment. Prayers. Cry in the desert.'[86] The number of dead as per official figures was 474 and 1,109 wounded on the Indian side. A stalemate followed which ended only when the international community under US leadership pressurized Pakistan to withdraw. While withdrawing, the Pakistani troops laid numerous anti-personnel mines. In the next six months, the Indian Army was able to defuse some 8,500 mines.[87]

The encounter in Kargil in many ways was an artillery duel similar to the battle of Somme in 1916. Like Douglas Haig at Somme,[88] General V.P. Malik, the Indian COAS at Kargil also wanted to destroy the enemy strong points by shelling with heavy artillery. The 400 pieces of 155-mm Bofors gun provided the Indian ground units an edge in artillery. Kargil led to the Indian Army acquiring snowmobiles for raising the infantry's mobility in the snowy terrain. And to increase firepower, the army is in the process of acquiring AK-47 rifles with under barrel grenade launchers and disposable flame throwers.[89]

In the late 1990s, Pakistan became an undeclared nuclear weapon state. And the Indian military feared that a deep penetration by the Indian ground forces into Pakistan might result in the nervous Pakistani leadership going for a nuclear strike. To prevent this General Sundarji had already chalked out a defence oriented doctrine for waging limited war under the nuclear shadow. He argued:

> The strategy of conventional defence consists of two parts. The first is a dissuasive part; a strong defensive position, which can extract a heavy toll from the attacker. However, when the attacker has choice of time and place of attack, he can mass enough forces to gain some success; this cannot be generally prevented by the defender. Nevertheless he can certainly make it very expensive for the attacker. The second part of the strategy is the almost axiomatic counter-offensive, at a time and place of the defender's choice; where he would do unto the attacker manifold what the attacker earlier did unto him. This threat of counter-offensive, and the certainty of heavy damage to the original attacker, is the deterrent part of the equation.[90]

Sundarji's plan of deterring the enemy's attack is partly derived from Kaul and Kalkat. As far as conventional defence is concerned, Sundarji's plan of being defensive in the first phase and then launching a counter-offensive in the second phase is similar to that school which takes from Clausewitz that defence is the stronger form of warfare.[91]

However, the prevailing doctrine of the Indian Army is to maintain a static linear defence all along the Pakistan border. This is in accordance with the political order that even an inch of land must not be lost.[92] India's current military strategy could be categorized as a defensive-defence doctrine. The aim is to protect the territorial integrity of India by merely deploying the military assets to physically push back the invaders to the international border. Bharat Verma, the editor of *Indian Defence Review* claims that this doctrine is an extension of the age-old fortress mentality of fighting from a fixed position. Verma goes on to say that because of long land borders and inhospitable terrain, it is impossible to guard every inch effectively. For instance, the 140-km. long Line of Control in the Kargil Sector alone would require seven divisions. Verma continues that a credible defensive posture necessitates that in case of a limited probe across the border by the enemy, i.e. a Kargil type scenario in the near future, the Indian armed forces should be allowed to cross the international border to punish the enemy.[93]

It is necessary to remember Sun Tzu who says: 'For if he prepares to the front his rear will be weak, and if to the rear, his front will be fragile. If he prepares to the left, his right will be vulnerable and if to the right, there will be few on his left. And when he prepares everywhere he will be weak everywhere.'[94] India cannot be strong all along the India–Pakistan border and so has to be prepared to launch limited counter-strikes across the border at points of its own choosing. In such a case the IDF's elite *sayarot* (reconnaissance units), which possess integral components of the infantry and armoured brigades, provide a viable model. These units, constitute the cutting edge of the IDF, and conduct both surgical combat operations as well as high-quality intelligence missions.[95]

CHINA AGAINST INDIA

In October 1950, about 40,000 Chinese troops crossed into Tibet. The Tibetan Army some 8,500 strong with only fifty artillery pieces, 250 mortars and 200 machine-guns was merely a police force. After a brief clash in which about 180 Tibetan soldiers were killed, 894 were captured and 4,317 surrendered, the Chinese occupied Chamdo. On 26 October 1951, the PLA entered Lhasa and a Chinese garrison of 20,000 troops was installed in the city, and in 1952, the Military Region of Tibet was established. The PLA was modernized gradually. In 1949, about 80 per cent of the PLA personnel were of peasant origin without any education. In Korea, about 30 per cent of the Chinese soldiers captured in 1951 were illiterate. Though literally spread with time, the PLA remained a infantry oriented force. In 1962, the strength of the PLA was 4.5 million. The combatant element numbered to about 2.5 million. Of the combatant element, about 90 per cent were infantry. An infantry division had about 14,000 soldiers. Most of the motorized infantry regiments were however deployed in the region opposite Formosa.[96]

By 1962, the Indian Army had expanded from 3,00,000 to 6,25,000 men. In October 1962, the PLA attacked the Indian Army at Ladakh and in the Kameng and Lohit Divisions of the North-East Frontier Agency (later known as Arunachal Pradesh). At NEFA, four Chinese divisions attacked four brigades of the Indian Army scattered along 600 miles. Just before the war, the 4th Indian Division was in-charge of defending NEFA with its two brigades. The 5th Brigade's headquarter was at North Lakhimpur and the 7th Brigade's headquarter was at Tawang. One battalion was deployed in Tawang, another at Bomdi La and the third one at Dirang Dzong. Brigadier John Dalvi commanded the 7th Brigade's three scattered battalions.[97]

Inadequate weapons and weak logistical infrastructure were the principal factors behind the Indian debacle—the Chinese had superior mortar and artillery. The Soviet influenced PLA regarded artillery as 'the god of modern warfare'. The 4.2-inch mortars of Dalvi's troops were out-ranged by the Chinese infantry mortars whose range was 7,000 yards. The 9th Punjab Regiment of the 7th Brigade lacked heavy weapons, and had no digging tools, no saws to cut logs for bunkers, no mines and no wires. The 2nd Rajput belonging to this brigade had only 50 per cent of the authorized digging tools. The 4th Grenadiers who arrived at Namka Chu did not have machetes to cut trees for bunkers. When the Chinese saw the pathetic attempts of the Indian soldiers to cut logs for constructing bunkers, they started deriding Dalvi's troops. The Chinese had large numbers of mechanical saws and so were able to construct bunkers and defences quite easily. Worse, was to follow. The 62nd and 65th Indian infantry brigades arrived in NEFA in November in summer uniforms.[98] This was a case of bad administration and inadequate preparation on part of the Indian Army's high command. The scenario was somewhat similar to the German high command sending troops in summer uniform to fight in Moscow in the winter of 1941.[99]

The logistical apparatus of the PLA had collapsed in Korea in 1950. As a result most of the soldiers went hungry and ate cold food—each personnel had two potatoes every day. The Chinese high command learnt from its mistakes and took care to ensure adequate logistical support during the Indian campaign. The Chinese had laid lateral roads in the Tsangpo Valley with feeders to the south which in some cases reached within a few miles of the McMahon Line. These were all-weather roads which were capable of taking the biggest military vehicles. The Chinese had a road capable of conveying vehicles weighing 7 tons running up to Le, a few kilometre north of Thag La. In contrast, India had only a solitary road capable of taking vehicle weighing only 1 ton to Tawang. Further, the road was damaged during the monsoon and required extensive repair. So, the Chinese were in a position to concentrate and maintain troops with heavy weapons more quickly and efficiently than the Indian Army. Finally, unlike the Indian troops, the Chinese soldiers were acclimatized for a large number of PLA forces were stationed in Tibet for years and many of them had fought the Khampa rebels. So, the PLA soldiers were physically attuned to living and fighting at high altitude and were suitably clothed and equipped.[100]

Faulty doctrine and defective tactics further aggravated the situation for the Indian Army. The various positions taken by different units of the 11th Infantry Brigade in the Walong Sector were not mutually supportive. On 14 November 1962, while the 6th Kumaon was on the West Ridge, the 4th Sikh was on the Maha Plateau, and the 3rd Battalion of the 3rd Gurkha Rifles was deployed on the Dakota Hill and Dong Plateau. If one of the units were attacked, the others being spread so widely and thinly would not have been able to come to the support of the sister units. The Indian high command decided to use tanks and the 7th Light Cavalry's Stuart tanks were ordered to move into Bomdila. The successful use of tanks at Zoji La in Kashmir during the 1947–8 War encouraged the Indian military pundits to use the armoured vehicles for retrieving the debacle in NEFA. However, the Kameng Sector in Arunachal Pradesh was much rougher than the Zoji La Pass. The highest point of the track in the Kameng Sector was 14,000 ft. at Se La and the track was narrow and rugged. There were constant dangers of road blocks caused by landslides and steep arrow bends which were impossible for the tanks to negotiate. It seemed that the officers higher up had taken the decision to use tanks without conducting any reconnaissance. In addition to the terrain, the Indian infantry also proved to be unreliable. On 17 November, 1962, the Chinese attacked Se La. The B Squadron of the 7th Light Cavalry found itself without infantry (62nd infantry Brigade) protection as the latter had already retreated. The 62nd Infantry Brigade had to retreat from Se La because the Chinese massed one division against it and it did not have sufficient artillery shells.[101]

The war lasted thirteen days. Of the 4,00,000 men deployed by the Indian Army against China, only 24,000 actually fought against the PLA. The Indian Army's total loss was about 9,743—1,423 killed, 3,078 wounded, 1,655 missing and 3,587 taken as prisoners.[102] The PLA's intrusion into India was a small affair compared to the PLA's campaign in Korea where in late 1950, China deployed 3,00,000 soldiers. But the tactics employed by the PLA against the Indian Army was somewhat similar to the one used against the US–British soldiers in Korea. The most favoured Chinese method of attack was to hit the enemy formations in the flank during the night—especially when the enemy's road-bound troops were retreating, the PLA constantly ambushed them.[103]

For the debacle against China, all the blame could not be laid on the shoulders of the India politicians. The *netas* and their minions, the *babus* (Indian Administrative Service officers) ought to be criticized for pursuing a faulty grand strategy. Certainly Nehru must be blamed for showing weak nerve. But for the lack-lustre performance of the Indian Army at the tactical and operational planes, the officer cadre cannot escape blame. The tactical value of reconnaissance and robust infantry–artillery cooperation were techniques that the Indian Army was yet to learn. However, the redeeming feature of the officer corps was that this body was willing to learn from past mistakes.

After the fiasco of 1962, the army officers were willing to redress the situation in case of a probable Chinese attack in the near future. Since the PLA trounced the Indian Army in the high Himalayas, the ability to fight successfully in the mountainous terrain became a topic of special interest for the officer corps. Most of the officers concluded that at the tactical plane, an aggressive stance would pay better dividend than a passive posture. This required radical reorganization and training of the combat formations.

Lieutenant-General G.G. Bewoor in 1968 pointed out the health hazards of mountain warfare.[104] As early as 1963, Major S.D. Parab emphasized that the troops should be better prepared to fight in cold mountainous environs. He argued that the soldiers should be trained to construct improvised brushwood and snow shelters though semi-permanent, prefabricated and portable huts should also be made available for constructing the battalion and brigade headquarters. He also made the suggestion of raising ski-trained paratroopers.[105] During the same year, some elements within the Indian Army toyed with the idea of using sleds in the high Himalayas. The sleds would provide mobility to the soldiers and could also be used for moving guns to the threatened points.[106]

Other officers emphasized the use of machines for acquiring mobility and firepower in the hilly terrain. One year after the humiliation in the hands of the Chinese, Brigadier Y.B. Gulati pointed out:

> Force is the product of mass and acceleration. Acceleration determines the rate at which the mass or numbers involved can be put into motion and the rate at which it can build up speed of the moving mass. A moving mass has momentum. The effectiveness of a military force may, therefore, be said to lie in its ability to build up and maintain momentum. . . . In mountainous terrain not suitable for tanks the armoured battle group should become a helicopter battle group, the armoured regiment providing the army element of an air force squadron and flying helicopters armed with guided rockets. The infantry battalion of the helicopter group would provide the holding power where and when necessary, the transport flights of a divisional air force helicopter squadron providing the air lift . . . the armoured regiment would hold light air transportable tanks (such as the present AMX tank).[107]

Major M.K. Puri in 1966 wrote that within the context of an aggressive defence, the army must be prepared to launch battalion-sized counter-attacks in the mountainous regions at night to recapture the territories overrun by the enemy, and the attacks to be supported by 81-mm mortars.[108] General J.N. Chaudhuri also concurred that the firepower of the Indian infantry division ought to be raised.[109]

The Indian Army is also aware that it might have to fight the PLA in the jungles in case Beijing decides to launch a limited offensive across north-east India. In 1965, Major K. Brahma Singh addressed this particular issue. He was for conducting mobile defence by raising the mobility of the infantry. In Singh's view, the Indian infantry carried a lot of unnecessary weight which reduced its mobility and efficiency. Since in the jungles motor transport is not possible, the infantry had to carry everything on its back. Singh said that the load could be reduced by using modern scientific techniques. For example the weight of the wireless sets could be reduced by greater use of aluminium. But, Singh's most radical recommendation was that the troops in the jungles be supplied with helicopters. Besides adding mobility, the helicopters could also increase the firepower and reach of the infantry by giving continuous close air support.[110]

An endeavour to study warfare within a broader perspective was also present. The Chinese armed forces were steeped in the principles of Sun Tzu. After the defeat in the hands of the Chinese, the Indian Army officers also showed a healthy respect for the theory of this ancient Chinese sage-cum-warrior. Major-General Har Prasad in 1966 deduced that the principles of surprise and concentration of forces are essential. He also evaluated the principle of flexibility of war plans, by analysing the historical example of the failed Imphal–Kohima offensive by the Japanese. When the Japanese forces faced strong opposition at Kohima, they should have gone to Dimapur. The failure of the Japanese Army to adapt their plans in fluid battle conditions, argued Har Prasad, resulted in the collapse of their offensive.[111]

In general Western analysts tend to disparage combat effectiveness of the Chinese military. Paul H.B. Godwin asserts that most of China's military technology is of the early 1970s vintage. It's industrial base is also regarded as three decades behind those of the Western industrial powers.[112] China therefore appears as a wooden Titan to Western strategists because they compare Chinese military effectiveness vis-à-vis America. But, China's military capacity vis-à-vis a medium Asian power like India is more than adequate. So India needs to take stock of doctrinal changes, recent reorganization of the Chinese military and its acquisition of new weapons systems.

From the late 1980s, the Indian military theoreticians feared the resurgence of Chinese military capacity. China is preparing a new doctrine of limited war shaped by the Confucian ethos which emphasizes limited aggression for maintaining order, a sort of conflict that might start because of unresolved territorial disputes along the borders. Following the modern *mantra* of waging local war under high technology conditions, the PLA is preparing to conduct limited offensives emphasizing mobility with modern weapons systems. According to this paradigm of warfare, the PLA will meet the enemy outside China's borders so is preparing to launch first in-depth punitive strikes against the enemy troops to destroy the latter's transportation, telecommunication hubs, etc. The central tenet of the new Chinese military theory is the creation of a firepower-heavy rapid response force. As part of this programme, in 1984, the Strategic Rocket Forces of the PLA was carved out of the Second Artillery Corps. In 1998, China came up with the 15th Airborne Army comprising of three brigades. In November 1995, a live fire exercise including night operations was held. From 1985 onwards, China is trying to transform the PLA into an organization composed of crack troops equipped with precision weapons capable of quick reaction. In pursuit of this policy, the Chinese high command is attempting to transform the infantry intensive PLA into a capital-intensive organization and several attempts have been made to reduce manpower. In 1987, the 4.238 million strong PLA was reduced to 3.325 million. By 1990, the PLA's manpower was further reduced to 3.199 million. By 2000, 121 PLA infantry divisions were reduced to 46.[113]

China has deployed about 1,50,000 troops in Tibet while India maintains six mountain divisions against China. This can be increased to thirteen divisions in case of a war. At present, the Indian XXXIII Corps guards the vulnerable Siliguri corridor, India's sole land link with north-east India. This corps with some 45,000 troops is also in-charge of defending Bhutan against Chinese incursions. The IV Corps with its headquarter at Tezpur protects Arunachal Pradesh.[114]

As regards fighting a possible limited war with China, Sundarji pointed out the limitations of India's doctrine of positional war and the difficulties posed by terrain and weather in conducting manoeuvres. Sundarji has written:

> The current doctrine for mountain warfare mandates positional warfare because of the difficulties of mobility and weather. Defences are

elaborate constructions, not only for ballistic protection from artillery shelling and direct fire weapons but also for protection from weather. The weather protection required is not only for comfort but for very survival, as most of our mountainous borders are in high altitudes ranging from four to six kilometers above sea level. Motorable road axes pose the biggest threat, because these can be exploited by an attacker for deep penetration utilizing it to bring forward his artillery support and logistics required to sustain a deep penetration.[115]

So, the strongest Indian defence is directed towards the motorable roads, while subsidiary defences guard the mule tracks that could be used by the enemy. However, Sundarji is not satisfied with passive defence. Sundarji says that any possible deep attack by the PLA would get support only from the light artillery and mortar transported by the animals. He is for strong counter-offensives supported by CAS. Compared to the Chinese operating their aircraft from the high altitude runways of Tibet, which would reduce their weapons load and range of action,

TABLE 4.7: MILITARY BALANCE BETWEEN CHINA AND INDIA: 1990–1

Nature and Strength of Military Units	China	India
Strength of the Army	23,00,000 (10,75,000 Conscripts)	11,00,000
Main Battle Tank	8,000 (6,000 T-54, the rest T-69 and others)	3,150 (800 T-55, 700 T-72, the rest Vijayanta)
Light Tank	2,000 (1,200 Type 63 amphibious and the rest Type 62)	100 PT-76
Armoured Personnel Carrier	2,800	450
Towed Artillery (Self Propelled Artillery, Multi-Barelled Rocket Launcher, Mortar, Anti-Tank Guided Weapon and Recoilless Gun)	14,500	3,860
Air Defence Gun (23-mm., 37-mm., 57-mm., 85-mm. and 100-mm.)	15,000	2,750

Source: *The Military Balance: 1990–91*, London: The International Institute for Strategic Studies, 1990, pp. 149, 161.

Sundarji asserts, the IAF would have the advantage of operating from many sea level airfields of east and north-east India.[116]

THE FACE OF FUTURE WARFARE

Towards the end of the 1990s, the majority of the Western military analysts accepted that the West is undergoing a sort of Revolution in Military Affairs. The US government's defence department describes RMA as a major change in the nature of warfare brought about by the innovative application of technologies combined with dramatic changes in military doctrine.[117] However, there is still some debate regarding the exact features of the ongoing revolution. According to the military theoreticians of the post-modern age, the core of RMA is the Information Revolution. Information Revolution in the military is part of the wider transformation of society. The information technologies are giving rise to what could be termed as a 'network society'.[118]

The American military sociologist Charles Moskos writes that the world is experiencing a shift from industrial to information-based economies. The application of information technologies in warfare promise greater military agility, precision and potency.[119] Mark McNeilly an American war theorist points out the importance of intelligence acquisition as a feature of Information War in the following words:

> To conquer, one must combine foreknowledge and deception. Learn everything possible about the enemy: troop levels, technological advancements, economic strengths, their leadership. This intelligence gives insight into not just the enemy's capabilities, but its intentions as well. Foreknowledge also means knowing one's own capabilities as well as the enemy's. Comparison of strengths and weaknesses will provide an understanding of the correlation of forces. Know the impact of terrain and weather on one's strategy and plans. Understand how they may affect the outcome of the battle. Build an intelligence infrastructure to provide this knowledge.[120]

Eliot A. Cohen asserts that rapid, violent and unpredictable changes are occurring in the realm of military technology. Very soon Cruise missiles and unmanned aerial vehicles will replace fighter planes and tanks. The emphasis will be on satellite imagery to provide quick information to the military user. And in the new set up, information warriors will become more important than the tankers and the pilots.[121]

The Chinese military theorists are also emphasizing RMA. In 1995, Chang Mengxiong had written:

Numerous facts show that we are in the midst of a new revolution in military technology in which electronic information technology is the central technology. This technology provides unprecedented applications for the development of new weaponry. Information acquisitions will be the main distinction of 21st century military forces. Military battles during the 21st century will unfold around the use of information for military and political goals.[122]

Mengxiong goes on to say that by 2020 the army would use information-intensified weapons like Precision Guided Munitions (PGMs) such as guided bombs, cluster bombs, artillery shells, cruise missiles, target-guided missiles and anti-radiation missiles. These are weapons that can acquire and use information from the targets to correct their trajectories. These smart weapons will be launched from outside the enemy's firepower network and these clever weapons will by themselves be able to identify and attack the targets. Mengxiong visualizes the future battlefield as full of robot soldiers, unmanned tanks and other unmanned information intensified combat platforms.[123]

The presence of very long range information intensified precision weapons, will make the future battlefield shapeless and formless. Colonel Xiao Zingmin and Major Bao Bin write:

The differences of front, rear, and side will be mitigated, the front and rear of the battlefield will be attacked simultaneously, and important facilities of the strategic rear area might be the first target of attack. The battlefront will not be fixed, the fight will be waged in all directions, and outer space will be a battlefield. All spaces will be full of intense combat. The expansion of land battlefields has greatly surpassed the firing range of guns and attacks by infantry and tanks. Bombers and tactical missiles can cover a fighting radius of several hundred or thousands of kilometers. Confrontation in outer space can be elevated to tens of thousands of meters or even kilometers high. The interlinking of battlefields far and near, or high and low in space, is unprecedented.[124]

The essence of winning an information-age war will be speed in dominating the battle space. Xiao Jingmin and Bao Bin assert that time is equivalent to force and speed is power. With the introduction of information technologies, operational activities in the inchoate battlefield will become faster.[125]

The Indian analysts also accept the RMA. The Indian military theoreticians accept the importance of enhancing war-fighting capability. As the retired Air-Commodore Jasjit Singh said: 'The synergy created by the interface of doctrine and technology, therefore holds the key to success in modern warfare.'[126] So the Indian military requires a new doctrine for the new form of warfare that is emerging in the information age. Akshay Joshi emphasizes the 'Information Revolution' as part of the RMA. Joshi points out:

> New tools and processes of waging war like Information Warfare (IW), Network-Centric Warfare (NCW), integrated Command and Control (C4ISR), system of systems, all powered by information technology, have led to the Revolution in Military Affairs (RMA). This is likely to broaden the parameters of our thinking about national security. . . . The military has to contend with the 5th dimension of warfare, information, in addition to land, sea, air and space.[127]

Joshi is obviously influenced by the American theorists who are over-enthusiastic about the role of Information Warfare within the context of RMA.

Most Indian analysts are dazzled by the rapid advancement in military hardware. Jasjit Singh writes: 'Given the nature of changes taking place in technology, technological, more than tactical surprise will increasingly hold the key to achieving asymmetry of military capabilities and operational results.'[128] In the near future, the RMA will involve the use of electromagnetic pulses, computer logic bombs and computer viruses.[129] Vinod Anand interprets the RMA as mostly a military-technical revolution based on the use of the microchip. And, these technologies, writes Anand, can only be fully exploited through jointness, but in practice the Indian Army is hopelessly out of date.[130]

India need not ape the RMA of the Western world in totality because the strategic–operational–tactical contexts in which the Indian Army operates or will operate are very different from the context in which the US Army will function. Eric R. Sterner rightly warns that post-RMA forces might face disadvantages in fighting certain kinds of war. He especially refers to terrorism, ethnic wars and the threat posed by the non-state actors.[131] In the near future Kargil-type scenarios might occur and this will require the Indian Army to remain as an infantry-heavy force.[132]

Nevertheless, it would be erroneous to completely ignore the RMA because certain capabilities of a post-RMA force will also certainly help to counter low intensity threats. For instance, the central theme of the RMA is cyber warfare. And cyber warriors are also prominent in conducting irregular warfare. To give an example, in September 2000, Israeli hackers created a website to jam the Hezbollah and Hamas websites in Lebanon. In response, the Palestinian and other supporting Islamic organizations called for a cyber *jihad*. The destruction of business sites with e-commerce capabilities caused 80 per cent dip in the Israeli stock exchange. The Palestinian hackers were also involved in defacing Indian websites. So governments have to use counter-hackers to track down hackers.[133] The crux is that a post-RMA force will enhance the Indian Army's cyber warfare capabilities for fighting the hackers as well as Information Age quasi-conventional wars under a nuclear umbrella.

The Indian Army is utterly unprepared to fight in a nuclear-biological-chemical (NBC) environment because the Indian Army possesses only a brigade's worth of NBC equipment. Finally the Indian Army remains an infantry-oriented force. For raising mobility, firepower and surveillance of its 350 infantry battalions, around Rs. 20 billion ($476 million) had been allocated in 1999 as part of the Ninth Army Plan that ended in 2002. It is planned to equip the infantry with Russian Dragunov 7.62-mm semi-automatic rifles and locally developed mortar fire control data computers. Each battalion is supposed to have five hand-held global positioning systems. But, in 1999 only 500 had been imported. According to the blue print of the Ninth Army Plan, each battalion will be equipped with six hand-held laser range finders. Some of these devices have been imported from United Kingdom. The artillery is demanding medium-range surveillance radars and thermal integrated observation equipment. In a way, the battle for Kargil was in many ways an artillery duel with markers for the future. The Indian Army fired about 1,50,000 rounds and the Indian artillery retaliation would have been more effective, if India's arsenal had Cannon-Launched Guided Projectiles. Such PGMs are more effective than ordinary conventional artillery shells. For instance three rounds of PGMs produce the same effect as 300 rounds of ordinary artillery shells.[134]

CONCLUSION

Independent India's Army like that of Pakistan Army has never fought *bitvas* (big bloody battles)[135] like Dien Bien Phu. The butcher's bill for India's successive wars with Pakistan and China is negligible compared to the casualties suffered by several other countries in conflicts in the post-Second World War era. Mass mobilization of soldiers was also absent in India's small wars with its neighbours. In 1950, during the Korean war, about 3,00,000 Chinese troops clashed with 4,40,000 Western soldiers operating under the United Nation's flag.[136] A small country like Vietnam between 1964 and 1969 lost 5,00,000 soldiers against the US armed forces.[137] During the 1980s, the Iran–Iraq War resulted in one million casualties.[138] Even the armour clashes between India and Pakistan are quite small by post-Second World War standards. The fiercest tank encounter between India and Pakistan occurred at Khem Karan, but it pales into insignificance when compared to the 'Valley of Tears' where during the 1973 War, the IDF destroyed 500 Syrian tanks and vehicles.[139] On 2 October 1973, the Syrians amassed 1,500 tanks (T-62s, T-55s etc.).[140] In the 18 days of the Arab–Israeli War in 1973, 500 Israeli tanks and 2,000 Arab tanks were destroyed. The Israeli and the Arab armed forces suffered 50 per cent material losses in less than two weeks.[141]

From 1947 onwards two schools of thought have dominated the Indian Army. The minority school as represented by Lieutenant S.G. Chaphekar, Brigadier Y.B. Gulati, Major Brahma Singh, General Krishnaswamy Sundarji, and civilian analysts like Ravi Rikhye, etc., are for a capital-intensive army capable of conducting mobile warfare. The objective is rapid penetration of the enemy's defence in depth by waging air–land warfare. But, the majority school including India's distinguished battlefield commanders like Lieutenant-General Harbakhsh Singh, Field-Marshal Sam Manekshaw, and other mid-level officers like Kalkat, etc., support the cause of a infantry dominated army geared for slow attritional campaigns by crumbling the enemy's defence incrementally through set-piece encounters. The majority school, i.e. the proponent of positional warfare is influenced by Monty's North African campaign and the British Indian Army's campaign in Burma during 1944–5. The British legacy should not be overemphasized as the Soviet military thinking had also shaped the contours of Indian military thought. The minority school's ideas regarding mobile warfare is shaped by the *panzer* campaigns that unfolded in Europe during World War II and by the Israeli *Blitzkrieg* against the Egyptians. It is inadequate to argue that the minority school is naïve as the economy of India would be incapable of maintaining a capital-intensive army. The reality is actually different. A lot of money that could be used for buying hardware goes towards pay and pension of the infantry. If the Indian government is able to maintain law and order with the aid of bureaucracy and police forces, then the Indian Army could be withdrawn from counter-insurgency operations and its infantry strength cut down significantly. Secondly, the linear defence against Pakistan could be replaced by mobile counter-attack groupings.

The strategy of conducting deep penetrations within Pakistan to give its army stunning blows and to capture a huge chunk of territory was not possible within the overall passive grand strategic framework of Nehruvian era. So, the army settled for inert defence. The Israeli-style pre-emptive attack emphasizing manoeuvre geared to create local superiority within a context of general numerical inferiority[142] remains an impossible task for both the Indian and Pakistani armies. The Pakistan Army's attempt to launch such operations both during 1965 and 1971 failed. To put simply, the Indian Army and its counterpart, the Pakistan Army, till 2006 were incapable of conducting deep battle characterized by quick penetration of the enemy lines by armoured formations and mechanized infantry supported by tactical air force. The objective in such fast-moving campaigns is to outflank and outmanoeuvre the enemy formations, encircling and finally annihilating them.

Both the Indian and Pakistani army theoreticians and the practitioners remain unclear about the concept of operational level of war. The operational level of war constitutes the intermediate level above tactics and below military strategy. For commanding armies, especially in manoeuvre warfare in a particular theatre, this concept is very important.[143] The Indian Army's theory of warfare has not been original but derivative in nature using both the theoretical tools of the Western military theories and history to justify the principles. The Indian military theoreticians imitate ideas and concepts from both the West and erstwhile USSR and the Indian Army's balance sheet as regards conduct of limited conventional war represents mediocrity. Clausewitz's phrase 'a genius for war' does not fit the Indian Army. Yet the Indian Army shows considerable expertise in coping with unconventional war.

NOTES

1. Lawrence Freedman, *Atlas of Global Strategy: War and Peace in the Nuclear Age,* London: Macmillan, 1985, p. 114.
2. Carl Von Clausewitz, *On War,* ed. and tr. Michael Howard and Peter Paret, 1976; rpt., Princeton, New Jersey: Princeton University Press, 1989, pp. 579–81, 601–2.
3. For the concept of limited war see, Robert O'Neill, 'Problems of Command in Limited Warfare: Thoughts from Korea and Vietnam', in Lawrence Freedman, Paul Hayes and Robert O'Neill, eds., *War, Strategy and International Politics: Essays in Honour of Michael Howard,* Oxford: Clarendon Press, 1992, p. 275.
4. General K. Sundarji, *Vision 2100: A Strategy for the Twenty-First Century,* Delhi: Konark, 2003, p. 74.
5. Stephen Hart, 'Montgomery, Morale, Casualty Conservation and "Colossal Cracks": 21st Army Group's Operational Technique in North West Europe, 1944–45', *JSS,* vol. 19, no. 4 (1996), pp. 133–4, 150; Andrew A. Weist, 'Haig, Gough and Passchendaele', in G.D. Sheffield, ed., *Leadership and Command: The Anglo-American Military Experience since 1861,* London/Washington: Brassey's, 1997, pp. 88, 90; Carter Malkasian, *A History of Modern Wars of Attrition,* Westport: Praeger, 2002, pp. 1, 3, 9.
6. Malkasian, *A History of Modern Wars of Attrition,* pp. 8, 10–11.
7. William S. Lind, *Maneuver Warfare Handbook,* Boulder/London: Westview, 1985, p. 125.
8. Brian Holden Reid, 'Introduction', in *JSS,* vol. 19, no. 4 (1996), p. 2.
9. Pradeep P. Barua, *The State at War in South Asia,* Lincoln/London: University of Nebraska Press, 2005, p. 295.
10. I accept Andrew Scobell's definition of military doctrine. He writes 'military doctrine is devised to prepare for the kinds of wars that the armed forces anticipate from the threat environment and national objectives defined by the security policy.' Andrew Scobell, *China's use*

of Military Force: Beyond the Great Wall and the Long March, Cambridge: Combridge University Press, 2003, p. 45.

11. The German paratroopers dazzled the world by their exploits in Holland in 1940 and in Crete in 1941. James Lucas, *Hitler's Enforcers: Leaders of the German War Machine, 1939–45,* 1996; rpt., London: Brockhampton Press, 1999, pp. 175–85. The best account of the Allied airborne forces in Operation Market Garden remains Cornelius Ryan's, *A Bridge Too Far,* 1974; rpt., London: Cornet Books, 1975.
12. For Hitler's rockets refer to Jozef Garlinski, *Hitler's Last Weapons: The Underground War Against the V1 and V2,* London: Julian Friedman Publishers, 1978.
13. Lieutenant S.G. Chapekar, 'A Frank Survey of India's Defence Problems–I', *JUSII,* vol. LXXVIII, no. 327 (1947), p. 240.
14. Lieutenant-Colonel C.B. Ponnappa, 'My Views on Post-War Forces', *JUSII,* vol. LXXVII, no. 326 (1947), p. 152.
15. Mohammad Ayub Khan, *Friends not Masters: A Political Autobiography,* Lahore: Oxford University Press, 1967, p. 43; Brigadier Chandra B. Khanduri, *Field Marshal K.M. Cariappa: A Biographical Sketch,* Delhi: Dev Publications, 2000, p. 58; Colonel H.C.B. Rogers, *Tanks in Battle,* 1965; rpt., London: Sphere, 1972, pp. 95, 124–59, 183, 194; General J.N. Chaudhuri, *Arms, Aims and Aspects,* Bombay: Manaktalas, 1966, pp. 16-17; Lieutenant-General L.P. Sen, *Slender was the Thread: Kashmir Confrontation, 1947-48,* 1969, rpt., New Delhi: Orient Longman, 1994, p. 296, Major-General Sukhwant Singh, *General Trends: India's War Since Independence,* vol. 3, 1982, rpt., New Delhi: Lancer, 1998, p. 5.
16. Sumit Ganguly, *Conflict Unending: India-Pakistan Tensions Since 1947,* New Delhi: Oxford University Press, 2002, p. 18; Bhaskar Sarkar, *Outstanding Victories of the Indian Army, 1947–71,* New Delhi: Lancer, 2000, pp. 12, 14; Lieutenant-General S.K. Sinha, *Operation Rescue: Military Operations in Jammu and Kashmir, 1947-9,* 1977, rpt., New Delhi: Vision Books, 2002, p. 20; C. Dasgupta, *War and Diplomacy in Kashmir: 1947-48,* New Delhi: Sage, 2002, p. 49.
17. Ganguly, *Conflict Unending,* pp. 17-18; Dasgupta, *War and Diplomacy in Kashmir,* p. 134; Sukhwant Singh, *General Trends,* p. 3.
18. Ganguly, *Conflict Unending,* p. 17; Major-General Jagjit Singh, *With Honour and Glory: Wars fought by India, 1947-99,* New Delhi: Lancer, 2001, pp. 14, 18. As usual there is disagreement among the various military officers regarding the number of casualties suffered by India. According to Brigadier Chandra B. Khanduri, in the First India-Pakistan war for Kashmir India lost 1,003 soldiers and 3,152 were injured. Khanduri, *Cariappa,* p. 49.
19. *The Military Maxims of Napoleon,* 1987; rpt., New York: Da Capo, 1995, pp. 64, 92, 152.

20. Chaudhuri, *Arms, Aims and Aspects*, p. 39.
21. Rajendra Singh, *Soldier and Soldiering in India: Twelve Essays*, 1954; rpt. New Delhi: Army Educational Stores, 1964, p. 5.
22. Ibid. p. 7.
23. Ibid. p. 4.
24. Chaudhuri, *Arms, Aims and Aspects*, pp. 4–5.
25. Lorne J. Kavic, *India's Quest for Security: Defence Policies, 1947-65*. Berkeley/Los Angeles: University of California Press, 1967, p. 37; Sukhwant Singh, *General Trends*, p. 11; Richard Overy, *Russia's War*, 1997; rpt., London: Penguin, 1998, pp. 58–9.
26. Major Harish Bhatia, 'Re-Organization of the Armed Forces', *JUSII*, vol. LXXXXV, no. 401 (1965), p. 209.
27. Lieutenant-Colonel Gautam Sharma, *Path of Glory: Exploits of 11 Gorkha Rifles*, Ahmedabad: Allied, 1988, p. 50.
28. Lieutenant-General Harbakhsh Singh, *War Despatches: Indo-Pak Conflict: 1965*, New Delhi: Lancer, 1991, p. 7; Roger Ford, *The World's Great Tanks*, Kent: Grange, 1997, pp. 89–93, 96–7, 99–101, 120–1; Khan, *Friends not Masters*, p. 62.
29. Brigadier Desmond E. Hayde, *The Battle of Dograi*, 1984; rpt., Dehra Dun: Natraj, 1991, pp. 11, 80; Khan, *Friends not Masters*, p. 67; Harbakhsh Singh, *War Despatches*, pp. 7, 20.
30. David M. Glantz, *Soviet Conduct of Tactical Maneuver*: Spearhead of the Offensive, London: Frank Cass, 1991, p. 76; Sukhwant Singh, *General Trends*, p. 8; Lieutenant-Colonel C.L. Proudfoot, *We Lead: 7th Light Cavalry, 1784–1990*, New Delhi: Lancer, 1991, p. 152; *The Indian Army*, New Delhi: Army Headquarter and Lancer, pp. 73, 75.
31. Harbakhsh Singh, *War Despatches*, pp. 14–15.
32. Sharma, *Path of Glory*, p. 50; Sukumar Biswas, *Three Weeks' War*, Calcutta: M.C. Sarkar & Sons Pvt. Ltd., 1966, p. 53. Most of the *Blitzkrieg* attacks started at dawn. The attack against Poland started on 1 September 1939 precisely at 4.40 a.m. Operation Barbarossa started on the early morning of 22 June 1941. J.F.C. Fuller, *The Second World War: 1939–45, A Strategical and Tactical History*, 1954; rpt., New York: Da Capo, 1993, pp. 50, 120; Lieutenant-General P.S. Bhagat, *The Shield and the Sword*, 1996; rpt., Delhi: Vikas, pp. 15, 18.
33. Chaim Herzog, *The War of Atonement*, 1975; rpt., Dehra Dun: Natraj, 1984, p. 270; G.D. Bakshi, 'Operational Art in the Indian Context: An Open Sources Analysis', *Strategic Analysis*, vol. 25, no. 6 (2001), p. 729; Bhagat, *Shield and the Sword*, p. 27; Richard Holmes, 'Kursk: 1943', in idem, Christopher Chant and William Koenig, eds., *Two Centuries of Warfare*, London: Octopus Books, 1978, pp. 370–2; Proudfoot, *We Lead: 7th Light Cavalry*, pp. 151–2.
34. Rupert Matthews, *Hitler: Military Commander*, London: Arcturus, 2003, pp. 148–50; Hayde, *The Battle of Dograi*, p. 11; Bhagat, *Shield and the Sword*, p. 18; Harbakhsh Singh, *War Despatches*, p. 98.

35. Biswas, *Three Weeks' War*, pp. 115–16; Lieutenant-General Phillip B. Davidson, *Vietnam at War, The History: 1946–75*, 1988; rpt., London: Sidgwick & Jackson, 1989, p. 257; *Indo-Pakistan War, 1965: A Flash-Back*, 1966, ISPR Publications: Rawalpindi, 2002, p. 14; Juneja, *Indo-Pak War*, p. 167.
36. Bakshi, 'Operational Art in the Indian Context', p. 728.
37. Brigadier Surinder Harsh, 'Desert Warfare in Our Context', *JUSII*, vol. 99, no. 417 (1969), pp. 397-403; Williamson Murray, 'Armored Warfare: The British, French, and German Experiences', in idem, and Alan R. Millett, eds., *Military Innovation in the Interwar Period*, Cambridge: Cambridge University Press, 1996, pp. 18–29, 34–49; Kenneth Macksey, *Tank versus Tank: The Illustrated Story of Armored Battlefield Conflict in the Twentieth Century*, 1999; rpt., London: Chancellor Press, 2001, pp. 158-9.
38. Major-General Gurcharan Singh Sandhu, *69th Armoured Regiment: 1968-93*, New Delhi: Lancer, 1996, pp. 11–12, 23–4, 29, 44; Major-General Sukhwant Singh, *India's War Since Independence: Defence of the Western Border*, vol. 2, 1981; rpt., New Delhi: Lancer, 1998, p. 277; A.M. Vohra, 'A Military Policy for the Nineties', *JUSII*, vol. CXX, no. 499 (1990), p. 13.
39. Lieutenant-General J.F.R. Jacob, *Surrender at Dacca: Birth of a Nation*, New Delhi: Manohar, 1997, pp. 65–70; Major-General Sukhwant Singh, *India's War Since Independence: The Liberation of Bangladesh*, vol. 1, 1980; rpt., New Delhi: Lancer, 1998, pp. 138–9; Davidson, *Vietnam at War*, p. 660.
40. Jacob, *Surrender at Dacca*, p. 60.
41. Gurcharn Singh Sandhu, *69th Armoured Regiment*, p. 45; Davidson, *Vietnam at War*, p. 559; Major-General Lachhman Singh, *Victory in Bangladesh*, 1981; rpt., Dehra Dun: Natraj, 1991, pp. 80, 85; D.R. Mankekar, *Pakistan Cut to Size*, Delhi: Hind Pocket Books, 1972, p. 51; Lieutenant-General A.A.K. Niazi, *The Betrayal of East Pakistan*, 1998, rpt., Karachi: Oxford University Press, 2000, p. 117.
42. Sukhwant Singh, *Defence of the Western Border*, p. 2.
43. Major-General D.K. Palit, *The Lightning Campaign: Indo-Pakistan War 1971*, New Delhi: Thomson Press, 1972, p. 77; Pervaiz Iqbal Cheema, *The Armed Forces of Pakistan*, 2002; rpt., Karachi: Oxford University Press, 2003, p. 79.
44. Palit, *The Lightning Campaign*, pp. 80–1.
45. Sukhwant Singh, *Defence of the Western Border*, p. 64; Palit, *The Lightning Campaign*, p. 79.
46. Sukhwant Singh, *Defence of the Western Border*, p. 3.
47. Lieutenant-General Depinder Singh, *Field-Marshal Sam Manekshaw: Soldiering with Dignity*, Dehra Dun: Natraj, 2002, p. 179.
48. Jacob is rightly scathing about the operational effectiveness of the *Mukti Bahini*. Jacob, *Surrender at Dacca*, p. 146. On the other hand Akhtar

Ahmed, an officer of the *Mukti Bahini* asserts in his memoirs that the war against Pakistan was almost won by his rag-tag organization. The Indian Army just chipped in to accept the surrender of Niazi. Major Akhtar Ahmed, *Advance to Contact: A Soldier's Account of Bangladesh Liberation War,* Dhaka: The University Press Limited, 2000, pp. 1–3, 6–7.

49. John A. English, *On Infantry,* 1981; rpt., New York: Praeger, pp. 186–7, 190; Field Marshal Lord Carver, 'Field Marshal Erich Von Manstein', in Correlli Barnett, ed., *Hitler's Generals,* London: Weidenfeld & Nicolson, 1989, pp. 221–44; Hayde, *The Battle of Dograi,* p. 86.
50. Sukhwant Singh, *Liberation of Bangladesh,* pp. 140–2.
51. Niazi, *Betrayal of East Pakistan,* p. 120.
52. Nigel Hamilton, *The Full Monty: Montgomery of Alamein, 1887–1942,* London: Penguin, 2001, pp. 696–780.
53. English, *On Infantry,* p. 158.
54. The *Luftwaffe* supplied Paulas with at least 100 tons of supplies every day. William Carr, *Hitler: A Study in Personality and Politics,* London: Edward Arnold, 1978, pp. 101–2. For the Stalingrad scenario refer to Antony Beevor, *Stalingrad: The Fateful Siege, 1942–43,* 1998; rpt., Harmondsworth: Penguin, 1999, pp. 239–351.
55. Jacob, *Surrender at Dacca,* pp. 73–4.
56. Lachhman Singh, *Victory in Bangladesh,* p. 2; Gunther E. Rothenberg, *The Anatomy of the Israeli Army,* London: B.T. Batsford Ltd., 1979, p. 123.
57. Sukhwant Singh, *Liberation of Bangladesh,* p. 137; Lachhman Singh, *Victory in Bangladesh,* pp. 2, 4.
58. Colonel Horst Zobel, '3rd Panzer Division Operations', and Lieutenant-General Karl Wilhelm Thilo, 'A Perspective from the Army High Command (OKH)', in David M. Glantz, ed., *The Initial Period of War on the Eastern Front: 22 June–August 1941,* 1993, rpt., London: Frank Cass, 1997, pp. 244, 306–7; John Keegan, *The Second World War,* 1989; rpt., London: Pimlico, 1997, p. 150.
59. Lachhman Singh, *Victory in Bangladesh,* p. 3.
60. Jacob, *Surrender at Dacca,* pp. 55–6.
61. English, *On Infantry,* pp. 203–4; Sukhwant Singh, *Liberation of Bangladesh,* p. 138; Rothenberg, *Israeli Army,* p. 123.
62. Jacob, *Surrender at Dacca,* pp. 130–48.
63. Overy, *Russia's War,* pp. 179, 181.
64. Sukhwant Singh, *Liberation of Bangladesh,* p. 214.
65. Major Yogi Saksena, 'Indo-Pak War, 1971: Some Lessons', *JUSII,* vol. CII, no. 428 (1972), pp. 255–6.
66. Ravi Rikhye, 'Rethinking Mechanized Infantry Concepts', *JUSII,* vol. CII, no. 427 (1972), pp. 163–4.
67. Herzog, *War of Atonement,* p. 270.

68. Len Deighton, *Blitzkrieg: From the Rise of Hitler to the Fall of Dunkirk,* 1979; rpt., London: Granada, 1981, pp. 168–220.
69. Ravi Rikhye, 'Tricap: New Dimensions in Armoured Warfare', *JUSII,* vol. CI, no. 425 (1971), pp. 319-20.
70. Samuel M. Katz, *Israeli Tank Battles: Yom Kippur to Lebanon,* London: Arms and Armour Press, 1988, pp. 30–1.
71. *MODAR: 1974–75,* p. 26.
72. Major-General O.S. Kalkat, 'Approach to Tactics', *JUSII,* vol. CII, no. 427 (1972), pp. 133–4, 136.
73. Ravi Kaul, 'Strategy and Weapons Systems', in idem, ed., *The Chanakya Defence Annual: 1973–74,* Allahabad: Chanakya Publishing House, n.d., p. 61.
74. P.A. Petersen, '*Perestroyka* and Planning Soviet Air Power', in Air-Commodore E.S. Williams, ed., *Soviet Air Power: Prospects for the Future, Perestroyka and the Soviet Air Forces,* London: Tri-Service Press, 1990, p. 54; English, *On Infantry,* p. 196; Malcolm Mackintosh, 'Continuity and Change in Soviet Strategic Thought', in Freedman, Hayes and O'Neill, eds., *War, Strategy and International Politics,* pp. 258, 260.
75. Rajesh Kadian, *India and Its Army,* New Delhi: Vision, 1990, p. 149; *MODAR: 1979–80,* p. 14; Glantz, *Soviet Conduct of Tactical Maneuver,* p. 58; Ross Masood Husain, 'Threat Perception and Military Planning in Pakistan: The Impact of Technology, Doctrine and Arms Control', in Eric Arnett, ed., *Military Capacity and the Risk of War: China, India, Pakistan and Iran,* Oxford: Oxford University Press, 1997, pp. 134–5.
76. Katz, *Israeli Tank Battles,* p. 111.
77. General K. Sundarji, *Blind Men of Hindoustan: Indo-Pak Nuclear War,* 1993, rpt., New Delhi: UBSPD; 1996, p. 44.
78. Colonel General Heinz Guderian had written in his memoir: 'It has been said, "only movement brings victory." We agree with this proposition. . . . Only fire can open the way to movement.' General Heinz Guderian, *Panzer Leader,* tr. from German by Constantine Fitzgibbon, 1952; rpt., London: Arrow Books, 1990, pp. 40, 44.
79. Sanjay Badri Maharaj, *The Armageddon Factor: Nuclear Weapons in the India-Pakistan Context,* New Delhi: Lancer, 2000, pp. 38–9; Bharat Karnad, *Nuclear Weapons and Indian Security: The Realist Foundations of Strategy,* Delhi: Macmillon, 2002, p. 676.
80. The Moscow counter-offensive of the Soviets in the winter of 1941 unfolded along a frontage of 1,000 km and the depth of operation was 300 km. The average rate of advance per day was between 6 to 9 km In the next year, the Stalingrad counter-offensive occurred over a 800-km front. The attack reached a depth of 150 km and the average rate of advance per day was 15 km. P.A. Zhilin, 'Policy and Strategy of the Soviet Union in the Second World War', in Arthur L. Funk, ed.,

Politics and Strategy in the Second World War: Germany, Great Britain, Japan, the Soviet Union and the United States. Manhattan, Kansas: Kansas State University, 1976, p. 60.

81. Karnad, *Nuclear Weapons and Indian Security*, p. 676.
82. Umer Farooq, 'Production under Way on Pakistan's Al-Khalid MBT', *JDW*, vol. 32, no. 13, 29 September 1999, p. 15; Rahul Bedi, 'India Delays Army MBTs', *JDW*, vol. 33, no. 21, 24 May 2000, p. 13; Christopher F. Foss, 'Enhanced T-90s Targets Asian Market', *JDW*, vol. 33, no. 22, 31 May 2000, p. 37; *Tenth Report, MoD, Standing Committee on Defence, Thirteenth Lok Sabha*, New Delhi: Lok Sabha Secretariat, 2001, p. 4; Sanjay Badri Maharaj, 'The Nuclear Battlefield: India vs Pakistan', *Indian Defence Review*, vol. 14, no. 2 (1999), 75–6; Baranwal, *SP's Military Yearbook: 1993–94*, pp. 117, 229; J. Baranwar, Editor-in-chief, *SP's Military Yearbook: 1993-94*, New Delhi: Guide Publications, 1993.
83. Lieutenant-General V.R. Raghavan, *Siachen: Conflict without End*, New Delhi: Viking, 2002, pp. 32, 66, 69; Pravin Sawhney, *The Defence Makeover: 10 Myths that Shape India's Image*, New Delhi: Sage, 2002, p. 13.
84. Khanduri, *Cariappa*, pp. 110–2; Vinod Anand, 'India's Military Response to the Kargil Aggression', *Strategic Analysis*, vol. 23, no. 7 (1999), p. 1056.
85. Samar Halarnkar, 'Prime-Time War', *India Today*, 14 June 1999, p. 33; *From Surprise to Reckoning: The Kargil Review Committee Report*, New Delhi, 15 December 1999, New Delhi: Sage, 2000, pp. 17, 19.
86. Ibid., pp. 20–1.
87. Satish K. Jain, 'Operation Safed Sagar', *IDR*, vol. 16, no. 3 (2001), p. 17; *From Surprise to Reckoning*, pp. 21–3.
88. Shelford Bidwell and Dominick Graham, *Fire-Power: British Army Weapons and Theories of War, 1904–1945*, 1982; rpt., Boston: George Allen & Unwin, 1985, pp. 82–7.
89. *MODAR: 1999–2000*, p. 21; Lieutenant-General R.K. Jasbir Singh, ed., *Indian Defence Yearbook: 2003*, Dehra Dun: Natraj, 2003, p. 89.
90. Sundarji, *Blind Men of Hindoostan*, p. 44.
91. Clausewitz in *On War*, p. 358 asserts, 'The latter [defence] increases one's own capacity to wage war, the former [offence] does not. So in order to state the relationship precisely, we must say that *the defensive form of warfare is intrinsically stronger than the offensive*.' (Italics in original.)
92. J. Baranwal (ed.), *SP's Military Yearbook: 1995*, New Delhi: Guide Publications, 1995, p. 95.
93. Bharat Verma, 'Pakistan: The Counter Strategy', *IDR*, vol. 14, no. 2 (1999), pp. 5, 7.
94. Quoted from Mark McNeilly, *Sun Tzu and the Art of Modern Warfare*, New York: Oxford University Press, 2001, p. 89.

95. Stuart A. Cohen, *Towards a New Portrait of the (New) Israeli Soldier*, Ramat Gan: Bar Ilan University, 1997, p. 85.
96. Edgar O'Ballance, *The Red Army of China*, London: Faber & Faber, 1962, pp. 199–201, 203; John Gittings, *The Role of the Chinese Army*, London: Oxford University Press, 1967, pp. 37–8, 145.
97. Neville Maxwell, *India's China War*, 1970; rpt., Bombay: Jaico, 1971, p. 295; Sen, *Slender was the Thread*, p. 297.
98. Proudfoot, *We Lead: 7th Light Cavalry*, pp. 143-4; Brigadier J.P. Dalvi, *Himalayan Blunder: The Curtain Raiser to the Sino-Indian War of 1962*, Bombay: Thacker & Co., 1969, pp. 307–8, 315; K.C. Praval, *The Red Eagles: A History of Fourth Division of India*, New Delhi: Vision Books, 1982, pp. 202, 211; Gittings, *Role of the Chinese Army*, p. 147.
99. Keegan, *Second World War*, pp. 164–9; Carr, *Hitler*, pp. 101–2.
100. Maxwell, *India's China War*, p. 301; Praval, *Red Eagles*, pp. 201–2; Gittings, *Role of the Chinese Army*, p. 133.
101. S.N. Prasad, *History of the Conflict with China: 1962*, www.timesofindia.com, pp. 196, 243–5; Proudfoot, *We Lead: 7th Light Cavalry*, pp. 144–5.
102. Brigadier H.S. Sodhi, *Top Brass: A Critical Appraisal of the Indian Military Leadership*, Noida: Trishul Publications, 1993, p. 36.
103. Edgar O'Ballance, *Korea: 1950–53*, Dehra Dun: Natraj, 1969, pp. 67, 72–3.
104. Lieutenant-General G.G. Bewoor, 'High Altitude Mountain Warfare', *Institute for Defence Studies and Analyses Journal*, vol. 1 (1968–9), p. 63.
105. Major S.D. Parab, 'Cold Weather Operations in Mountainous Terrain', *JUSII*, vol. LXXXXIII, no. 392 (1963), pp. 259, 263.
106. Lieutenant Andrew W. Furlan, 'The Effects of Cold Weather on Infantry Weapons', *JUSII*, vol. LXXXXIII, no. 393 (1963), p. 394.
107. Brigadier Y.B. Gulati, 'A Size and a Shape for the Army', *JUSII*, vol. LXXXXIII, no. 391 (1963), pp. 100, 108–9.
108. Major M.K. Puri, 'Local Counter Attacks in the Mountains', *JUSII*, vol. LXXXXVI, no. 402 (1966), pp. 51, 53, 56.
109. Chaudhuri, *Arms, Aims and Aspects*, p. 17.
110. Major K. Brahma Singh, 'Infantry Mobility in the Jungle', *JUSII*, vol. LXXXXV, no. 401 (1965), pp. 223–7.
111. Major-General Har Prasad, 'The Principles of War: Lessons from History to Remember', *JUSII*, vol. LXXXXVI, no. 404 (1966), pp. 175–88.
112. Paul H.B. Godwin, 'Military Technology and Doctrine', in Chinese Military Planning: Compensating for Obsolescence', in Arnett, ed., *Military Capacity*, pp. 40–1, 57–9.
113. A.K. Sachdev, 'Modernization of the Chinese Air Force', *Strategic Analysis*, vol. 23, no. 6 (1999), pp. 992–3; Srikanth Kondapalli, 'Towards a Lean and Mean Army: Aspects of China's Ground Forces

Modernisation', *Strategic Analyses*, vol. 26, no. 4 (2002), pp. 462–3; Srikanth Kondapalli, *China's Military: The PLA in Transition,* New Delhi: Knowledge World, 1999, pp. 60–2, 76, 248, 299; Sawhney, *The Defence Makeover*, p. 48.

114. Sawhney, *The Defence Makeover*, pp. 45, 48; Badri Maharaj, *Armageddon Factor*, p. 22; *MODAR: 1974–75*, p. 5.
115. Sundarji, *Vision 2100*, p. 213.
116. Ibid. pp. 212–14.
117. Mungo Melvin and Stuart Peach, 'Reaching for the End of the Rainbow: Command and the Revolution in Military Affairs', in Gary Sheffield and Geoffrey Till, eds., *The Challenges of High Command: The British Experience*, Basingstoke: Palgrave Macmillan, 2003, p. 177.
118. Julian Reid, 'Foucault on Clausewitz: Conceptualizing the Relationship Between War and Power', *Alternatives*, vol. 28, no. 1 (2003), p. 5.
119. Charles Moskos, 'Towards a Postmodern Military?', in Stuart A. Cohen, ed., *Democratic Societies and Their Armed Forces: Israel in Comparative Context,* London: Frank Cass, 2000, pp. 5, 6.
120. McNeilly, *Sun Tzu,* p. 95.
121. Eliot A. Cohen, 'A Revolution in Warfare', *Foreign Affairs*, vol. 75, no. 2 (1996), pp. 37–54.
122. Chang Mengxiong, 'Weapons of the 21st Century', in Michael Pillsbury, ed., *Chinese Views of Future Warfare,* 1997, rpt., New Delhi: Lancer, 1998, p. 249.
123. Ibid., pp. 249–51.
124. Colonel Xiao Jingmin and Major Bao Bin, '21st Century Land Operations', in Pillsbury, ed., *Chinese Views of Future Warfare,* p. 312.
125. Ibid.
126. Air Commodore Jasjit Singh, 'Strategic Framework for Defence Planners: Air Power in the 21st Century', *Strategic Analyses*, vol. 22, no. 12 (1999), p. 1807.
127. Akshay Joshi, 'A Holistic View of the Revolution in Military Affairs (RMA)', *Strategic Analyses*, vol. 22, no. 11 (1999), p. 1743.
128. Jasjit Singh, 'Strategic Framework for Defence Planners', p. 1807.
129. Vice-Admiral Nayyar et al., *National Security: Military Aspects*, New Delhi: Rupa, 2003, p. 65.
130. Vinod Anand, *Joint Vision for the Indian Armed Forces*, Delhi Papers, no. 16, New Delhi: Institute for Defence Studies and Analyses, 2001, p. 74.
131. Eric R. Sterner, 'You Say You Want a Revolution (in Military Affairs)', *Comparative Strategy: An International Journal*, vol. 18, no. 4 (1999), pp. 300, 306.
132. Kaushik Roy, 'The Battle for Kargil: Post-Modern War in South Asia', in Kanti Bajpai, Afsir Karim and Amitabh Matoo, eds., *Kargil and*

After: Challenges for Indian Policy, New Delhi: Har-Anand, 2001, p. 101.

133. Colonel Patrick D. Allen and Lieutenant-Colonel Chris C. Demchak, 'The Palestinian-Israeli Cyberwar', *Military Review,* vol. LXXXIII, no. 2 (2003), pp. 52–8.
134. Anand, *Joint Vision,* pp. 75–6; *MODAR: 1999–2000,* p. 20; Rahul Bedi, 'Indian Army Plans Boost for Infantry Battalions', *JDW,* vol. 31, no. 8, 24 February 1999, p. 28; Karnad, *Nuclear Weapons & Indian Security,* p. 681.
135. A Russian concept meaning an encounter on a grandiose scale resulting in huge losses and major consequences. Sergei Kudryashov, 'The Soviet Perspective', in Paul Addison and Jeremy A. Crang, eds., *The Burning Blue: A New History of the Battle of Britain,* London: Pimlico, 2000, p. 78.
136. O'Ballance, *Korea,* p. 71.
137. Davidson, *Vietnam at War,* p. 402.
138. Freedman, *Global Strategy,* p. 113.
139. Katz, *Israeli Tank Battles,* p. 41.
140. Ibid., p. 10.
141. Theodore G. Stroup, Jr. and Leonard Wong, 'Re-establishing the Force: The Revolution in Military Affairs in the Human Resource and Leadership Systems', in Cohen, ed., *Democratic Societies and Their Armed Forces,* pp. 152–3.
142. Mork A. Heller, *Continuity and Change in Israeli Security Policy,* Oxford: Oxford University Press, 2000, p. 13.
143. Peter Browning, *The Changing Nature of Warfare: The Development of Land Warfare from 1792 to 1945,* Cambridge: Cambridge University Press, 2002, p. 195.

FIVE

Indian Army and Counter-Insurgency

The Indian subcontinent is characterized by a high degree of pluralism. India has eighteen languages, 200 dialects, a dozen ethnic and seven religious groups. These in turn are fragmented into numerous sects, castes and sub-castes.[1] In such a multi-ethnic society, intrastate conflicts are inevitable. Hence, from its very independence, India has been experiencing civil wars. The centrifugal tendencies continue to operate in the twenty-first century insurgencies, and the Indian Army, being the state's last line of defence, has been brought in frequently to stabilize the scenario. From the India–Myanmar border in the east upto the Line of Control (LOC) in Kashmir, the Indian military units have been deployed repeatedly for guarding the state against the armed rebels. In India, the insurgents have been supported by foreign powers. For the foreign powers, patronizing insurgencies offer a technique to keep the Indian Army engaged. Combat with the insurgents can be categorized as counter-insurgency war. The insurgents are also known as terrorists or guerrillas and action against them can be termed as counter-terrorism or counter-guerrilla war. Such wars are fought against foreign-sponsored internal enemies and not against the regulars of a foreign state. Hence, these conflicts are also categorized as unconventional war. Unconventional war may also be referred to as low-intensity warfare or small war because such warfare do not involve large-scale use of tanks, aeroplanes and artillery. Nevertheless, unconventional warfare has the characteristics of being attritional protracted warfare.

CONCEPTUALIZING COUNTER-INSURGENCY WARFARE

Paul Collier, a World Bank official and Nicholas Sambanis assert that democracies with a poor economy and many ethnic groups are characterized by civil wars.[2] Collier and Anke Hoeffler write that internal rebellion occur either due to greed or grievances. The latter refers to opposition to perceived or actual injustice. Economic inequalities along with ethnic divisions and religious polarizations further fuel opposition.[3] Patrick M. Regan claims that identity wars last longer than ideological conflicts. He continues that the duration of civil wars is directly proportional to the extent of ethno-linguistic fractionalizations.[4] Marta Reynal-Querol argues that religion is the most important dimension of ethnicity. And religiously divided societies are more prone to intense conflict than linguistically divided societies. This is because religious identity is fixed and non-negotiable.[5] Reynal-Querol's statement has special validity as regards the late-twentieth century insurgencies in India.

Jean Dreze, an economist has theorized on unconventional warfare with special reference to India. He says: 'Most armed conflicts today are 'internal' wars, rather than inter-state disputes. Internal wars are socially divisive and undermine the integrity of the nation state. This creates a vicious circle of violence and social disintegration.'[6] Dreze continues that such conflicts far from promoting participation of the underprivileged in the mainstream exacerbate their marginalization. Also, war conditions often undermine earlier achievements in fields such as democratic rights, gender equality and ethnic harmony.[7] General K.S. Sundarji, former Chief of Army Staff claimed that in the coming decades the use of unconventional force (such as proxy war by sponsored insurgency, subversion and terrorism including narco-terrorism) would be used by the nation states as an instrument of policy against neighbouring enemy states to a greater degree.[8]

A group of retired Indian military officers rightly claim that insurgency in the Indian context includes terrorism plus guerrilla warfare. Guerrillas unlike the terrorists do not attack common people. The guerrillas confront the state's security forces only.[9] Violence directed by the terrorists against civilians is a case of displaced or redirected aggression.[10] In this case anger generated by other sources (i.e. the state machinery) is displaced or redirected to other targets (for example civilians). The Indian military officers categorize three types of insurgencies: secessionist, restorational and reformist. The secessionist movement aims to establish a new autonomous political entity in place of the existing political framework. The objective of the restorational movement is to transform the existing social structure in accordance with the principles of past religious or political systems. The restorational movements are backward looking and reactionary in nature. And the reformist movement tries to bridge the gulf between the haves and have nots without upsetting the existing system.[11] Rohan Gunaratna divides the objectives of the insurgents into three categories: irredentism, (i.e. reunification of the two regions), separatism (for example establishment of the Tamil Ealam in Sri Lanka) and autonomy.[12]

The Indian doctrinaires have forgotten about the revolutionary insurgencies. Brigadier-General Samuel B. Griffith (a retired officer of the United States Marine Corps) writes about the communist inspired guerrilla warfare in the following words:

> A revolutionary war is never confined within the bonds of military action. Because its purpose is to destroy an existing society and its institutions and to replace them with a completely new state structure, any revolutionary war is a unity of which the constituent parts, in varying importance, are military, political, economic, social, and psychological. For this reason, it is endowed with a dynamic quality and a dimension in depth that orthodox wars, whatever their scale lack. This is particularly true of revolutionary guerrilla war, which is not susceptible to the type of superficial military treatment frequently advocated by antediluvian doctrinaires.[13]

Griffith's definition could well apply to the restorational movements dominated by Islamic ideology in recent times in Kashmir as well as the secessionist movements occurring in north-east India.

It would be worthwhile to compare the counter-insurgency doctrine of the Indian Army with the unconventional war doctrines of the other armies. The French Army while fighting insurgencies in Vietnam and Algeria evolved *guerre revolutionnaire* which is described as a rigid military response to unconventional warfare.[14] In contrast, the principal elements of British Army's counter-insurgency doctrine are political primacy, flexibility and coordination of civil-military responses at every level.[15] Gerald Templer, British commander of the forces in Malaya during the early 1950s, was able to defeat the Communist insurgents. According to Templer, shooting to kill the insurgents constitutes 25 per cent of the counter-insurgency operations. The rest, 75 per cent, involves winning the hearts and minds of the people in the disturbed zone.[16] And the British Army emphasizes that during policing operations, the military officers must be given training in handling public relations.[17] Brigadier M.J. Strudwick of the British Army warned in 1993 that civil rights are a major issue and often become the principal reason for popular support behind the insurgencies. Once the security forces infringe human rights, organizations like Amnesty International start intervening and thus obstruct the government's counter-insurgency operations.[18] The South African Army focuses on psychological actions. The security personnel are advised, trained and instructed to maintain good relations with the local population.[19]

After the Vietnam War, the American Army accepted Templer's emphasis on political initiatives and 'proper treatment' of the civilians in the insurgency prone areas. One of the American proponents of the counter-insurgency doctrine Wray R. Johnson writes:

> The application of purely military measures may not, by itself restore peace and orderly government because the fundamental causes of the condition of unrest may be economic, political or social. . . . The solution of such problems being basically a political adjustment, the military measures to be applied must be of secondary importance and should be applied only to such an extent as to permit the continuation of peaceful corrective measures.[20]

Johnson continues, that at heart insurgencies are not military problems but political, economic and social problems. Therefore the military measures against the guerrillas that includes the tenets to find, fix and destroy them in a firefight are inadequate. The political, economic and social aspects of insurgencies too require to be addressed. The non-military programme aims to win the hearts and minds of the populace and to construct grass root level political institutions.[21] For this, writes Peter M. Dunn, the security forces require a thorough knowledge of the needs, customs and beliefs of the people.[22] The post-modern name for a pacification campaign is peacekeeping operations. In peacekeeping operations, the actual use of force is mostly replaced by mere threats of using force and thus forcing the insurgents to negotiate.[23]

The Indian analysts accept the necessity of fighting the battle for the hearts and minds of the people and political primacy in counter-insurgency warfare. Major Vivek Chadha while analysing the nature of counter-insurgency operations emphasizes the primacy of political initiatives over the purely military aspects.[24] In a similar vein, Major-General V.K. Shrivastava says that since in the counter-insurgency operations there is neither a visible enemy nor a definite territory to capture, such operations demand the application of 'minimum force'.[25] Nevertheless, the role of force though minimized cannot be eliminated completely.

Lieutenant-General S.K. Sinha was one of the foremost counter-insurgency theoretician of the Indian Army. When he was a Brigadier in 1970, he attempted to construct a framework of anti-insurgency operation, emphasizing the military dimension. In Sinha's words, a handful of skilled guerrillas could tie down a disproportionately large number of security forces. According to his calculation the ratio is 1:20.[26] In 1964, Major-General P.S. Bhagat claimed that a worthwhile guerrilla force must have an effective strength of 2,000 in an area with a population of 1,00,000. Bhagat argues that the counter-insurgency forces are required to be 15–20 times stronger than the guerrillas as the army has to fight the guerrillas as well as protect the people.[27] So, guerrilla warfare is relatively cheap to wage, though it is rather expensive to counter.[28] Nevertheless, too much emphasis on statistics might prove to be problematical. In 1952, in Malaya, the British superiority in practice amounted to only 2.5:1. Still the British were successful in eliminating the guerrillas.[29]

Despite accepting overall political supremacy at the theoretical plane, at the executive level, the Indian military officers demand total control. As part of the counter-insurgency operations, most of the Indian military officers demand unified control over the coercive apparatus of the state. For instance Colonel Bhaskar Sarkar writes that in the disturbed areas while conducting counter-insurgency, the state's paramilitary forces and the police should be subordinated to the concerned military authorities. Sarkar points out the purely military aspects of counter-insurgency operations involving unified forces (police, paramilitary and the military units) for the following tasks: (a) Killing and capturing the maximum number of insurgents. (b) Containing the insurgents. (c) Progressively reducing the area of influence of the insurgents.[30]

Colonel Harjeet Singh writes that the troops engaged in counter-insurgency operation should follow the 'ten commandments'. They are as follows: no rape, no molestation, no torture resulting in death and maiming, respect human rights, fear God and uphold *Dharma*, no meddling in civil administration, carry out civic actions, develop a model for media interaction and use it as a force multiplier, no military disgrace (loss of arms, surrender, etc.) and competence in platoon/company actions.[31]

It is interesting to note that out of the ten commandments only two commandments (the last two) deal with military aspects. So, while Templer says that shooting concerns 25 per cent of counter-insurgency operation, in Harjeet Singh's model shooting constitute only 20 per cent of the anti-insurgency campaign. And Singh points out that the military actions are going to be small unit decentralized operations. Proper behaviour of the troops especially with respect to the women of the disturbed areas is emphasized. About 40 per cent of the counter-insurgency campaign, according to Harjeet Singh, are related with ensuring human rights of the populace. The morale and behaviour of the soldiers are maintained by focusing on their ethical code of conduct, i.e. religious duty of the troops. Finally Singh tells us that the troops should under no circumstances disrupt the civil administration of the area. And to win the affection and confidence of the people the troops must conduct civic actions to upgrade the local populace's conditions, a trend which is emphasized in both the Latin American and South African armies fighting insurgencies. The Latin American armed forces emphasize civil development projects like building roads, clinics and courteous treatment of the civilians to win their support while fighting insurgent.[32] Similarly, the South African troops while combating the insurgent are supposed to aid the local agriculture and irrigation projects.[33]

Unlike Harjeet Singh, Bhaskar Sarkar gives less importance to the issue of public support to the insurgents. According to Sarkar, in a disturbed area, only 10 per cent support insurgency, another 10 per cent are against it and the 80 per cent remaining populace are indifferent. However, Sarkar does not underestimate the battle for winning 'the hearts and mind of the people.' Influenced by the Vietnamese General Vo Ngyuen Giap's writings, Sarkar states that the security forces must respect the people and help them in distress.[34] An Indian police officer Ved Marwah accepts that a successful terrorist movement cannot function without the support of the populace.[35] It is also important to remember the father of modern guerrilla warfare Mao Tse-Tung who asserts: 'Because guerrilla warfare basically derives from the masses and is supported by them, it can neither exist nor flourish if it separates itself from their sympathies and cooperation.'[36] So, the objective of the counter-insurgency forces is to strike at the contact between the masses and the insurgents.

Hence, the counter-insurgency campaign should also involve the use of propaganda to win the 'Information Battle'. While fighting a low intensity war, the British priority is to isolate the insurgents from the support of the populace. This is done through propaganda such as showing films, distributing leaflets, etc.[37] Greater interaction between the civilian population of the disturbed area and the counter-insurgency forces, is necessary as part of the state's campaign to win the affection of the populace. It is necessary for the personnel of the counter-insurgency forces to learn the local language and familiarize themselves with the customs and religious practices of the various segments of the populace. This will help in breaking down the wall of mistrust between the populace and the state's forces created by rebel propaganda.[38] In Vietnam, since the American intelligence officers had to rely on interpreters, they could not establish rapport with the local population. Hence, the United States military authorities failed to acquire adequate intelligence about the ground situation[39] and generate friendly bonds with the locals.

Both the proponents of the guerrilla warfare and the doctrinaires of counter-insurgency warfare note the importance of media and use of new technology in the overall framework of their campaign. Urban guerrilla warfare or terrorism involves attacks on banks, abduction of businessmen and assassination of political leaders, etc. Such activities compared to the actions of the guerrillas in the rural area generate greater media publicity.[40] Carlos Marighela, the urban guerrilla warfare theorist writes that for the urban guerrillas, media is the vital weapon. For bank raids, street ambushes, spectacular kidnappings, etc., the media provide propaganda.[41]

Several Indian military officers have noted the use of media as a force multiplier by the insurgents. Brigadier R.K. Nanavatty had written in 1993:

> The new forms of warfare thrive on publicity. The 'enemy' seeks to manipulate the media. The media falls prey to the lure of sensationalism and unwittingly endangers individuals, state and public security. The government lacks an effective and modern apparatus to disseminate *'Public information'* despite exclusive control over the electronic means of mass communications.[42]

One Indian police officer writes: 'The information technology has greatly enhanced the ability of the groups to spread terror and thus achieve the desired emotional response that made this particular mode of violence a psychologically effective weapon.'[43] He continues that the global media by highlighting terrorist activities furthers the objectives of the terrorists to terrorize the members of the civil society.[44]

For gaining an advantage over the insurgents, the Indian Army requires to wage what could be categorized as 'information warfare'. Akshay Joshi emphasizes the 'Information Revolution' as part of the Revolution in Military Affairs. He asserts:

> In the Information age, using information, communication and media technologies in pursuance of national aims and foreign policy objectives has to supplement the use of force. It will be wrong to ignore the basic realities of international politics in the information age, that is the supremacy of the mind over muscle.[45]

Joshi further states:

> We have enough material and technical expertise to launch a propaganda blitz using modern information and communication technologies. The first step is to formally include information warfare as part of our national security strategy. There is tremendous patriotism and talent in the media, armed forces, government, and research community to launch and win the information war.[46]

SHADOW WARRIORS FROM THE PERIPHERY

Unlike criminals, rebels possess a definite political objective. Scott Gates writes that in contrast to the criminal gangs, an unconventional army is able to engage the government militarily.[47] To acquire the designation of a 'non-state army', the rebel force must number more than 1,000 men and must cause more than 100 fatal casualties to the security forces.[48] Gates says that rebel armies are characterized by a high degree of cohesion and primary group solidarity.[49] Primary bonding and group cohesion are the products of indoctrination and a rigorous discipline. About the shadow warriors' ideological motivation, Griffith writes:

> In the United States and Britain we go to considerable trouble to keep soldiers out of politics, and even more to keep politics out of soldiers. Guerrillas do exactly the opposite. They go to great lengths to make sure that their men are politically educated and thoroughly aware of the issues at stake. A trained and disciplined guerrilla is much more than a patriotic peasant, workman, or student. . . . His indoctrination begins even before he is taught to shoot accurately, and it is unceasing. The end product is an intensely loyal and politically alert fighting man.[50]

Along with the United States and Britain, Griffith might have added India's policy of maintaining apolitical soldiers. As regards discipline, the right hand man of Ho Chi Minh, General Vo Nguyen Giap writes: 'The Vietnam People's Army practices a strict discipline. . . . Point two of its Oath of Honour requires: "The fighter must rigorously carry out the orders of his superiors and throw himself body and soul into the immediate and strict fulfillment of the tasks entrusted to him."'[51] Giap's interpretation of discipline was very much similar to the *Wehrmacht's* definition of 'inhuman' discipline for its personnel.[52]

Despite the presence of intense ideological indoctrination, strong group solidarity and a rigorous disciplinary mechanism, the insurgent armies cannot be successful without external support. Foreign financial aid and arms are vital for conducting insurgencies. China supplied the North Vietnamese with weapons from 1949 onwards.[53] The Provisional Irish Republican Army raised most of their funds to procure firearms in the USA but waged their violent campaign in Northern Ireland and mainland Britain.[54] The Afghan *mujahideens* were able to resist the Soviet Red Army due to the generous funding by the Central Intelligence Agency of USA. In 1980, USA provided $30 million to the Afghan guerrillas to fight the Soviets and in 1981–2, $50 million was provided. The American aid rose by $30 million in 1983 and then by another $40 million in 1984. In 1985, it was

$250 million, in 1986, it was $470 million and in 1987 it was $630 million.[55] Let us analyse the genesis of insurgent armies in India and their financial structures.

KASHMIR

The biggest insurgency problem for independent India remains Kashmir. The Kashmir insurgency could be traced back to the immediate aftermath of Partition. The Muslims of Kashmir in particular and the population in general are not homogeneous. The Buddhists are in the Leh region and the Shia Muslims inhabit the Kargil region. The Muslims of the Pakistan occupied Kashmir are also known as Mirpuris. They are similar to the Punjabi Muslims of the Potowar region of Pakistan. In Jammu, one-third of the population is Muslim. At present about 45.7 per cent of the 2,22,336 sq. km. of the area of Jammu & Kashmir is with India, 35.1 per cent is with Pakistan and 19.2 per cent is under China.[56]

In 1947, when the Pathan raiders came to Kashmir, their standard tactics were sniping and ambushing army columns. After 1971, the Pakistan Army's officers went to China for getting training on guerrilla operations. When they came back, they raised a militia for operations in Kashmir. Since the late 1980s, ethno-religious secessionist insurgency has been troubling the Indian state in Kashmir. Islamabad's strategy is to conduct low-intensity warfare and pose the threat of an escalation of violence with nukes waiting in the shadows. The aim is to internationalize the Kashmir issue. If successful, this would ultimately weaken India's hold over the Kashmir Valley and the protracted war in Kashmir would become a bleeding wound for India.[57]

Nobody is sure about the exact number and size of the *tanzeems* (Islamic militant outfits) operating in Kashmir at any moment. One can only speculate. As evident from Table 5.1 the militant groups in Kashmir could be divided into three categories: those who demand self-determination for Kashmir, those who demand merger with Pakistan and finally those *tanzeems* whose aim is to establish the 'rule of *Allah*' throughout the world. The combat strength of the militant outfits vary from company size to that of a regimental battle group. In 1993, there were about 120 militant outfits in Kashmir, the Jammu and Kashmir Liberation

TABLE 5.1: PRINCIPAL INSURGENT GROUPS OPERATING IN KASHMIR IN 2003

Name of the Organization	Equipment	Finances	Strength	Objectives
Al Barq	100 AK series rifles	Rs. 7,00,000 per month	350 (including 20 foreigners)	Liberation of Kashmir
Al Jihad Force			125 (including 25 foreign mercenaries)	Liberation of Kashmir
Ikhwan-ul-Musalmeen	250 AK series rifles, 20 rocket launchers, 50 hand grenades, 10 LMGs.		225	Self determination of Kashmir
Hizb-ul-Mujahideen	1,550 AK series rifles, 55 LMGs, 90 rocket launchers, 2,000 hand grenades, 500 mines plus 1,000 pistols and revolvers.		1,400	Merger with Pakistan
Al Fateh Force	30 machine guns and rocket launchers	Rs. 3,00,000 per month	35	Merger with Pakistan
Al Mujahid Force			125	Merger with Pakistan
Al Umar Mujahideen	20 AK series rifles and 5 pistols	Receives Rs. 3,00,000-4,00,000 per month from ISI plus local collection (extortion and protection money)	700	Merger with Pakistan
Dukhtaran-e-Millat			1,000	Merger with Pakistan
Harkat-ul-Ansar	50 LMGs, 50 rocket launchers, 400 AK-47, 10 sniper rifles and 100 pistols.	Grant from ISI	800 Pakistan/ Afghanistan trained militants (including 350 foreigners)	Establishment of Islam in the world

Source: Sreedhar K. Santhanam, Sudhir Saxena and Manish, *Jihadis in Jammu and Kashmir: A Portrait Gallery,* New Delhi: Sage, 2003, pp. 55–136.

Front (JKLF) being the oldest. While Jamaat-i-Islami enjoys support in the rural areas, the Hizb-ul-Mujahideen, Al Omar Mujahideen has bases in the urban areas. After 1993, most of the terrorist acts have been committed by Hizb-ul-Mujahideen rather than the JKLF. In the mid-1990s, the Hizb-ul-Mujahideen's strength was about 2,500. In the last decade of the twentieth century, the ISI of Pakistan created a new terrorist outfit named Al Faran which included criminals released from the jails of Pakistan. According to one author, in 2002, about 4,000 militants attached to twenty-two insurgent groups were operating in Kashmir. Most of the militants belonged to Hizb-ul-Mujahideen, Harkat-ul Ansar and Lashkar-e-Taiba. And another 3,000 trained militants were awaiting entry into Kashmir from the POK. In addition, 5,000 more were trained in the various *madrasas* (religious seminaries) in the North-West Frontier Province of Pakistan.[58]

Various organizations like the Jamiat Ulema Pakistan of Maulana Fazlur Rahman are active in Islamizing and motivating the Afghan refugees to conduct *jihad* and gain martyrdom. General Zia-ul-Haq started the programme to Islamize Pakistan in the 1980s. Not only did he provide funds to the *madrasas* but made the certificates given by the *madrasas* equivalent to those provided by the universities. Saudi Arabia provides financial support to the *madrasas*. According to one estimate there were about 40,000–50,000 *madrasas* in Pakistan in 2003. From the 1970s, the Jamat-e-Islami has been propagating communal hatred in the *madrasas* run by them.[59]

A significant chunk of the insurgents fight to attain *shahada* (martyrdom) in Kashmir. Most of the Palestinian suicide bombers also emphasize the wish to become *shaheed*. Interrogation of the captured terrorists operating in Kashmir show that many had joined only to wage *jihad*. A few others stated the sense of adventure as a motivating force. Again, for many holding an assault rifle made them feel powerful and they also acquired respect and fear from the common people. Most of them were between fifteen and twenty-five years of age. Defection is rare among the terrorist groups as most defectors and even dissidents are frequently killed or maimed.[60]

Training of the terrorists includes instructions in the use of small arms and explosives along with intelligence collection and indoctrination in the group's cause. The ISI recruits Pakistan Army's retired Junior Commissioned Officers (JCO) and the Non-

Commissioned Officers (NCO) to train the militants. In the early 1980s, the US special forces gave training to the ISI especially in the fields of communications and sabotage so that they could train the Afghan *mujahideens* in unconventional warfare against the Soviet troops in Afghanistan. And the ISI is now using its US-transmitted knowledge to train the militants operating in Kashmir. From 1986 onwards, thousands of ISI-trained Kashmiri insurgents were sent to Afghanistan for further training in order to fight more effectively in the Kashmir Valley. In 1993, Pakistan diverted over a thousand *mujahideens* into Kashmir—militants who had experience of combat against the Soviet Army. After the collapse of the Taliban, more *mujahideens* have moved into Kashmir. Between 1989 and 2000, the percentage of foreign militants in Kashmir rose from a mere 6 per cent to over 60 per cent. Over 90 per cent of the foreign militants were from Pakistan and the rest came from Afghanistan, Sudan, etc. After the Soviet withdrawal from Afghanistan, about 10,000 Pakistanis received military training in Afghanistan for conducting *jihad*. Between 1990–4, the camps in the tribal areas of North-West Frontier Province of Pakistan, and in the Afghan provinces of Zawar and Paktia were training personnel of the Lashkar-e-Taiba and the Jamaat-i-Islami. In the 1990s, infiltration and ex-filtration of the guerrillas occurred primarily through the Pir Panjal and Keran and the jungles of Tithwal and Kupwara provided sanctuaries to the guerrillas.[61]

The Al Qaeda is the symbol of transnational terrorism with its network in thirty-four countries. The Al Qaeda used to aid many *tanzeems* in Kashmir with money and weapons till the beginning of Operation Enduring Freedom in 2001. The Harkat-ul-Ansar (Mujahideen) was in close touch with the Taliban and Al Qaeda. In 2003, the Indian intelligence agencies calculated that 75 per cent of the funding of terrorist activities in Kashmir was done by the ISI. Another 15 per cent came from abroad and the remaining 10 per cent from within the Kashmir Valley. Looting is used for raising money within Kashmir. For example, on 20 December 1989, the branch of the Union Bank of India on Residency Road of Srinagar was looted. Numerous such examples can be given. Rather than pillage and plunder, foreign funding constitutes the financial backbone of the terrorist organizations. In 2000, the ISI was spending $3.3 million per month on the Kashmiri militants from the revenue generated out of the drug trade. Afghanistan and the North-West Frontier Province of Pakistan are the biggest producer of opium in the world. In 1999, out of 6,000 metric tons of illicit opium produced in the world, about 4,600 metric tons were produced in Afghanistan. Pakistan's production of opium came to about 65 metric tons. Pakistani heroin along with Columbian marijuana are the most cherished drugs in Europe and in the USA. One kg. of heroin from the Golden Crescent costs Rs. 1 lakh in South Asia. By the time, the consignment reaches the USA, the price soars up to Rs. 1 crore. In 1992, Pakistan earned US $1.5 billion from export of heroin.[62]

Some of the *tanzeems* like Lashkar-e-Taiba also receive funds from the Muslim industrialists of Saudi Arabia, Kuwait and Pakistan. The Lashkar-e-Taiba enjoys support of the lower-middle classes in Pakistan. Petty traders, merchants and lower level government employees from Punjab and the North-West Frontier Province of Pakistan attend the congregations called by the Lashkar-e-Taiba and provide donations. The Tehrik-ul-Mujahideen leaders till 2002 visited Pakistan's Embassy in Kathmandu. The Kathmandu network was used for distributing funds received through *hawala* transactions for a number of *tanzeems*. A few handicraft emporiums were also opened in Kathmandu which provided cover for financial transactions involving the insurgents. Many Pakistani businessmen function as ISI's conduits for large scale *hawala* transactions and use their contacts in the USA, UK, Saudi Arabia and Bangladesh to facilitate such transactions.[63]

Along with money, the ISI also provides arms to the *mujahideens* operating in Kashmir. The ISI use smugglers to carry explosives and arms across the international border and reward them with a couple of kilograms of heroin, overruling the officers of the Narcotics Bureau. There is a lot of resentment among the Narcotics Bureau officers against the ISI because of this practice. The North-West Frontier Province of Pakistan has a flourishing arms industry. In 2001, the annual turnover of the trade amounted to $1 billion and about 50,000 people were involved in this illegal arms industry. According to another estimate, in 2002 there were 60,000 shops dealing with arms and they employed about 70,000 people.[64] Products from this industry to some extent equip the insurgents operating in Kashmir.

Thanks to their foreign sponsors, the insurgents do not suffer from any technological inferiority via-à-vis the security forces. The *mujahideens* in Afghanistan used 107-mm and 122-mm rockets and RPG-7s against the Soviet helicopters and aircraft in Afghanistan.[65] In the 1970s, the most favourite weapon of the guerrillas in Rhodesia was land mines.[66] The insurgents in Kashmir too use land mines operated by remote control to make the movement of security forces through the highways unsafe.[67] And they target both the civilians and the security forces in the countryside as well as in the cities. From 1994, the Israeli cities suffered from an upsurge in attacks by suicide bombers identified with the Islamist Hamas movement.[68] In 2000, Islamic suicide bombers appeared both in Chechnya and Kashmir.[69] On 1 October 2001, a suicide attack occurred on the Jammu & Kashmir Assembly. The Talibanized Jaish-e-Mohammad took responsibility for the attack that left twenty-nine people dead. In May 2002, the suicide bombers started attacking various army camps in Rajouri, Jammu and Udhampur. In a pre-dawn *fidayeen* attack on the Kaluchak camp near Jammu about thirty-two people (mostly women and children of army families) were killed. In 2002, ten *fidayeen* attacks took place. Between 1989 and 1997, the militants in Kashmir organized 1,900 kidnappings and in 700 of those cases, the captives were killed. From 1989, the terrorist outfits started killing Kashmiri Pandits in order to induce them to leave the Kashmir Valley. This is part of the militants' ethnic cleansing programmme, i.e. to transform Kashmir into the land of the 'pure' by driving out the infidels. The Muslims have economic grievances against the Pandits. Being better educated the latter hold more jobs disproportionate to their numbers. Further, the terrorists conceive the Pandits as stooges of the government. In the last decade between 1,00,000 to 2,50,000 Hindus had been forced to leave the Kashmir Valley. Groups like Lashkar-e-Taiba are also against the Sufi or mystical syncretistic Islam prevalent in Kashmir which generates the culture of *Kashmiriyat*, a culture more or less secular in nature. The insurgents also force the Muslim women of Kashmir to follow the fundamentalist way of life, i.e. wearing *burqa* and not sporting modern hairstyles, etc., in an attempt to destroy the *Kashmiriyat* culture, for the objective of Lashkar-e-Taiba is to fight against democracy and to establish an Islamic *Caliphate*.[70]

NORTH-EAST INDIA

The north-east region of India comprises of the seven states (often called the seven sisters) of Assam, Arunachal Pradesh, Manipur, Meghalaya, Mizoram, Nagaland and Tripura. The relation between the 'sisters' is anything but cordial. The whole region covers an area of 2,60,000 sq. km. with a population of thirty-two million (1991 census). The north-east has a 4,500 km. long international border, but is connected to the Indian mainland by a 22-km. land corridor in Siliguri in West Bengal. This tenuous land link is known as the 'chicken's neck'. Lack of proper surface communication between the north-east and the rest of India is hurting the economic development this region which in turn boosts insurgencies in that region. There are about thirty insurgent groups representing various ethnic groups and tribes operating in this region.[71]

In the 1940s, the Indian Nagas numbered 2,00,000 while about 4,00,000 Nagas were in Burma especially in the Kachin State. There is no single tribe called the Nagas. About thirteen tribes constitute the category known as the 'Nagas'. The most populous tribe among the Nagas is the Konyak. They numbered 63,000 in 1960. The Rengma and the Zeliang constitute the smallest tribes of the Naga federation. Each of the last two named tribes numbered 5,000 in 1960. The Nagas in quest for independence formed the underground 'Federal Government of Nagaland' in 1956. Angami Zapu Phizo slipped into East Pakistan in late 1956 and then moved into West Pakistan. From 1958 onwards, training camps for the Naga rebels were set up in the Chittagong Hill Tracts of East Pakistan. Towards the end of the 1950s, Laldenga, an ex-Havildar initiated insurgency in the Mizo Hill District. In the 1960s, the ISI in East Pakistan also started giving assistance to the Mizo rebels. Between 1966 and 1969, the personnel of the National Socialist Council of Nagaland (NSCN) went through Kachin State to China to as well as to procure weapons. In 1969, Pakistan and China set up a Coordination Bureau in order to coordinate the training, arming and funding of the insurgent movements in north-east India. Hence, terrorist outfits like the ANVC in Meghalaya and the NSCN of Nagaland in comparison to the police possess more sophisticated arms like the AK-47, and AK-56 rifles and grenades.[72]

Scott Gates writes that leaders of the ethnic groups attempt to inculcate a sense of membership and solidarity among them in order to construct an ethnic identity that would hold the rebel units together. And ethnically based groups are less likely to suffer from desertion.[73] Phizo attempted to do exactly the same thing. His army was organized on a tribal basis and the aim was to use tribal kinship affinities to mobilize insurgent manpower. Each brigade raised from a tribe was commanded by a brigadier. Each brigade had one or more battalions, and each comprised of two or three companies of four sections of eleven men each. As different brigades had separate tribal identity this caused disunity among the various Naga tribes. After 1956, Phizo's reorganized his army shifting from a tribal to a more bureaucratic organization frame. The region was divided into four geographical commands and a Commander-in-Chief headed the army with four commands under him. Each command had several battalions which were raised from the several tribes under a particular command. A battalion was composed of three companies and each company had two platoons and one platoon had two sections each with twenty-eight men. In addition, there were volunteer parties, courier parties plus women's volunteer organizations. The personnel belonging to the latter organization not only functioned as nurses but also occasionally looted grain and pigs from the rural markets. Phizo's army was quite combat effective. In 1962, just before the border war between India and China erupted, the 11th Brigade of the 4th Division was engaged in fighting the Naga rebels and could not be deployed in the NEFA against the PLA in the North-East Frontier Agency. The Naga rebels could move fast in difficult terrain and these units enjoyed complete freedom from logistical constraints. The Nagas rebels unlike the Indian soldiers did not need rations supplied by the road-bound convoys and could eat anything while fighting in the jungles. The dense jungles within the states and on the boundaries with the neighbouring states favours the north-east militants and provide them with ample hideouts. The jungles along with inadequate surface communications make it difficult for the security forces to concentrate and operate effectively.[74]

The huge influx of Muslims from Bangladesh after the 1971 War into north-east India has generated fear among the inhabitants in this area that they are fast becoming a minority. According to one scholar, the north-east receives between 500 to

1,400 Bangladeshi immigrants every day.[75] The people of the north-east having failed to get support against immigration from New Delhi have vented their frustration through insurgency against the Indian state. Ironically, Bangladesh, the root cause of all trouble has begun providing sanctuary and support to the insurgent movement like the United Liberation Front of Asom (ULFA) operating in Assam. The ULFA's transformation into an organized militant group started in 1983 under the tutelage of the NSCN. In 2003, about 3,000 militants belonging to the ULFA and the Bodos were hiding in the jungles of south Bhutan. In that region, the ULFA, National Democratic Front of Bodoland (NDFB) and the National Socialist Council of Nagalim (Isaac-

TABLE 5.2: PRINCIPAL INSURGENT GROUPS OPERATING IN NORTH-EAST INDIA IN 2003

Name of Insurgent Group	Date of Origin	Cadre Strength	Objective	Area of Operation
United Liberation Front of Asom	1979	2,500	Independent Assam	Assam
National Democratic Front of Bodoland	1986	3,500	Independent Bodoland in areas north of river Brahmaputra	
United Nationalist Liberation Front	1990	1,500	Independent Socialist Manipur	Imphal Valley and Myanmar-Manipur border
People's Liberation Army*	1979	1,500	Liberated North-east	Imphal Valley
Hynniewtrep National Liberation Council	1992	200	To make Meghalaya the exclusive homeland of the Khasi tribe	Khasi hills of Meghalaya
National Liberation Front of Tripura		1,500	Independent Tripura	Tripura
National Socialist Council of Nagaland	1980	2,000	Formation of Greater Nagaland	Nagaland, Tirap and Changlang districts of Arunachal Pradesh and four districts of Manipur (Ukhrul, Senapati, Chandel and Tamengong)

Note: *Not to be confused with the Peoples Liberation Army of China.
Source: Saba Naqvi Bhaumik, 'A Million Mutinies', *Outlook*, 20 October 2003, p. 42.

Muviah) or NSCN (I-M) has set up about fifty training camps. The ULFA uses the contacts of the NSCN to establish warehouses and liaison offiçes in Thailand for procuring arms. Through the NSCN, the ULFA has also established linkages with China for procuring arms and ammunition which are imported through Bangladesh and Myanmar. Weapons are bought in the arms *bazars* of Bangkok, and are transported into Bangladesh in fishing vessels. After landing around Cox's Bazar, the weapons are carried by porters to the insurgent training bases in the Chittagong Hill Tracts and then handed over to the insurgents in Manipur. Most of the insurgent groups finance themselves from the revenue generated by the narcotics trade and taxes collected forcibly from the populace. Every government employee is forced to pay 25 per cent of his income as tax for the Republic of Nagalim. From the late 1980s, the ULFA started extorting large sums from tea plantations, businessmen, etc.[76]

Sri Lanka

In Sri Lanka, the Sinhalese constitute 74 per cent and the Tamils comprise 18 per cent of the populace. The rest are Muslims. In northern Sri Lanka and in the Jaffna region, the Tamils are pre-dominant. Most of the Tamils are Hindus while the Sinhalese are Buddhists. In 2000, the total population of Sri Lanka was nineteen million. The Liberation Tigers of Tamil Ealam (LTTE) receives funds from the expatriate Tamil communities in Canada, the USA, Australia, and South Africa and substantial amount of fund is raised overseas by the Tamil Rehabilitation Organization. During the 1980s, it was calculated that monthly contributions to the LTTE from the expatriate Tamil workers came to about US $1 million. In 1984, the Indian Army trained the Tamil separatist groups in the art of guerrilla warfare. By 1985, around 5,000 armed Tamil guerrillas were operating in Sri Lanka. They were armed with AK-47 and G-3 rifles. The LTTE fought the Indian Peace Keeping Force with booby traps and improvised explosive devices in the 1,300 sq. km, of the Jaffna peninsula where 50 per cent of Sri Lanka's Tamil populace live. Most of the militants operated in groups of ten to twelve. They used to ambush IPKF's patrols and convoys. The LTTE personnel were highly motivated and most carried cyanide capsules so that they could not be taken prisoner and tortured for information. The LTTE noticed the combat effectiveness of the Indian units were reduced by 50 per cent when their commanding officer was killed. Hence, the LTTE targeted the unit leaders. After 967 days, the IPKF lost 1,200 men against the LTTE and was withdrawn from Sri Lanka in 1990.[77]

THE CENTRE STRIKES BACK

There is a general reluctance to arm the police heavily, as this might result in greater number of human rights violations.[78] Once accustomed to the use of heavy firepower it may find difficult to revert to its original doctrine of minimal use of force, after the problem of insurgency is over. Hence, the Indian Army is used in counter-insurgency campaigns. The situation to an extent is similar to that faced by Israel. The Israeli Defence Force has to intervene because the Palestinian insurgents armed with rocket-propelled grenades, light anti-armour and anti-aircraft missiles are more than a match for the Israeli police.[79]

The Indian Army's counter-insurgency activities is very extensive. In 1994, only 42,000 Russian troops were engaged in peacekeeping duty throughout the ex-Soviet Union.[80] In Mozambique, the Portuguese deployed 60,000 troops in 1974.[81] By comparison, more than 2,00,000 Indian soldiers are always engaged in counter-insurgency duty. The Indian experience could be compared with the Soviet experience in Afghanistan, where in 1987 Moscow deployed more than 1,00,000 troops.[82] However, the Indian experience is not as extensive as the French experience in Algeria where in 1956, about 4,70,000 French military personnel were serving.[83] Between 1979 and 1987, there were 13,310 Soviet soldiers killed, 35,478 were wounded and 311 were missing in action in Afghanistan.[84] These figures should be compared with the losses suffered by the Indian Army in the counter-insurgency campaigns. Table 5.3 shows that unconventional war had proved more costly than conventional war for the Indian Army. Between 1986 and 1996, 2,467 soldiers died and 14,359 were wounded in various counter-insurgency operations. According to Bharat Karnad, between 1992–2002, counter-insurgency resulted in 61,000 civilian and military casualties and had cost India Rs. 45,000 crore.[85] According to another calculation in the last decade, insurgency in Kashmir resulted in the death of 25,000 persons.[86] The severity of the counter-insurgency operation in Kashmir can be gleaned from the following figures. In 1995, 202 security personnel were killed and the tally for the militants was 1,308. In 1996, 1,209 militants were killed, 2,567 were arrested, and 655 surrendered and 185 security forces were killed. In Assam, between 1991 and 2004, 2,000 militants and 769 security personnel were killed. During the same period, 12,000 militants surrendered.[87] However, it is important to note that both France and USSR, unlike India, conducted pacification campaigns in foreign countries. India probably tops the list as regards the scope and intensity of internal pacification duties.

The use of army in pacification campaigns is quite common in South Asia. The Pakistan Army is also used heavily in aid to civil life. In the 1970s, Afghanistan assisted the Baluchis to fight Pakistan. The Baluchis consider the Punjabi Muslims who dominate the Pakistan Army and administration as foreign colonialists. Between 1973–7, about 80,000 Pakistani soldiers were deployed in Baluchistan to tackle the Baluchi separatists—and 3,300 soldiers and 5,300 Baluchis died in the ensuing confrontation. In September 1974 about 15,000 armed Baluchi tribesmen fought with the Pakistan Army and in 1978, for six months 45,000 Pakistani soldiers were deployed in Sindh.[88]

TABLE 5.3: CASUALTIES SUFFERED BY THE INDIAN ARMY IN VARIOUS CONVENTIONAL WARS AND IN UNCONVENTIONAL WARFARE

War	Killed	Wounded	Missing	Total
India–Pakistan War (1947–8)	1,103	3,152	–	4,255
India–China War (1962)	1,521	548	1,729	3,798
India–Pakistan War (1965)	2,902	8,622	361	11,855
India–Pakistan War (1971)	3,630	9,856	213	13,699
Low-Intensity Warfare (1972–96)	3,187	15,299	–	18,486
Total	12,343	37,477	2,303	52,093

Source: Colonel Harjeet Singh, *Doda: An Insurgency in the Wilderness*, New Delhi: Lancer, 1999, p. 245.

The French counter-insurgency operation in Algeria between 1954 and 1958 failed because of the flawed French strategy of holding a dialogue with the Algerian nationalists only after achieving pacification.[89] Dialogue and punitive military action should occur side by side in a counter-insurgency campaign because cooperation of the local inhabitants is required to achieve military success over the insurgents. The Indian Army was successful in eliminating the Pakistan sponsored guerrillas in Kashmir when the local populace cooperated with the security apparatus of the state. When in August 1965, the Pakistani guerrillas entered Kashmir through unguarded tracks, the local populace reported their presence to the security units and even guided the military patrols to trace the guerrilla hideouts in the jungles.[90]

However, civilian cooperation has not been forthcoming from 1990 onwards. The Indian military operates under extremely hazardous conditions in Kashmir because it is impossible to distinguish a militant from a civilian. The local press is hostile, and house to house searches conducted by the army further alienate the local population. Major Chadha writes that one of the crucial ingredients of counter-insurgency operation is area domination. But area domination should be people friendly and should not involve random searches at odd hours in the night without any specific information.[91] A positive step on part of India is that from the late 1990s, in place of the army more and more Border Security Force (BSF) units have been inducted,[92] who have more knowledge of the local populace. While the BSF and the Central Reserve Police Force (CRPF) conduct counter-insurgency operations in the urban areas especially in Srinagar, the military units are deployed along the LOC and in the difficult mountainous region.[93] The paramilitary forces conduct policing to prevent intrusion of insurgents and arms from Pakistan and the Indian Army's units function as a second line of defence and provide back up forces during emergencies. The scenario has some similarities with the counter-insurgency operations conducted by the security forces in Punjab. While fighting the Khalistani insurgents, the army performed 'cordon off' duties and the commando element of the Punjab police pacified the countryside.[94]

Unified command is one of the principal ingredients of successful counter-insurgency operation, but, one that this is lacking in India. In 2000 in Kashmir, there were about 2,00,000 military and paramilitary personnel—thirty Rashtriya Rifles battalions, fifty BSF battalions, Indo–Tibetan Border Police and Assam Rifles battalions plus regular units of the three corps of the army.[95] According to an estimate, there are over 4,50,000 security personnel deployed in Kashmir. There is no combined command over them—the paramilitary units are under the Home Ministry and the army is under the Defence Ministry—and cooperation between the two is weak.[96] The myriad intelligence agencies that operate without any overall coordinating authority creates not only confusion among the forces operating in the Kashmir Valley but also sends contradictory signals to the political bosses at Delhi.[97] Till now the counter-insurgency tasks have been inefficiently coordinated by both the Defence and Home Ministries and the result has been *ad hoc* response characterized by lack of cooperation and coordination. India needs to set up a temporary unified command for conducting low-intensity warfare. The staff of this command should be drawn from defence, police, diplomatic corps as well as civilian bureaucrats. This command should also include individuals from the civilian society–journalists, academics and social workers—acting in an advisory capacity.[98]

For peacekeeping operations, the West now emphasizes jointness having learnt its lesson in Bosnia-Herzegovina between 1996 to 1999 where operations were hampered by lack of co-operation among the different components of the multinational force like the police, military, and the humanitarian agencies.[99] In the past, India's security forces had been able to achieve some sort of jointness. In 1957, when Lieutenant-General S.P.P. Thorat took over the Eastern Command, he created the Naga Hill Force under Major-General R.K. Kochar. It included one mountain artillery regiment, seventeen infantry battalions, fifty Assam Rifles platoons, thirty-three platoons of Assam Police and fifteen police companies.[100] Lieutenant-General P.N. Kathpalia writes that presence of police officials with troops are necessary to overcome misunderstanding between the soldiers and the local populace.[101] When the guides of Ealam People's Revolutionary Front (EPRF, a local organization which cooperated with the IPKF) were incorporated within the IPKF's companies, the latter were able to engage the LTTE as the EPRF had intimate knowledge of the terrain.[102] On 14 September 1991, the Indian Army launched Operation Rhino against the ULFA, a joint operation involving close interaction between the army, paramilitary, state government and various intelligence agencies. Because of civilian cooperation it was possible for the army to use civilian jeeps, a technique essential for surprising the insurgents.[103]

Joint forces including army units and paramilitary forces and police have to be deployed while fighting insurgents. Since the paramilitary and the army have separate organizational culture, difference and tension are bound to crop up. Some Western theorists argue that in cases where diverse organizations are operating, the culture of cooperation can be generated through drinking, eating and partying together. Informal gatherings can help people of different security organizations to familiarize with each other and to create an atmosphere of trust which generate an ethos of solidarity and a sense of team spirit—a necessity for effective crackdown on the elusive insurgents.[104] From the late 1970s onwards, the *Bundeswehr's* personnel have been trained for international peacekeeping missions. The training of the soldiers aim to raise their social awareness and knowledge about the context of international security policy under which they operate as part of the alliance. Their education is geared to encourage creativity, flexibility, critical rationalism, the capability of getting information and organizing it, the capacity to communicate and exchange information, evaluation of one's behaviour and its impact on others, the readiness to take responsibility, criticism and lastly the capacity to inspire corporate unity.[105]

Despite the demand of the Indian Army officers to have complete autonomy in implementing counter-insurgency plans, political and judicial interference are necessary for checking excessive repression by the military personnel. For example The Armed Forces Special Powers Act 1958 excludes legal redress against the military personnel. But, in September 1982, the Imphal High Court restricted the powers of the armed forces under this Act. The court argued that the people arrested by the military engaged in counter-insurgency tasks in Manipur should be handed over to the police.[106] While political control should be there, the temptation to exercise micro-management of the security forces by the politicians and the bureucrats ought to be avoided. The American military units in Vietnam were hampered by too much interference in tactical operations—ranging from restrictions on targets, bomb loads and routes of aircraft to the complex Rules of Engagement for the troops in the field.[107]

In 1989, the Indian 8th Mountain Division conducted anti-insurgency duties in Nagaland. In March 2000, the 21st Mountain Division of the Indian Army was deployed in lower Assam to check the activities of the NDFB.[108] It is to be noted that the mountain divisions were originally raised to fight China but instead remains engaged in counter-insurgency duties. Thus, counter-insurgency tasks are degrading the conventional deterrent capacity of the Indian Army. Apart form this there are other adverse effects of low-intensity warfare on the militaries. The IDF commanders worry that policing duties against the *Intifada* are harming the former's preparation for modern conventional warfare. After all, combat in the cities with the Palestinians equipped with mortars and bombs is not helping the IDF in preparing for manoeuvre warfare under high technological conditions that it might have to fight against the conventional armies of the Arab states.[109]

Protracted counter-insurgency campaign has a negative impact on the psychology of the Indian soldiers. Sundarji writes: 'Fighting an insurgency these days appears to be a full-time occupation in "peace". In such a low intensity conflict you cannot tell friend from foe, all being in civilian garb, when the other guy always has the initiative to strike first, and the soldier is mostly reactive. Also, in popular mythology the soldier is always the goon who commits atrocities, and that tells on his self-esteem and consequently degrades his morale.'[110] In August 2002, some incidents of 'fragging' were reported. Deployment for a period stretching upto nine months has a negative effect on the mental health of the soldiers and is also adversely affecting the officer corps. There are many cases of middle-level officers committing suicides and there have been instances of them shooting senior officers.[111] General Shankar Roychowdhury (COAS from November 1994 to September 1997) writes in his autobiography:

Looking somewhat resplendent in my Cavalry blues amid the gathering of sombrely clad Riflemen, my interaction with unit level officers, particularly battalion commanders from active operational areas in Kashmir and the north-eastern states, was interesting and informative. . . . Based on one of the issues discussed that evening, I initiated action the very next morning: to revise the tenure of infantry battalions in active counter-insurgency areas from three years to two, as the existing three-year tour of combat duty did not take into account the progressive erosion of individual and unit fighting capacity due to combat fatigue resulting from harsh operational conditions in Kashmir and the North East.[112]

The General Officer Commanding the XV Corps admitted in June 1997 that the newly raised mixed Rashtriya Rifles and the special forces units lacked regimental feeling and group cohesiveness. And cases of soldiers inflicting self-casualties are increasing.[113] The appeal to patriotism, which in anthropological terminology is known as 'fictive kinship',[114] is useless in fighting unconventional warfare, yet regimental *esprit de corps* is vital in conducting low-intensity warfare. Sundarji writes:

> Wanting to be a soldier in this milieu seems ultimate in madness. All theories of people being motivated by 'King and Country' and pure abstract patriotism are highly inadequate. While idealism does have a part to play, the real motivation in my view, is the urge not to let one's buddy down, and not to let the team—the battalion or regiment down. It is in building up this regimental spirit, so vital in war that activities like acquiring regimental trophies play a significant part.[115]

Aggressive patrolling and successful ambushes of the insurgents depends on leadership of the junior officers leading the combat team. General Roychowdhury says about the counter-insurgency operation in Kashmir: 'This war was essentially an affair of small units where the junior leader was the supreme commander. These young officers and jawans, mostly in the one-to six-years service bracket—Second Lieutenants, Lieutenants, Captains and junior Majors as well as their riflemen, sowars, gunners and sappers.'[116] Roychowdhury emphasizes the role of *Augtragstaktik* in such small unit actions.

> These youngsters were capable of far more than what the Army generally allowed them to handle, so, wherever I went, I stressed in the unit and formation commanders the absolute necessity of devolving initiative and responsibility to the young officers, JCOs and NCOs. The nature of the war dictated the operational necessity of giving leeway to junior commanders to act on their own, but a regular and more than

somewhat conventional Army like ours sometimes tended to forget these things.[117]

Sundarji writes: 'In good leadership the theme has to be one of "do as I do" and not "do as I say".[118] After all, 'leading by example' is all the more essential in low-intensity war.

The IDF engages in company-size encounters with the insurgents. And most of the counter-insurgency operations in India can be categorized as a 'Corporal's War' as a great deal of responsibility is placed on the NCOs. They lead patrols, guards camps, lay ambushes, etc. The shortage of more than 13,000 commissioned officers forces the Indian Army to rely more and more on the NCOs especially in the counter-insurgency operations which are mostly conducted by small teams. Pilot leadership courses for junior and senior NCOs have been initiated and a Junior Leader's Academy has been established at Bareilly.[119] No doubt the well-trained and well-experienced NCOs would serve the army well in the event of a conventional campaign characterized by manoeuvre warfare. This is indeed a plus point in the army's involvement in counter-insurgency warfare.

Counter-insurgency is generally characterized by dispersed operations.[120] The US Army came to grief in Vietnam because instead of resorting to small unit tactics, it focused on big unit 'search and destroy' sweeps.[121] The Indian Army also faced serious trouble when the IPKF numbering 60,000 and equipped with T-72 tanks and infantry combat vehicles (BMP-1) tried to conduct 'big sweeps' in the American style in Sri Lanka,[122] and it retreated, too, in 'American style' from Sri Lanka.

In the campaign against the communist Hukbalahap during 1946–54, the Filipino Army did well as it launched self-contained and self-supporting battalion combat teams designed for small unit operations rather than large unit sweeps of the Vietnam style. The small fighting units emphasized on subtlety rather than predictability.[123] When the Rhodesian Army engaged in counter-insurgency duties in the 1950s, about 50 per cent of the training was in the form of small unit operations, and it was quite successful in rooting out the insurgency.[124]

Similarly, the Indian Army was successful when it launched company-level operations under the leadership of the havildars against the LTTE. On 23 May 1989, an Indian Army battalion experienced in counter-insurgency duties launched Operation Kontakkarankukum against the LTTE in the Vavuniya Sector. Omantai was the staging camp for the east-west movement of the LTTE. One company was detailed to open roads and protect convoys. It was also entrusted to gather intelligence. The battalion headquarter with one company was located at Puliyankulam. Another company, i.e. the C Company, carried out extensive area reconnaissance. Between 21–3 June 1989, the C Company, less a platoon, was ordered to carry out a search and destroy mission and lay multiple ambush along the probable routes of the LTTE. The main ambush was laid by two platoons of the C Company backed by two 81-mm. mortars. In the Vietnam style, Indian helicopters were standing by to provide immediate evacuation to the wounded soldiers.[125] Small-unit tactics proved effective also in Assam against the ULFA. During Operation Rhino in September 1991, the army deployed fifteen brigades within an area of 3,000 sq. km. on both the banks of the Brahmaputra. Quick Reaction Teams consisting of eighteen to twenty soldiers were set up for search and cordon duties.[126]

Success in counter-insurgency operations depends on nature being used as an ally. In 1957, F.L. Freemantle with the 3rd Battalion of the 5th Gorkha Rifles who was deployed in Nagaland provides a glimpse of the tactical level of operations in the following words: 'Tactically, it was best to operate on dark nights or in pouring rain to surprise the hostiles and nullify their intimate knowledge of the terrain.'[127] Bikash Basu argues that most of the encounters between the insurgents and the security forces would occur in the jungles. Probably he has north-east India in mind. He writes about the morale of the counter-insurgency forces operating in the jungle areas:

> The jungle terrain poses geomedical problems for men and pack animals deployed for military actions. . . . The armed personnel also suffer from the psychological effect of isolation and loneliness due to restrictions imposed by the density of jungle foilage on visibility, mobility and concentration of manpower. The lack of battle efficiency in the jungle terrain demoralizes the armed personnel seriously. The proliferation of microorganisms, insects, bacteria, etc., in the jungle terrain is at the highest level on account of the combination of high temperature and

high humidity. The military evaluation of the geomedical problems of the operational theatres in jungle terrain can help the organization of remedial and preventive measures.[128]

Freemantle who had fought in the late 1950s in Nagaland as a battalion commander paints the face of low-intensity campaigns in the jungle: 'Leeches, which infest Nagaland, are another hazard especially when one is forced to stay absolutely still for long periods when laying an ambush. We all suffered but I think the record was held by Raj Tiagi who once had to remove over thirty from his person.'[129]

Vertical envelopment of the insurgents is another feature of counter-insurgency warfare. In 1966, the Portuguese at Angola used Alouette helicopters for carrying troops as well as for machine gunning the insurgents.[130] But, such 'fire and movement' tactics did not achieve much success. In 2000, Russia used SU-25 aircraft against the Chechen rebels.[131] After 1965, the Indian Army used helicopters for inserting and de-inducting troops in the insurgency areas of Nagaland, Manipur and Mizoram.[132]

According to one Indian analyst, the new technologies especially in the field of transportation, microelectronics and explosives generated as part of the RMA greatly facilitates terrorist activities.[133] This is probably not true because the counter-insurgency forces could also use the latest technologies. In 1999, the Indian Army planned to import about 150 vehicle-mounted radars and 500 manportable radars (EL/M-2128/9) from Israel to use with the infantry battalions. The EL/M-2129 could be used both for detecting movement of the insurgents as well as hostile enemy regulars. The EL/M-2129 can also detect vehicles and helicopters up to 10 km.[134] The Indian military officials want to deploy Unmanned Aerial Vehicles in the mountainous region bordering China and Pakistan, especially in the militant-infested areas of Kashmir and the north-east for surveillance and identification of the targets.[135] The UAVs will also be of use in detecting hostile troop movement in case of a conventional war. These systems will enhance the Indian Army's capability both as regards counter-insurgency and conventional warfare.

However, technological modernization to a significant extent is hampered because most of the capital that ought to be invested in buying hardware is being used up in maintaining the infantry required for combating the terrorists. After all, counter-insurgency is a manpower intensive task. The Arjun Singh Committee recommended reduction of manpower in the Indian Army from 12,50,000 to 9,50,000. This reduction in manpower would enable the army to go for modernization. However, in 1993, about six divisions were engaged in counter-insurgency operations and the recommendation of the Arjun Singh Committee was not implemented.[136]

Bharat Karnad says that a guerrilla is best hunted down by another guerrilla.[137] Some officers and civilian theoreticians accept that special combat tactics and organizational mechanisms are required for fighting the guerrillas in difficult terrain.[138] Most of the retired Indian military officials are of the opinion that special forces with special knowledge of local conditions, religious and cultural susceptibilities of the disturbed areas should be used to counter the insurgents.[139] Most of the training camps and ammunition dumps of the insurgents are located along the LOC in Pakistan and along the Bangladeshi side of India–Bangladesh border and so remain out of bounds for the Indian troops.

From the 1950s, the IDF followed the policy of cross-border retaliation known as 'reprisal raids',[140] but the Indian Army is not allowed to mount such raids along the borders of Pakistan and Bangladesh. It must be noted that such raids would not automatically result in Pakistan launching nuclear weapons as Pakistan is well aware that India's nukes could hit back. So, there exists space for a limited sub-conventional war under the nuclear shadows. Raids against Bangladesh are also necessary to check infiltration across the border. Because of India's superiority in conventional weapons, Dacca could not hope to escalate a crisis into a full scale conventional war. Major-General Afsir Karim writes that India should raise special forces such as the British SAS for attacking terrorist training camps in POK.[141] In the 1960s, the latter were famous for laying ambushes with electronically detonated Claymore mines. With the SAS team, General Walter Walker conducted cross-border raids for 'teaching' the Indonesians a lesson. The depth of each raid was between 10,000–20,000 yards inside enemy territory.[142] In Afghanistan, in the 1980s, small *Spetsnaz* units with sniper rifles equipped with night sights and silencers were used by the Soviets to ambush the *mujahideens* after dark.[143]

Till there is proper organization of special forces, police forces remain vital for fighting insurgencies. The police forces need to be strengthened in the disturbed areas for conducting joint operations efficiently with the army. Along with the police, organization of local village guards can also prove helpful in fighting the insurgents. Such a voluntary force of peasant villagers was raised in 1957 in Nagaland to defend their villages against the insurgents.[144]

One of the principal factors behind young persons joining the insurgencies is unemployment. During 1947, most of the raiders in Kashmir were soldiers who were demobilized from the Sepoy Army in the immediate aftermath of World War II. When these demobilized troops were denied employment in the Maharaja of Kashmir's Army, they turned to looting. Thus, begun the first insurgency in Kashmir.[145] Many potential insurgents, as well as those who surrendered, are bought into the fold of the state by incorporating them within the security agencies. A Naga Regiment with two battalions was raised especially with the objective of co-opting such people. The Nagas also join the Assam Regiment. Those Naga insurgents who surrender are also taken into the police and in the BSF.[146] On 23 November 1970, the Deputy External Affairs Minister Surendra Pal Singh told the Lok Sabha that in the last two years 2,189 Naga insurgents had surrendered. Of them, 329 were absorbed in the Nagaland Armed Police, seventy were recruited in the Naga Regiment, ten into the BSF and ten into the CRPF.[147] New Delhi is also able to coopt some Kashmiri insurgents by promising money and employment. These 'counter-insurgents' work in concert with the security forces to hunt down their former insurgent comrades. One such organization of the counter-insurgents is headed by Kukka Parray, a former insurgent leader.[148] Even after joining the state apparatus, some of these ex-insurgents might function as double agents for the insurgent groups. Despite this risk, this policy should be followed on a larger scale both in Kashmir and in the north-east.

CONCLUSION

Every action has an equal and opposite reaction. The rise of globalization on one hand results in the emergence of multiple local conflicts on the other hand. Rakesh Gupta defines localization as the rise of ethnic identities not foreseen before globalization.[149] Ethnicity and religion rather than secular ideology (like Marxism) and economic factors are the key reasons behind the emergence of extremism. One could partly explain it due to the rise of religiosity all over the world,[150] as a reaction to globalization. So, in recent times, Islam has replaced Marxism as the ideological force behind the insurgencies. The rebel armies to an extent are mirror images of conventional armies. At times the rebel armies are more tightly knit with stronger ideological motivations than the states' armies. Whatever may be the level of popular support for the insurgents, without external sponsors, insurgencies cannot succeed against the state. Transnational terrorism is however not unique to South Asia. Lebanon and Syria support armed guerrillas against Israel.[151] A nexus has developed between the terrorists, drug mafia and the criminals. Not only the LTTE and the Kashmiri militants use drug money to buy weapons in South-East Asia. The *Kosovar* Albanian terrorist also used drug money to buy weapons in Italy which were then sent to Kosovo.[152]

Pacification campaigns in India have been quite bloodless by world standards. The Algerian War for instance cost 2,50,000 Algerian and French lives.[153] None of the counter-insurgency campaigns in India have resulted in such horrendous casualties. In 1986, the Soviet-sponsored Kabul government resettled 3,00,000 Afghans from the eastern provinces of Afghanistan to west Afghanistan in an attempt to wean away the people from the *mujahideen* liberated zones.[154] Resettlement as a state policy has never been tried in India. However, command has been a problem not only at the grand strategic level that involves the politicians but also at the operational plane for the Indian Army. This in turn harms combat effectiveness of the troops. Technology is no panacea as far as counter-insurgency is concerned. There is no proverbial silver bullet. In the Indian context, the presence of a quite sophisticated counter-insurgency theory and absence of excessive use of military force to some extent explain the Indian state's ability to control the insurgencies within tolerable limits. The next chapter shifts the focus of Indian military activity from land to air.

NOTES

1. Satish Kumar, 'Sources of Democracy and Pluralism in India', in Vice-Admiral K.K. Nayyar and Jorg Schultz, eds., *South Asia Post 9/11: Searching for Stability*, New Delhi: Rupa, 2003, p. 74.
2. Paul Collier and Nicholas Sambanis, 'Understanding Civil War: A New Agenda', *Journal of Conflict Resolution*, vol. 46, no. 1 (2002), p. 11.
3. Paul Collier and Anke Hoeffler, 'On the Incidence of Civil War in Africa', *Journal of Conflict Resolution*, vol. 46, no. 1 (2002), pp. 14–15. One might add that internal wars are due to an amalgam of greed and grievances on part of the rebels. However, the percentage of greed and grievances varies according to local situations.
4. Patrick M. Regan, 'Third-Party Interventions and the Duration of Intra-State Conflicts', *Journal of Conflict Resolution*, vol. 46, no. 1 (2002), pp. 58, 63.
5. Marta Reynal-Querol, 'Ethnicity, Political Systems, and Civil Wars', *Journal of Conflict Resolution*, vol. 46, no. 1 (2002), pp. 29, 32.
6. Jean Dreze, 'Militarism, Development and Democracy', in M.V. Ramana and C. Rammanohar Reddy, eds., *Prisoners of the Nuclear Dream*, New Delhi: Orient Longman, 2003, p. 282.
7. Ibid., p. 296.
8. General K. Sundarji, *Vision 2100: A Strategy for the Twenty-First Country*, Delhi: Konark, 2003, p. 37.
9. Vice-Admiral K.K. Nayyar et al., *National Security: Military Aspects*, New Delhi: Rupa, 2003, pp. 213–14.
10. For the concept of redirected or displaced aggression see David P. Barash and Charles P. Webel, *Peace and Conflict Studies*, New Delhi: Sage, 2002, p. 134.
11. J. Baranwal, Editor-in-chief, *SP's Military Yearbook 1995*, New Delhi: Guide Publications, 1995, p. 47; Nayyar et al., *National Security*, p. 213.
12. Rohan Gunaratna, 'Transnational Terrorism: Support Networks and Trends', in K.P.S. Gill and Ajai Sahni, eds., *Faultlines: Writings on Conflict and Resolution*, vol. 7, New Delhi: Bulwark Books and the Institute for Conflict Management, 2000, p. 6.
13. Brigadier-General Samuel B. Griffith, 'Introduction', in idem and Major Harris-Clichy Peterson, eds., *Guerrilla Warfare*, 1962; rpt., London: Cassell, 1963, p. 9.
14. John Pimlott, 'The French Army: From Indochina to Chad, 1946–84', in Ian F.W. Beckett and John Pimlott, eds., *Armed Forces and Modern Counter-Insurgency*, London: Croom Helm, 1985, p. 46.
15. John Pimlott, 'The British Army: The Dhofar Campaign, 1970–75', in Beckett and Pimlott, eds., *Armed Forces*, p. 20.
16. Jeffrey Grey, 'Malaya, 1948–60: Defeating Communist Insurgency', in

Major-General Julian Thompson, ed., *The Imperial War Museum Book of Modern Warfare: British and Commonwealth Forces at War, 1945–2000,* 2002; rpt., London: Pan Macmillan, 2003, pp. 81–2.

17. Raffi Gregorian, *The British Army, the Gurkhas and the Cold War Strategy in the Far East: 1947–54,* Basingstoke: Palgrave, 2002, p. 55.
18. Brigadier M.J. Strudwick, 'A National Plan to Contain Militancy and Insurgency', *National Defence College Journal,* vol. 15 (1993), p. 50.
19. Francis Toase, 'The South African Army: The Campaign in South West Africa/Namibia since 1966', in Beckett and Pimlott, eds., *Armed Forces,* p. 211.
20. Wray R. Johnson, *Vietnam and American Doctrine for Small Wars,* Bangkok: White Lotus Press, 2001, p. 20.
21. Ibid., pp. 87, 91.
22. Peter M. Dunn, 'The American Army: The Vietnam War, 1965–73', in Beckett and Pimlott, eds., *Armed Forces,* p. 80.
23. David N. Solomon, 'The Soldierly Self and the Peacekeeping Role: Canadian Officers in Peacekeeping Forces', in Jacques Van Doorn, ed., *Military Profession and Military Regimes: Commitments and Conflicts,* The Hague/Paris: Mouton, 1969, p. 59.
24. Major Vivek Chadha, *Company Commander in Low Intensity Conflict,* New Delhi: Lancer, 1997, p. 121.
25. V.K. Shrivastava, 'Indian Army: The Challenge Ahead', *Strategic Analysis,* vol. 25, no. 4 (2001), p. 497.
26. Brigadier S.K. Sinha, 'Counter Insurgency Operations', *JUSII,* vol. 100. no. 420 (1970), p. 262.
27. Major-General P.S. Bhagat, *Forging the Shield: The Defence of India and South East Asia,* Calcutta: The Statesman, p. 85.
28. S. Kalyanaraman, 'Conceptualizations of Guerrilla Warfare', *Strategic Analysis,* vol. 27, no. 2 (2003), p. 173.
29. Ian F.W. Beckett and John Pimlott, 'Introduction', in Beckett and Pimlott, eds., *Armed Forces,* p. 9.
30. Colonel Bhaskar Sarkar, *Tackling Insurgency and Terrorism: Blueprint for Action,* Delhi: Vision Books, 1998, pp. 95, 102.
31. Colonel Harjeet Singh, *Doda: An Insurgency in the Wilderness,* New Delhi: Lancer, 1999, Appendix D.
32. F.A. Godfrey, 'The Latin American Experience: The Tupamaros Campaign in Uruguay, 1963–73', in Beckett and Pimlott, eds., *Armed Forces,* pp. 120–1.
33. Toase, 'The South African Army', in Beckett and Pimlott, eds., *Armed Forces,* p. 211.
34. Sarkar, *Tackling Insurgency and Terrorism,* pp. 17, 111.
35. Ved Marwah, *Uncivil Wars: Pathology of Terrorism in India,* 1995; rpt., New Delhi: Indus, 1996, p. 109.

36. Mao Tse-Tung, 'Guerrilla Warfare', in Griffith and Peterson, eds., *Guerrilla Warfare*, p. 33.
37. Pimlott, 'The British Army', in idem. and Beckett, eds., *Armed Forces*, p. 21.
38. Indian Defence Review Research Team, 'OP TOPAC: The Kashmir Imbroglio', *IDR*, vol. 14, no. 2 (1999), p. 33.
39. S.P. MacKenzie, *Revolutionary Armies in the Modern Era: A Revisionist Approach*, London: Routledge, 1997, p. 160.
40. Kalyanaraman, 'Conceptualizations of Guerrilla Warfare', p. 181.
41. Godfrey, 'The Latin American Experience', in Beckett and Pimlott, eds., *Armed Forces*, p. 117.
42. Brigadier R.K. Nanavatty, 'Use of Armed Forces in Aid to Civil Authority: Problems and Prospects', *National Defence College Journal*, vol. 15 (1993), p. 13. Italics in original.
43. Ashwini Kumar, 'Latest Trend in Terrorism', in V.D. Chopra, ed., *Rise of Terrorism and Secessionism in Eurasia*, New Delhi: Gyan Publishing House, 2001, p. 73.
44. Ibid.
45. Akshay Joshi, 'Information Warfare in Kargil Operations', *IDR*, vol. 14, no. 2 (1999), p. 37.
46. Ibid., pp. 36–7.
47. Scott Gates, 'Recruitment and Allegiance: The Microfoundations of Rebellion', *Journal of Conflict Resolution*, vol. 46, no. 1 (2002), p. 112.
48. My definition of unconventional armies like all definition is open to criticism. I am influenced by S. Brock Blomberg and Gregory D. Hess's, 'The Temporal Links between Conflict and Economic Activity', *Journal of Conflict Resolution*, vol. 46, no. 1 (2002), p. 77.
49. Gates, 'Recruitment and Allegiance', p. 115
50. Griffith, 'Introduction', in Griffith and Peterson, eds., *Guerrilla Warfare*, p. 10.
51. *The Military Art of People's War: Selected Writings of General Vo Nguyen Giap*, edited and with an introduction by Russell Stetler, 1970; rpt., New York/London: Monthly Review Press, 1971, p. 112.
52. Omer Bartov, *The Eastern Front, 1941–45: German Troops and the Barbarisation of Warfare*, 1985; rpt., Basingstoke: Macmillan, 2001, pp. 30, 144.
53. Pimlott, 'The French Army', in idem. and Beckett, eds., *Armed Forces*, p. 51.
54. Gunaratna, 'Transnational Terrorism', in Gill and Sahni, eds., *Faultlines*, vol. 7, p. 4.
55. Shri Prakash, 'Transformation of Pakistan and Afghanistan as Spring-Boards of Cross-Country Terrorism in the Region', in Chopra, ed., *Rise of Terrorism*, pp. 114–15.
56. Jyoti Trehan, 'Violence in J&K: Complexities and Pathways', in Gill and

Sahni, eds., *Faultlines*, vol. 7, p. 152; Owen Bennet Jones, *Pakistan: Eye of the Storm,* New Delhi: Viking, 2002, p. 107; Marwah, *Uncivil Wars*, pp. 35–6.

57. Singh, *Doda*, p. 56; Smruti S. Pattanaik, 'Pakistan's Kashmir Policy: Objectives and Approaches', *Strategic Analysis*, vol. 26, no. 2 (2002), p. 202; R. Rama Rao, 'Peace in the Subcontinent', *JUSII*, vol. 103, no. 429 (1973), pp. 5–6; Major-General F.L. Freemantle, *Fred's Foibles*, New Delhi: Lancer, 2000, p. 17.
58. Pattanaik, 'Pakistan's Kashmir Policy, p. 224; General Shankar Roychowdhery, *Officially at Peace: Reflections on the Army and its Role in Troubled Times*, New Delhi: Viking, 2002, p. 57; G.M. Shah, 'Militancy in Jammu and Kashmir: A Study of Actors, Objectives and Strategies', in Chopra, ed., *Rise of Terrorism*, p. 148; Sawhney, *The Defence Makeover*, p. 147; J. Baranwal, ed., *SP's Military Yearbook: 1993–94,* New Delhi: Guide Publications, 1993, p. 101; Sumita Kumar, 'The Role of Islamic Parties in Pakistani Politics', *Strategic Analysis*, vol. 25, no. 2 (2001), p. 277.
59. Marwah, *Uncivil Wars*, p. 52; S.K. Lambah, 'Democracy in Pakistan', in Nayyar and Schultz, eds., *South Asia Post 9/11*, p. 157; S.V.R. Nasr, 'Islam, the State and the Rise of Sectarian Militancy in Pakistan', in Christophe Jaffrelot, ed., *Pakistan: Nationalism Without a Nation?* New Delhi: Manohar, 2002, p. 92; Vinod Anand, 'Pakistan and Taliban Role in Chechnya Conflict', in Chopra, ed., *Rise of Terrorism*, p. 211; Arvind R. Deo, 'Pakistan's Unending Search for a Viable Foreign Policy', in Vice-Admiral K.K. Nayyar, ed., *Pakistan at the Crossroads,* New Delhi: Rupa, 2003, p. 23.
60. Shitanshu Mishra, 'Exploitation of Information and Communication Technology by Terrorist Organisations', *Strategic Analysis*, vol. 27, no. 3 (2003), p. 442; Lieutenant-General Y.M. Bammi, 'The Evolution of the Pakistani Army and the Likely Military Strategy Against India from 2002 to 2010', in Nayyar, ed., *Pakistan at the Crossroads*, p. 291; N.S. Jamwal, 'Terrorists' Modus Operandi in Jammu and Kashmir', *Strategic Analysis*, vol. 27, no. 3 (2003), p. 384; Assaf Moghadam, 'Palestinian Suicide Terrorism in the Second Intifada: Motivations and Organizational Aspects', *Studies in Conflict and Terrorism*, vol. 26, no. 2 (2003), p. 70; Mariam Abou Zahab, 'The Regional Dimension of Sectarian Conflicts in Pakistan', in Jaffrelot, ed., *Pakistan: Nationalism Without a Nation?*, p. 124.
61. Brigadier Chandra B. Khanduri, 'The Face of Proxy War in J&K', in Chopra, ed., *Rise of Terrorism*, p. 121; Zahab, 'Regional Dimension of Sectarian Conflict', and Saeed Shafqat, 'From Official Islam to Islamism: The Rise of *Dawat-ul-Irshad* and *Lashkar-e-Toiba*', in Jaffrelot, ed., *Pakistan: Nationalism Without a Nation?*, pp. 120, 138; Trehan, 'Violence in J&K', p. 155; Baranwal, ed., *SP's Military Yearbook: 1993–*

94, p. 101; Lieutenant-General V.K. Sood and Pravin Sawhney, *Operation Parakram: The War Unfinished,* New Delhi: Sage, 2003, p. 34; Sreedhar K. Santhanam, Sudhir Saxena and Manish, *Jihadis in Jammu and Kashmir,* New Delhi: Sage, 2003, p. 260; Mishra, 'Exploitation of Information and Communication Technology', pp. 442–3.

62. Major-General Afsir Karim, 'Terrorism and National Security Concerns', and Kalim Bahadur, 'Pakistan and Afghanistan as Sources of Terrorism in the Region', in Chopra, ed., *Rise of Terrorism,* pp. 102, 110; Kshitij Prabha, 'Narco-Terrorism and India's Security', *Strategic Analysis,* vol. 24, no. 10 (2001), pp. 1877–9; Ian Talbot, 'Does the Army Shape Pakistan's Foreign Policy?', in Jaffrelot, ed., *Pakistan: Nationalism without a Nation,* p. 313; Marwah, *Uncivil Wars,* p. 74; Jamwal, 'Terrorists' Modus Operandi', p. 389; Sood and Sawhney, *Operation Parakram,* p. 39.
63. Santhanam et al., *Jihadis in Jammu and Kashmir,* pp. 226, 265–6; Shafqat, 'The Rise of *Dawat-ul-Irshad* and *Lashkar-e-Toiba*', p. 132.
64. Pattanaik, 'Pakistan's Kashmir Policy, p. 223; Ajay Darshan Behera, 'The Supporting Structures for Pakistan's Proxy War in Jammu and Kashmir', *Strategic Analysis,* vol. 25, no. 3 (2001), p. 397; Rammohan, 'Disquiet on the Western Front', p. 259; Baranwal, ed., *SP's Military Yearbook: 1993–4,* p. 101.
65. Edgar O'Ballance, *Afghan Wars, Battles in a Hostile Land: 1839 to the Present,* 2002; rpt., Karachi: Oxford University Press, 2003, p. 145.
66. Ian F.W. Beckett, 'The Rhodesian Army: Counter-Insurgency, 1972–79', in Beckett and Pimlott, eds., *Armed Forces,* p. 180.
67. Major-General Afsir Karim, 'Confrontation in Kargil', *IDR,* vol. 14, no. 2 (1999), p. 12.
68. Mark A. Heller, *Continuity and Change, in Israili Security Policy,* Adelphi Paper, no. 335, Oxford: Oxford University Press, 2000, p. 24.
69. Jean-Luc Racine, 'Pakistan and the "India Syndrome": Between Kashmir and the Nuclear Predicament', in Jaffrelot, ed., *Pakistan: Nationalism Without a Nation?,* p. 215.
70. Karim, 'Terrorism and National Security Concerns' and K. Warikoo, 'Islamist Extremism and Terrorism in Kashmir', in Chopra (ed.), *Rise of Terrorism,* pp. 102, 128–9; Jones, *Pakistan,* p. 85; Sood and Sawhney, *Operation Parakram,* pp. 31, 41; Praveen Swami, 'A Message Loud and Clear', *Frontline,* 15 August 2003, p. 36.
71. Nitin A. Gokhale, 'Borderliners: Shades of Pacifism and Fatigue on a Canvas of Violence', *Outlook,* 20 October 2003, p. 56; Wasbir Hussain, 'Cross-Border Human Traffic in South Asia: Demographic Invasion, Anxiety and Anger in India's North-east', in Gill and Sahni, eds., *Faultlines,* vol. 7, p. 107; Sreeradha Datta, 'What Ails the North-East:

An Enquiry into the Economic Factors', *Strategic Analysis*, vol. 25, no. 1 (2001), pp. 73, 77.

72. Binalakshmi Nepram, *South Asia's Fractured Frontier: Armed Conflicts, Narcotics and Small Arms Proliferation in India's North-East*, New Delhi: Mittal, 2002, pp. 87, 91; Sushil K. Pillai, 'The Invisible Country: Ethnicity and Conflict Management in Myanmar', in Gill and Sahni, eds., *Faultlines*, vol. 7, pp. 57–8; Sharma, *Path of Glory: Exploits of 11 Gorkha Rifles*, Ahmedabad: Allied, 1988, p. 79; Gurudas Das, 'India's North-East Soft Underbelly: Strategic Vulnerability and Security', *Strategic Analysis*, vol. 26, no. 4 (2002), pp. 539, 541; Major-General Afsir Karim, 'Pak-Sponsored Terrorism', in Nayyar, ed., *Pakistan at the Crossroads*, p. 377; Lieutenant-General S.P.P. Thorat, *From Reveille to Retreat*, New Delhi: Allied, 1986, pp. 179–80.
73. Gates, 'Recruitment and Allegiance', pp. 115, 124.
74. Sarkar, *Tackling Insurgency and Terrorism*, p. 52; Freemantle, *Fred's Foibles*, p. 70; Neville Maxwell, *India's China War*, 1970; rpt., Bombay: Jaico, 1971, p. 295; Thorat, *From Reveille to Retreat*, pp. 182–3.
75. Kanti P. Bajpai, *Roots of Terrorism*, New Delhi: Penguin, 2002, p. 138.
76. Jaideep Saikia, 'Autumn in Springtime: The ULFA Battles for Survival', in Gill and Sahni, eds., *Faultlines*, vol. 7, p. 31; Jaideep Saikia, 'Changing Contours of Separatism', *Aakrosh*, vol. 6, no. 18 (2003), pp. 55–6; Bibhu Prasad Routray, 'Growing Tentacles of Insurgency in Arunachal Pradesh', *Peace and Conflict: An IPCS Bulletin*, vol. 5, no. 9 (2002), p. 4; Brigadier R.S. Grewal, 'Ethno Nationalism in North Eastern India', *JUSII*, vol. CXXXIII, no. 552 (2003), pp. 268, 271; Roychowdhury, *Officially at Peace*, p. 85; Dinesh Kotwal, 'The Contours of Assam Insurgency', *Strategic Analysis*, vol. 24, no. 12 (2001), pp. 2223–4; Kalyan Chaudhuri, 'Bhutan's Resolve', *Frontline*, 15 August 2003, p. 49; *Gokhale*, 'Borderliners', p. 56; Hussain, 'Demographic Invasion, pp. 108–112.
77. Nayyar et al., *National Security*, p. 237; M. Maroof Raza, *Low Intensity Conflicts: The New Dimension to India's Military Commitments*, Meerut: Kartikeya Publications, 1995, pp. 27–9, 34, 36; Chadha, *Company Commander*, p. 126; V. Suryanarayan, 'Ethnic Cauldron in Sri Lanka', in Chopra, ed., *Rise of Terrorism*, p. 291; V.S. Sambandan, 'A Rebellion in the East', *Frontline*, 26 March 2004, p. 115; N.S. Jamwal, 'Counter Terrorism Strategy', *Strategic Analysis*, vol. 27, no. 1 (2003), p. 60; Kingsley Wickremasuriya, 'Community Oriented Policing and Conflict Management: The Jaffna Experiment', in Gill and Sahni, eds., *Faultlines*, vol. 7, p. 86; Baranwal, ed., *SP's Military Yearbook: 1995*, p. 49; Rajesh Rajagopalan, *Fighting like a Guerrilla: The Indian Army and Counterinsurgency*, New Delhi: Routledge, 2008, p. 99.
78. Jamwal, 'Counter Terrorism Strategy', p. 72.

79. Heller, *Continuity and Change*, p. 29.
80. Pavel K. Baev, *The Russian Army in a Time of Troubles*, London: Sage, 1996, p. 132.
81. Ian F.W. Beckett, 'The Portuguese Army: The Campaign in Mozambique, 1964–74', in Beckett and Pimlott, eds., *Armed Forces*, p. 142.
82. O'Ballance, *Afghan Wars*, p. 182.
83. Martin Thomas, 'Order before Reform: The Spread of French Military Operations in Algeria, 1954–58', in David Killingray and David Omissi, eds., *Guardians of Empire: The Armed Forces of the Colonial Powers c. 1700–1964*, Manchester/New York: Manchester University Press, 1999, p. 208.
84. O'Ballance, *Afghan Wars*, p. 183.
85. Bharat Karnad, *Nuclear Weapons and Indian Security: The Realist Foundations of Strategy*, New Delhi: Macmillan, 2002, p. 497.
86. S. Sanyal, 'One Man's Terrorist: Victims' Perspectives on Terrorism', in Gill and Sahni, eds., *Faultlines*, vol. 7, p. 140.
87. Sushanta Talukdar, 'Peace at a Premium', *Frontline*, 5 November 2004, p. 28; General K.V. Krishna Rao, *In the Service of the Nation: Reminiscences*, New Delhi: Viking, 2001, pp. 533–9; Jones, *Pakistan*, p. 87; Harjeet Singh, *Doda*, p. 245.
88. Christophe Jaffrelot, 'Introduction, Nationalism without a Nation: Pakistan Searching for its Identity', in idem., ed., *Pakistan: Nationalism Without a Nation?*, pp. 24, 29; Jones, *Pakistan*, p. 134; Stephen P. Cohen, *The Pakistan Army*, 1984; rpt., Karachi: Oxford University Press, 1993, pp. 113–14.
89. Thomas, 'Order Before Reform', in Killingray and Omissi, eds., *Guardians of Empire*, p. 201.
90. R.D. Palsokar, 'The Essentials of Guerrilla War', *JUSII*, vol. 103, no. 429 (1973), pp. 36, 41.
91. Chadha, *Company Commander*, p. 123.
92. Vernon Hewitt, *Towards the Future: Jammu and Kashmir in the 21st Century*, Cambridge: Granta, 2001, p. 2.
93. Sawhney, *The Defence Makeover*, p. 158.
94. Karnad, *Nuclear Weapons and Indian Security*, p. 496.
95. Khanduri, 'Proxy War in J&K', in Chopra, ed., *Rise of Terrorism*, p. 123.
96. Sawhney, *The Defence Makeover*, p. 147. According to another estimate, some 6,00,000 Indian Army, police and paramilitary personnel were stationed in Jammu and Kashmir in 1998. Amelie Blom, 'The "Multi-Vocal State": The Policy of Pakistan on Kashmir', in Jaffrelot (ed.), *Pakistan: Nationalism without a Nation?*, p. 306.
97. Indian Defence Review Research Team, 'OP TOPAC', p. 21.

98. For a narrow interpretation regarding the requirement of a distinct body for fighting low-intensity warfare, see, Brigadier D.S. Chauhan, 'Low Intensity Conflict in India: Organization and Resources Required to Meet this Threat', *National Defence College Journal*, vol. 15 (1993), pp. 150–1.
99. Erwin A. Schmidl, 'The Evolution of Peace Operations', in idem, ed., *Peace Operations Between War and Peace*, London: Frank Cass, 2000, p. 18.
100. Thorat, *From Reveille to Retreat*, p. 182.
101. Lieutenant-General P.N. Kathpalia, 'Internal Security and CI Operations in Urban Areas', *IDR*, vol. 2, no. 1 (1987), p. 53.
102. Chadha, *Company Commander*, p. 126.
103. Pravin Sawhney, 'Operation Rhino: A Case Study', *IDR*, January 1992, pp. 71, 74.
104. Efrat Elron, Boas Shamir and Eyal Ben-Ari, 'Why Don't They Fight Each Other? Cultural Diversity and Operational Unity in Multinational Forces', in Stuart A. Cohen, ed., *Democratic Societies and their Armed Forces: Israel in Comparative Context*, London: Frank Cass, 2000, p. 126.
105. Hans Eberhard Radbruch, 'From Scharnhorst to Schmidt: The System of Education and Training in the German Bundeswehr', *Armed Forces and Society*, vol. 5, no. 4 (1979), pp. 619–20.
106. *Amnesty International Report 1983*, London: Amnesty International Publications, 1983, pp. 197–8.
107. Dunn, 'The American Army', in Beckett and Pimlott, eds., *Armed Forces*, p. 84.
108. Saikia, 'Autumn in Springtime', in Gill and Sahni, eds, *Faultilines*, vol. 7, p. 29; Rao, *Reminiscences*, p. 308.
109. 'Special Document: Mahmud Abbas' Call for a Halt to the Militarization of the Intifada', *Journal of Palestine Studies*, vol. 32, no. 2 (2003), p. 75; Stuart A. Cohen, *Studying the Israel Defense Forces: A Changing Contract with Israeli Society*, Ramat Gan, The BESA Centre: Bar-Ilan University, 1995, p. 14.
110. General K. Sundarji, *Of Some Consequences: A Soldier Remembers*, New Delhi: HarperCollins, 2000, p. 91.
111. Major Praveen Badrinath, 'Psychological Impact of Protracted Service in Low Intensity Conflict Operations (LICO) on Armed Forces Personnel: Causes and Remedies', *JUSII*, vol. CXXXIII, no. 551 (2003), p. 39.
112. Roychowdhury, *Officially at Peace*, pp. 16–17.
113. Sawhney, *The Defence Makeover*, p. 160.
114. Barash and Webel, *Peace and Conflict Studies*, p. 124.
115. Sundarji, *Of Some Consequences*, p. 91.

116. Roychowdhury, *Officially at Peace*, p. 46.
117. Ibid.
118. Sundarji, *Of Some Consequences*, p. 168.
119. Rahul Bedi, 'Indian Army Plans Boost for Infantry Battalions', *JDW*, vol. 31, no. 8, 24 February 1999, p. 28; Robin Neillands, 'Cyprus, 1955–59: EOKA', in Thompson, ed., *British and Commonwealth Forces at War*, p. 179.
120. For a detailed analysis of tactical and organizational aspects of counter-insurgency operations in the Indian subcontinent refer to Kaushik Roy, 'Goliath Against David: Militaries Against the Militias', *Contemporary India*, vol. 1, no. 2 (2002), pp. 138–48.
121. MacKenzie, *Revolutionary Armies*, p. 168.
122. Raza, *Low Intensity Conflicts*, p. 34.
123. Beckett and Pimlott, 'Introduction', in Beckett and Pimlott, eds., *Armed Forces*, pp. 6–7.
124. Beckett, 'The Rhodesian Army', in Beckett and Pimlott, eds., *Armed Forces*, p. 176.
125. Chadha, *Company Commander*, pp. 125, 127, 129.
126. Sawhney, 'Operation Rhino', pp. 74–5.
127. Freemantle, *Fred's Foibles*, p. 70.
128. Bikash B. Basu, 'Military Evaluation of Jungle Terrain: A Study in Military Geography', *JUSII*, vol. 99, no. 416 (1969), p. 288.
129. Freemantle, *Fred's Foibles*, p. 70.
130. Beckett, 'The Portuguese Army', in Beckett and Pimlott, eds., *Armed Forces*, p. 144.
131. Georg Mander, 'Sukhoi Pushes Again for Reform', *JDW*, vol. 33, no. 21, 24 May 2000, p. 18.
132. Brigadier Vijay K. Nair, 'Employment of Military Helicopters: Part II, The Indian Experiences and Compulsions', *IDR*, January 1992, p. 103.
133. B.K. Shrivastava, 'The United States and International Terrorism in South Asia', in Chopra, ed., *Rise of Terrorism*, p. 33.
134. Rahul Bedi, 'India Buys Radars from Israel', *JDW*, vol. 31, no. 8, 24 February 1999, p. 5.
135. Rahul Bedi, 'India Set to Put Nishant UAV to the Real Test', *JDW*, vol. 31, no. 7, 17 February 1999, p. 29.
136. Baranwal, ed., *SP's Military Yearbook: 1993–94*, p. 117.
137. Karnad, *Nuclear Weapons and Indian Security*, p. 497.
138. For innovative tactical and strategic policies that ought to be followed by the security forces against the insurgents see Kaushik Roy, 'Mars Defeated? Conventional Militaries in Unconventional Warfare', *Contemporary India*, vol. 2, no. 2 (2003), pp. 79–101.
139. Nayyar et al., *National Security*, p. 214.
140. Cohen, *Studying the Israeli Defense Forces*, p. 7.
141. Karim, 'Confrontation in Kargil', p. 14.

142. Walter Walker, 'Brunei and Borneo, 1962–66: An Efficient Use of Military Force', in Thompson, ed., *British and Commonwealth Forces at War*, pp. 215, 218.
143. O'Ballance, *Afghan Wars*, p. 144.
144. Rao, *Reminiscences*, p. 310.
145. Shah, 'Militancy in Jammu and Kashmir', in Chopra, ed., *Rise of Terrorism*, p. 138.
146. Rao, *Reminiscences*, pp. 310–11.
147. S.K. Ghosh, 'Insurgency in South and South-East Asia, 1969–70', *Institute of Defence Studies and Analyses Journal*, vol. 3 (1970–1), pp. 452–3.
148. Sumit Ganguly, 'The Islamic Dimensions of the Kashmir Insurgency', in Jaffrelot, ed., *Pakistan: Nationalism Without a Nation?*, p. 189.
149. Rakesh Gupta, 'Towards a Political Economy of Intra State Conflicts', in Chopra, ed., *Rise of Terrorism*, p. 65.
150. K. Santhanam, 'Sources of Terror: India', in Nayyar and Schultz, eds., *South Asia Post 9/11*, p. 31.
151. Heller, *Continuity and Change*, pp. 22, 27.
152. V.D. Chopra, 'Introduction', in idem, ed., *Rise of Terrorism*, pp. 19, 20.
153. Barash and Webel, *Peace and Conflict Studies*, p. 165.
154. O'Ballance, *Afghan Wars*, p. 154.

SIX

Reach for the Sky: The Indian Air Force

In the new millennium, the Indian Air Force represents one of the strongest aerial war machines within Afro-Asia. In the last fifty years, it evolved from its predecessor the Royal Indian Air Force (RIAF). The RIAF started as an ancillary organization to the Royal Air Force (RAF). In World War II, the RIAF saw limited action against the Japanese in Burma. After Independence, the IAF grew slowly but steadily. Meanwhile it had to encounter the armed forces of Pakistan in several limited wars. After 1990, the IAF emerged as an instrument of deterrence. However, the genesis of the IAF was a tortuous process. From 1947, the military establishment had been divided over the issue whether the IAF was going to be a tactical air force or a strategic deterrent. At present, the debate is regarding the role of IAF in nuclear war and star war. Let us have a flashback at the historical evolution of the IAF.

THE IAF IN KASHMIR: 1947–1949

With Independence, India got six and half squadrons and Pakistan acquired three squadrons.[1] Air Marshal Thomas Elmhirst was the first commander of independent India's air force. After his appointment in August 1947, he met Jawaharlal Nehru who was not oblivious of the importance of air power in military affairs. In 1940, Nehru while witnessing the successful Nazi *Blitzkrieg* had written: 'In modern warfare armies have become secondary factors. The air arms counts for most.'[2] Elmhirst describes his first meeting with the Indian prime minister in the following words:

Mr. Nehru had told me at my 'contract' interview that in the first year . . . approximately twenty million pounds, was all that the country could afford. The Financial Experts at Air Headquarters had to work out what that would mean in terms of pay, food, new aircraft, buildings, signals organization, and supplies and spares. Luckily, I had the experience under Lord Trenchard at London Air Ministry in the twenties of watching him lay the foundation of RAF, and remembered his policy of 'Training and Accommodation' first, followed by reequipping squadrons or forming new ones later.[3]

Elmhirst continues: 'I had a bit of luck in a hint from an old friend Sam Elsworthy, who was closing down the old RAF Base at Karachi. "Had I use for fifty new Spitfires in cases there?" I said, put them on the first ship to Bombay. They gave a great start to two squadrons and the Flying School.'[4] The Spitfires were made by Vickers Armstrongs Limited of the United Kingdom. Their operational range was 740 km. Each Spitfire was armed with two 20-mm. cannons, two .50 machine-guns and two 250-lbs. bombs.[5]

In the meantime, Pakistan-trained raiders invaded Kashmir. On 26 October 1947, the Army Headquarter decided to despatch one battalion by air to Srinagar. The aim was to concentrate at least one brigade in Kashmir Valley before the onset of the winter and on 27 October troops were flown from Palam Airport. About 300 men were sent on the first day and from the next day, these flights were organized from Safdarjang Airport. Some thirty Dakotas were mustered to airlift the 1st Sikh on the morning of 28 October. The Douglas Aircraft Company of the USA constructed the Dakotas. Each Dakota carried thirty-two men. Though its maximum speed was 340 km. per hour, it had an impressive range of 2,430 km. The airfield at Srinagar was the pivot of the whole campaign. Just as the failure of the Allied troops to hold on to Maleme airfield in May 1941 allowed the *Luftwaffe* to fly in troops and capture the island of Crete, the failure of the Pakistani raiders to capture the airfield at Srinagar enabled the IAF to insert troops and wrecked Pakistan's plan of annexing Kashmir. At that time the only airfield available in the plains of Jammu was the Maharajah of Jammu & Kashmir's football field, not far away from Jammu city. It was a grass covered field and became soft and slushy after the winter showers. Most of the Indian ground formations scattered in various points within Jammu & Kashmir mostly depended on air supply. There was no land link with the Indian garrison at Poonch. Brigadier Pritam Singh, the garrison commander constructed an airstrip close to the town. And the IAF transported a battalion. Around mid March 1948, the Pakistanis used howitzers for shelling the airstrip of Poonch and India responded by sending 25-pounder guns which were carried by the Dakotas. Credit is due to the dare devil Mehar Singh for landing the aircraft at night in order to avoid enemy shelling which was carried out during daylight. The IAF's airlift capabilities not only aided the Indian Army to defend Kashmir but also Ladakh. On 1 June 1948, six Dakotas carried a company of the 2nd Battalion of 4th Gurkha Rifles to Leh.[6]

Besides ferrying troops, the IAF was also used for strafing the insurgents. This trend was a continuation of the British policy of using air power for subduing insurgencies along the trans-Indus frontier. The RAF and the RIAF were used to strafe the Pathan tribes along the North-West Frontier.[7] In 1947, India negotiated for the delivery of eight–nine Tempest-II from the RAF. In the same year, Pakistan ordered twenty-four Tempest fighter-bombers from Britain. The Tempests continued in the IAF and in the Pakistan Air Force till 1953. The Tempests were both single seat fighters as well as fighter-bombers. Its speed was about 440 miles per hour at an altitude of 15,000 ft. and had an operational range of 800 miles. The Tempests were armed with either four 20-mm. cannons or carried two 1,000 lb. bombs under the wings. In addition, the IAF also deployed Supermarine Spitfires. The IAF's Tempests and Spitfires operating from Ambala and Amritsar provided firepower support to the Indian troops battling at the outskirts of Srinagar. The Spitfires strafed the Pakistani raiders on the outskirts of the airfield at Badgam. When the Pakistanis attempted to construct an airfield near Gilgit, the IAF struck. In August 1948, Number 7 Squadron's Tempests and Spitfires of the Advanced Flying School destroyed the Gilgit airfield along with the wireless station and the newly constructed military barracks. The IAF also provided aerial reconnaissance to the ground forces. While the Harvards and the Austers conducted air observation of the hostile forces, the Tempests escorted them. The Austers were slow machines with a maximum speed of 250 miles per hour and were used for observations for the artillery and guided cannon fire to the targets. The Harvards and the Austers like the Fiesler Storchs of the *Luftwaffe* were also used for conveying army officers to strategic locations.[8]

During emergencies, some IAF officers were able to display innovative capabilities. The IAF had no bombers for bombing the raiders. Mehar Baba came up with an innovative solution. Baba had passed from the RAF College at Cranwell in 1936. Baba argued that the Dakotas should be transformed into bombers. Baba himself flew a Dakota-turned-bomber on a raid to Domel bridge which was the lifeline of the raiders and heavily protected. Nevertheless, Baba pressed home his attack and destroyed the bridge.[9] This was probably the first time in the history of aerial warfare that transport aircraft had been transformed into ad hoc bombers.

Many Tempests were lost to the Bofors 40-mm guns which the Pakistanis had set up near Tithwal. The PAF possessed thirty-two Dakotas, thirty-five Tempests, twenty-nine Harvards, sixteen Tiger Moths and three Austers with 200 officers and 2,112 men. Nevertheless, the PAF did not dare to challenge the IAF. Initially, the PAF used its Dakotas for dropping supplies in daytime. But, when the IAF's fighters conducted regular sorties over the skies of Kashmir, the PAF's Dakotas resorted to nighttime supply operations.[10]

One positive impact of the war was that Elmhirst was allowed to go abroad for buying aircraft for the IAF. In his own words: 'In that year [1948], with some crores of rupees in my pocket, I went to London and with the very personal help of Lord Tedder, obtained fifty new Spitfire aircraft and a few of the first production of the De Havilland and Vampire jet fighters. At this date no Air Force in the world, other than the British and the United States, was equipped with jet fighters. It was a good start.'[11] The speed of a Vampire was 880 km. per hour and each was armed with four 20-mm cannon and eight 3-inch 60 lb. rockets or 2,000 lb. bombs.[12] The year 1948 also witnessed the foundation of the Technical Training Establishment at Bangalore for producing engineer officers and skilled mechanics for the IAF.[13]

In the aftermath of the First India–Pakistan War, the Indian government poured a lot of ink over the question of reorganization of the defence forces. Noble laureate P.M.S. Blackett, Nehru's Defence Advisor, to a large extent shaped the doctrine and force structure of the IAF. Blackett was against India establishing a long-range bomber force required for strategic bombing. Blackett told Nehru that if India used bombers to bomb enemy cities then the enemy would also retaliate on Indian cities. And this would ruin both the countries. Blackett advised Nehru to maintain a tactical air force consisting of fighters and fighter-bombers which at best could only rake the enemy's ground formations. Blackett also emphasized to Nehru and the Defence Minister Baldev Singh that a tactical air force would cost less than a strategic air force consisting of long-range heavy bombers.[14] So, both the 'humane factor' and economic dimensions set the stage for shaping the IAF as a tactical air force.

A decorated Italian combat aviator of the First World War, Amedeo Mecozzi, stressed the need for military aviation to cooperate with the army. This is categorized as tactical air power. Mecozzi rejected Giulio Douhet's theory of strategic bombing which emphasizes bombing the enemy's cities and industrial infrastructure.[15] A tactical air force is geared for providing support to the ground forces. The support could be Close Air Support (i.e. CAS), Battlefield Air Interdiction (BAI) and Air Interdiction (AI). AI is conducted to destroy, neutralize and delay enemy ground and naval forces before they can be brought to bear against friendly forces. In other words AI occurs far behind the battlefields. But, CAS operations are designed to provide responsive, sustained and concentrated firepower in close integration with the fire and manoeuvre of the friendly surface forces. The proponents of CAS are not interested in the application of air power beyond the immediate ground battle zone, and regard airpower as merely a desirable adjunct of the ground forces. They consider airpower as merely flying artillery for disabling and destroying hostile ground forces and as flying cavalry for reconnaissance of the enemy's ground units. BAI lies between CAS and AI and refers to aerial destruction of the enemy ground formations which are at the point of entering the battlefield for engaging the friendly ground forces.[16]

At the polar opposite of CAS, BAI and AI lies the theory of strategic bombing. Douhet (1869–1939) was an Italian airpower theorist who first developed the strategic bombing theory. His masterpiece, *The Command of the Air* (1921), predicts that air fleets would become the dominant feature of future war, far out weighing the armed forces of a nation. The aeroplane, he argues has a tremendous potential for bombing over land and sea. Douhet focuses on the linkages between industrial might and modern warfare and argues that military prowess is dependent on industrial and commercial might. Industrial preparation at home during peacetime would be required to produce the necessary equipment for the air force and keep it supplied. The military capacity of a country depends on its industrial infrastructure and on the mental toughness of the civilians who sustain the nation in arms. Douhet advocates carpet bombing for wiping out the enemy's industry and for causing collapse of civilian morale. He says that the grand air fleets would be sent to attack arsenals, ports, magazines, industrial plants, and banks for inflicting immense, irreparable, and decisive damage on the enemy. Douhet recommends that the first raids should be designed to win air superiority by bombing the enemy's aircraft, aeronautical installations, and plants for aeroplane production and then continuous raids should be launched on the enemy's home front so that its morale is broken.[17]

Nehru agreed with Blackett's advice that India required a tactical air force and the financial and strategic factors were the principal imperatives that shaped the IAF. Nehru argued that the IAF required interceptors and fighter-bombers, a view he expounded on 25 March 1954, in Parliament:[18]

> We do not think in terms of offensive war at a distance. . . . In the air do we require long-distance bombers to go far afield and bomb distant countries? Well, from the point of view of defence, we do not. From the point of view of practical approach to the problem, they would be out of place. One long distance bomber could be replaced by or would cost as much may be as, twenty, thirty, or fifty normal aircraft and the normal aircraft are much more useful to us in defence than one long-distance bomber. We may not have even the industrial background for the long-distance bomber.[19]

Nevertheless, Nehru was concerned about the threat posed by the PAF as can be seen from his note to the Defence Secretary dated 21 February 1954:

> I do not think that it will be an easy matter for Pakistan to build up an adequate trained personnel for a large increase of its Air Force and, more particularly, to man the later types of aircrafts, which require special training. But it is quite possible that some of the later type of

aircraft, which are better than ours, might come into the possession of the Pakistan Air Force. Even a few such aircraft would be a danger to us, though not perhaps as great as one might be led to expect. . . . I think that we should: (1) keep as fully informed as possible of developments in Pakistan, more especially, in regard to the Air Force there.[20]

Nobody at that time knew what exactly was going to be the future of air forces as the military and political establishments all over the world were in a turmoil. The advent of surface to air radar-controlled guided missiles encouraged the pundits all over the world to assert that the age of manned aircraft had come to an end. The 1957 British Defence White Paper argued that in the near future missiles would replace fighter aircraft.[21] Taking into account the wider perspective, one could say that despite the shortcomings in the rearmament programme, the Indian political and military leaders were prudent enough not to pursue any drastic course.

THE PASSIVE AIR FORCE: 1962

Until 1952, all the fighter aircraft of the IAF were of British origin. Then, a decision was taken to diversify the sources of military supply and Nehru was willing to buy aircraft from both the East and the West. In 1954, Nehru said:

I think that we should . . . investigate possibilities of purchasing better types of aircraft wherever they might be available. In doing so, we need not exclude any country, though we should proceed a little cautiously in this matter. In the final analysis, there is no reason why we should not purchase aircraft from the USSR, if they are adequate for our purpose, fit in with our establishments and training, and are not wholly beyond our capacity, financial or otherwise. In purchasing aircraft from the USSR, one has to remember that we cannot easily shift over to completely new types which should involve upsetting our present day arrangements. We should certainly, therefore enquire from various countries, to begin with, the UK and European countries. I would just at present avoid the USA, till we know more about their aid to Pakistan . . . Sweden is a country which might well be approached in this connection.[22]

Two squadrons of French MD Ouragon fighters (Indian name is Toofani) were purchased though, Britain continued to remain one of the chief suppliers of the IAF. Between 1957 and 1960, the IAF ordered 160 Hawker Hunter Mark 56 aircraft from Britain. Jordan and Lebanon also bought Hunters for equipping their air forces. In addition, the IAF also followed the policy of modernizing its equipment. The Spitfires were phased out in 1959, and from the mid 1950s, new aircraft like Canberra, Mystere and Gnat were inducted. The Canberra was a medium bomber which the IAF used for training, high altitude photographic reconnaissance and bombing. The Mystere was a single-seat fighter fitted with cannon, rockets and bombs and achieved a speed of 700 miles per hour. It was used for interception as well as ground attack. And the Gnat possessed a good rate of climb and small turning radius which made it a good interceptor.[23]

In the 1950s, the doctrine of the IAF was to prepare for a limited war against Pakistan, but ironically fate ordained that the next confrontation would be with China. In 1950, the People's Liberation Army Air Force (PLAAF) was equipped with MIG-15s by USSR. Over the Yalu River in Korea, the MIG-15 interceptors proved superior to the F-84 Thunderjets and F-80s (Shooting Stars) of the United States Air Force. The same decade also witnessed the induction of TU-2s and IL-10s from USSR to China. The PLAAF also acquired MI-4 transport helicopters and IL-28 light Jet bombers and the TU-16 Badger bombers.[24]

In 1962, war erupted between India and China. When the Indian XV Corps became conscious of its inferior artillery, it requested the Western Command for CAS on 31 October. The IAF and India's political masters were afraid to throw India's aerial machines against Red China. Probably, the Indian security managers were afraid of the light jet bombers and long range bombers of the PLAAF. And the IAF had no long-range bombers at that time to threaten China's heartland. But, it could be argued that the PLAAF's bombers might have failed to penetrate the fighter defence of India. The Chinese had six airfields in Tibet, but because of the elevation, aircraft operating from Tibet were able to carry less fuel and weapon loads. Hence, the potential capability of the PLAAF to bomb the Indian airfields and cities was quite low.[25] In contrast the IAF had many airfields in West Bengal almost at sea level. So, in terms of payload and range the IAF was in a favourable situation vis-à-vis the PLAAF aircraft that would operate from the airfields of Tibet. There was no clear political directive regarding the use of air power against China and Nehru's Cabinet probably lost nerve. But equally, the IAF chiefs were also unwilling or unable to articulate their views clearly and decisively. The Chief of Air Staff of the IAF should have pressed the issue of using aircraft in the conflict far more vigorously or could have threatened to resign over this issue. The lost opportunity for the IAF is summed up by Air Vice Marshal Harjinder Singh who in his memoirs says: 'If we had used our fighter-bombers to support the Army I do not think that the Chinese could have effectively retaliated; and the morale effect on a demoralized front—particularly in NEFA—might have been crucial.'[26]

Still the IAF by functioning as a medium of transport not only maintained the army units in inhospitable terrain but also increased the ground forces' firepower indirectly. Two troops of AMX-13 light tanks were airlifted by AN-12 aircraft to Chusul by 26 October 1962. These tanks were deployed at the base of the Gurung Hill and were able to inflict significant casualties on the PLA in Ladakh. Even before the Chinese attack, the 14th Jammu & Kashmir Militia was airlifted from Srinagar to Leh. As the PLA trounced the Indian Army, the IAF became nervous. The Chief of Air Staff demanded transport aircraft (C 119 and Caribous) from Canada and USA.[27] In the meantime Mao stopped the war.

The passivity of the IAF in 1962 will always remain a black blot in its history. However, the war with China was a blessing in disguise for after 1962 India's political establishment decided to raise the strength of the Indian armed forces. The programme was to raise the strength of the IAF from fifteen to forty-five squadrons (each squadron had between twelve to eighteen aircrafts). The plan was to induct supersonic combat aircraft, but expansion was a tortuous process. The IAF wanted French Mirage or at best F-104 Starfighters to counter the F-104s which Pakistan had acquired from USA. The IAF was against acquiring the MIG-21 from USSR, but, Krishna Menon, the Defence Minister put his heart into the MIG-21 deal. The MIG-21 had a higher rate of climb (45,000 ft. per minute) compared to the F-104 (36,000 ft. per minute) but the MIG-21 was not an all weather fighter. While the MIG-21 was only an interceptor, the F-104 was a fighter-bomber, equipped with cannons, bombs, etc. Further, the F-104 being equipped with advanced electronic gadgets and radar was a more sophisticated plane compared to the MIG-21. However, even if the IAF's view prevailed, the USA was unwilling to provide

India with F-104s and in the end Krishna Menon was able to push the MIG deal through.[28] The 1965 War with Pakistan came when the IAF was in the throes of expansion.

AERIAL ATTRITION: 1965

On the afternoon of 1 September 1965, the Chief of Army Staff General J.N. Chaudhuri requested the Chief of the Air Staff Arjan Singh for CAS. The army needed air support to check the advancing Pakistani armour at the Chamb sector. Both the service chiefs then went to the Defence Minister, Y.B. Chavan who asked Arjan Singh, 'Can you do it?'.[29] Arjan Singh replied: 'We just can. If you give orders now, we may be able to attack before night.'[30] In late afternoon at 4 p.m., the Defence Minister's order reached the Air Headquarter. By 5.19 p.m., twelve Vampires from No. 220 Squadron based in Pathankot were airborne. The pilots were tense from waiting for the call to go in action. They were flying westward in late afternoon. Not only were the Indian pilots blinded by the setting sun but their aircraft were also clearly visible in the sky. The advantage lay with PAF and the Pakistani anti-aircraft gunners and in the encounter that followed 4 Vampires were lost. Only one pilot was able to bale out. Nevertheless, the Vampires drew blood by attacking the tanks, guns and vehicles of the advancing Pakistan Army. Some Vampires also functioned as night fighters. Besides the Vampires, the other workhorse of the IAF was the Mystere IV. In addition, the IAF also possessed Gnats and Hunters.[31]

Stung by the IAF's attack on the advancing Pakistani ground forces, an angry and arrogant PAF came up to challenge the Indian aerial raiders. Though the IAF was numerically stronger than the PAF, the latter had much more sophisticated aircraft at its disposal. The core of the PAF was composed of 100 Sabre jets and twelve Lockheed Starfighters (F-104). The Pakistani Sabres were armed with heat seeking air to air Sidewinder missiles.

The F-86 Sabre jet, a much-hyped American fighter had made its debut during the Korean war (1950–3). In the aerial battles over the Yalu River of Korea, the Sabres flying at 760 km. (i.e. 475 miles) per hour downed several MIG-15s and MIG-17s. The superior hydraulic flight controls gave the F-86 an advantage over the MIG-15 as regards manoeuvreing, and the Sabre pilots in Korea were able to establish 10:1 kill ratio over the MIG-15s.[32]

On 6 September in the late afternoon as the sun was setting, the PAF attacked the Indian airbases at Adampur, Halwara, Pathankot, Srinagar and Jamnagar in the Western theatre. A series of dogfight between the Sabres and Hunters took place. Over Halwara, the Hunters were able to shoot down two Sabres. As early as 4 September, four Sabre jets had clashed with four Gnats over Akhnur. The Gnats were armed with only conventional guns. One Gnat closed an a Sabre and fired from a distance of 600 yards. The Gnat continued to fire till it had reached a striking distance of only 300 yards from the Sabre. Finally the Sabre burst into flames. This somewhat restored confidence among the Indian pilots. Generally the tactics followed by the IAF against the Sabres was to send the Mysteres forward with the Gnats providing high altitude escort. When the Sabres descended to attack the Mysteres, the Gnats pounced upon the Sabres from high altitudes. However, the combat air patrol flown by the IAF was not always able to stop penetration by the PAF. At Pathankot, the PAF was able to destroy some of the newly acquired MIGs on the ground. India's radar installations at Porbander in Gujarat and at Ferozepur in Punjab were also attacked.[33]

In retaliation, the IAF also attacked the bases of the PAF, the most important being Sargodha in Pakistani Punjab. When on 7 September at 5.50 a.m. 6 Mysteres paid a visit to Sargodha, the Pakistanis were more alert than the IAF was at Pathankot air-base. One Starfighter scrambled and it was able to down one of the Mysteres. However, the other Mysteres were able to even the loss by getting the Starfighter. Overall, the PAF lost one F-104 and one F-86 in air combat over Sargodha and the IAF lost five aircraft.[34]

An example of AI is the attack by the Mysteres of No. 31 Squadron on 9 September on a Pakistani train at the Raiwand railway station. This train was carrying tanks to the front. The use of Canberra bombers on 10 September to bomb the Pakistani armour concentration in Khem Karan sector during the battle of Asal Uttar was an instance of BAI by the IAF.[35] On the Indian side, the Austers as usual carried out vital non-combat duties like battlefield surveillance and casualty evacuation.[36]

The Soviet military theorists calculated that about 50 per cent of the armoured fighting vehicles are destroyed by military aviation during battle.[37] The Indian theorists also accept that the chief characteristic of the CAS is its armour killing capacity.[38] And the IAF was geared primarily for that function. Air Marshal Arjan Singh, who was Chief of Air Staff during the 1965 War explains the IAF's operational doctrine in the following words:

> I just want to mention here that the thinking in the Air Force during my time was that the enemy Air Force should be disabled to the extent that it cannot effectively interfere with our Army, Navy and Air Force formations, road and rail traffic and other vital targets. That can only be ensured with a favourable air situation. It is essential that before the land battle starts, the air effort should be largely shifted to the land battle by attacking troop concentrations, trains, tanks, vehicles, guns, radars, etc. just behind the enemy battle positions. That will pay good dividends and achieve success in the shortest time possible, at the minimum cost in men and materials to us.[39]

Compared to the passive performance of the PAF over Kashmir during 1947–8, its posture during the 1965 War was quite aggressive. Several factors caused this transformation. The PAF entered the 'Jet Age' in the 1950s, and from the mid 1950s onwards, was modernized through American assistance. In 1956, it acquired F-86 Sabres. The B-57s came in 1959, the F-104 Starfighters in 1961, and the C-130Bs came in 1963.[40] The Pakistani official claim is that by 9 September, the IAF had lost fifty aircraft compared to the PAF's six planes.[41] One Indian commentator calculated that in this war, while the IAF lost thirty-five planes, the PAF lost seventy-three.[42] According to Air Marshal M.S. Chaturvedi, against the thirty-three aircrafts lost by the IAF, the PAF lost seventy-three.[43] The true figures will never be established in the midst of contradictory claims and counter-claims.

After the war ended and the dust settled down, the IAF started evaluating its own performance. Arjan Singh admitted the IAF's command had bungled at Pathankot. Arjan Singh says: 'After the 1965 War, I asked two Air Commodores and a couple of Group Captains to put in their resignation, as well as some other people. I told them that they could continue receiving their pension if they resigned; otherwise I would have them dismissed as they had not done well and could not continue in service. . . . You have to maintain discipline. If someone has done wrong, punish him, if he has done well, reward him.'[44]

The IAF also learnt a few lessons. The MIG-21s being interceptors were found ineffective in conducting low-level ground attacks. In contrast, low-level attacks were carried out effectively by the Sabres. At the tree top level, the K-13 heat seeking missiles of the MIG-21s were affected by ground heat. And the MIG-21s had no cannons for air to ground strikes. After the 1965 conflict, the Soviet 30-mm. gun (which had a very high rate of fire—3,400 rpm) was fitted to the MIG-21.[45] Between 1962 and 1965, the interceptors of the IAF marked aside for air defence practiced mainly medium-and high-level combat manoeuvre. The pilots carried on exercises at 40,000 ft. above the sea level, but during the war most of the engagements occurred within 10,000 ft. of sea level.[46] At this time the commercial pilots received much greater pay than the combat pilots, which did affect the morale of the pilots. Arjan Singh set up the Air Force Academy at Hyderabad to train future pilots, and took steps to improve the service conditions and flying allowances of the pilots.[47] All these steps were necessary because the future aerial duel between India and Pakistan would occur on a bigger scale. Till 1965, aerial combat between India and Pakistan was concentrated mostly over the Punjab,[48] but the next war witnessed the widening of the air campaign.

GAINING AIR SUPERIORITY: 1971

On a grey dawn of 22 June 1941, the opening day of Operation Barbarossa, the *Luftwaffe* destroyed the bulk of the Soviet Air Force (SAF). Within a week, about 3,000 Soviet aircraft were destroyed and most of them while still on the ground.[49] Thus, was born the notion of pre-emptive aerial strike. History repeated itself in the early hours of 5 June 1967, when the Israeli Air Force's Mirage fighter-bombers arrived at low height to escape radar detection over the nine Egyptian airfields. Within hours, the neatly parked Egyptian MIG-17s were burning hulks of metal.[50] In three hours, the Israeli Air Force was able to destroy 300 out of 400 planes of the Egyptian Air Force.[51] The PAF was influenced by the Israeli style of attack and decided to implement it against the IAF. The PAF like the Israeli Air Force boasted of some Mirages, though the core of the PAF consisted of Sabres and B-51 bombers.[52]

At 5.45 p.m. on 3 December 1971, a pre-emptive air strike was launched by the PAF on a number of Indian airfields—Srinagar,

Avantipur, Pathankot, Jodhpur, Ambala and Agra. Later during the night of a full moon, a second wave of Pakistani aircraft came to deliver a second blow. The PAF's aim was to damage the runways in the first strike to prevent the IAF from getting airborne, and a second wave to finish off the IAF's stranded craft. However, not a single IAF aircraft was hit. The PAF's plan failed because the Pakistanis forgot that their pilots lacked the elan, confidence and training of the *Luftwaffe* crews of 1941 or that of the Israelis. Nor were the Indians as incompetent as the Egyptians! The PAF's doctrine was faulty. Instead of concentrating all their aircraft against the radar installations and airfields of the IAF, the PAF tried to conduct both pre-emptive strike as well as deep interdiction raids simultaneously. On 3 December, the PAF attacked not only the twelve Indian airfields but also bombed the railway junctions of Godra Road, Jammu and Barmer.[53] Instead of dissipating their effort, the PAF should have concentrated on a pre-emptive strike against the IAF's ground installations, and once it was disabled, the PAF could have attacked strategic targets deep inside India at its leisure. Again instead of a concentrated thrust against some of the airfields, the PAF made a mistake of using only two or three aircraft to raid each Indian airfield. And the tonnage of bomb dropped by them was inadequate and characterized by poor marksmanship. The Pakistani air strike was also unsuccessful because unlike the Egyptian air force the IAF was not caught napping nor were its planes parked in the open in rows like the Egyptian aircraft. The dispersal of the IAF's aircraft was planned in great detail. Most of the aircraft were in concrete pens and in less than twenty-four hours, the IAF hit back to gain air superiority.[54]

In 1970, the Gnat's role as interceptor was taken over by the MIG-21s. The IAF had eight squadrons of MIG-21 but still retained eight squadrons of Gnats. Before the 1971 War, the IAF pilots were well trained on the two seater Hawker Hunter-T Mark-66D trainers acquired from Britain. The IAF started raiding the PAF's airfields at Chanderi, Shorkot, Sargodha, Muri, Mianwali, Karachi, Rawalpindi and Lahore. The IAF did not escape unscathed for the Pakistani airfields were well guarded by anti-aircraft guns, and they accounted for six IAF aircraft over West Pakistan and five over East Pakistan. To guard the IAF's ground installations against sneaking raids by the PAF, the IAF's

MIG fighters maintained combat air patrol over the Indian airfields. Once the pre-emptive strike failed, the PAF aimed to provide an air umbrella over the Pakistani ground forces in West Pakistan concentrating to attack India.[55]

Even before gaining air superiority, the IAF was ordered to the task of CAS for the embattled ground forces. The IAF used Hunter squadrons and the six squadrons of SU-7 for ground attack. As in the 1965 War, at Chamb, the situation was threatening for the Indian land forces. While the fighter-bombers attacked the Pakistani troop concentration during day, the Canberra medium bombers conducted bombing raids at night. The IAF bombed Pakistani troop concentrations in the forests north-west of Poonch and thus disrupted the enemy's plan of action. Mehar Baba's legacy of converting transport aircraft into bombers lived on in the IAF. The Soviet AN-12 transport aircraft were converted into bombers, and were used to drop bomb at Kahuta on 5 December. However, the bombing was neither very effective nor heavy as only forty tons of bombs were dropped. On 6 December, Air Marshal M.M. Engineer, heading the Western Air Command directed the IAF to attack the railway communications of southern Pakistan.[56] This meant that the IAF was shifting its war plan from CAS to AI.

Over the skies of East Pakistan, the IAF had a field day. When fighting started on the night of 3 December, the IAF bombed Tezgaon and Chittagong. From the morning of 4 December, the IAF conducted raids on all the airfields of East Pakistan, and gained air superiority over East Pakistan within forty-eight hours from the commencement of the conflict, because the PAF was hopelessly outnumbered. After gaining air superiority the IAF attacked the river borne crafts thus disrupting the enemy movements across the waterways of East Pakistan. Air Marshal H.C. Dewan, in-charge of the Eastern Air Command also directed the IAF to attack Pakistani troop concentrations at Rajendraganj, Jamalpur, etc. The MIGs with their 57-mm. rockets attacked concentration of Pakistani ground forces. The IAF's air superiority made possible airborne landings on part of India. On 10 December, helicopter borne troops were landed on the west bank of the Meghna River, opposite Ashuganj, about 60 km. from Dacca. One MI-4 helicopter unit with 14 helicopters took part in this operation.[57]

In the 1980s, two American airpower theorists John Boyd and John Warden developed the theory of strategic paralysis. This theory argued that the enemy leadership must be attacked physically and psychologically.[58] This marked a shift from Douhet's dictum of carpet bombing of the enemy's civilians concentrated in the cities. Though the IAF did not possess a theory of strategic paralysis of the enemy, it did implement a similar plan of action in East Pakistan. On 14 December 1971, Indian intelligence picked up the information that the Governor of East Pakistan Dr Malik was holding a meeting in the Government House at Dacca at 12 p.m. that day. The IAF's Headquarter had picked up the message at 11.45 Dacca time. Immediately, the IAF Headquarter informed the Eastern Air Command's Headquarter over telephone: 'Please put on a strike on Dacca as soon as you can after 12 O'clock at Government House.' The staff officers at the Headquarter in Shillong got hold of guide maps of Dacca. These maps were on the scale of about four inches to a mile. These maps enabled the Headquarter at Shillong to direct the pilots with pinpoint accuracy. The fighters took off from Gauhati, and the MIGs standing by at Shillong were also ordered to scramble. When the MIGs were airborne they were given the exact location of the Government House and instructions regarding how to recognize it. At 12.20 (Dacca time) the MIGs appeared over the Government House at Dacca. It was exactly 35 minutes from the time the IAF Headquarter at Delhi had picked up the phone to talk with the Eastern Air Command's Headquarter at Shillong. Destruction rained over the Government House. Malik immediately reached for his pen and wrote his resignation on a piece of paper. Then he went into an air raid shelter. Malik's action in turn further demoralized an already shaken General A.A.K. Niazi, commander of the Pakistani Army units in East Pakistan.[59] It was a classic case of using real time information by the IAF.

Next the IAF decided to carry their psychological warfare further. Niazi and his staff were cooped up at Dacca University campus. Between the afternoon of 14 December and the evening of 15 December, the IAF fired 1,500 rockets and several thousands rounds of heavy ammunition on the Dacca University campus. The university building was heavily damaged. P.C. Lal, the Chief of Air Staff at that time, says that this attack on this

TABLE 6.1: LOSS OF PAKISTANI AND INDIAN AIRCRAFT DURING THE 1971 WAR

Date	Number of aircraft lost by the Indian Air Force	Number of aircraft lost by the Pakistan Air Force
3 December	–	4
4 December	11	29
5 December	6	14
6 December	5	5
7 December	X	1
8 December	5	20
9 December	4	X
10 December	6	3
11 December	1	–
12 December	2	3
13 December	–	–
14 December	–	–
15 December	–	–
16 December	3	6
17 December	3	8
Total	45	94

Notes: X denotes no loss.
'–' denotes figures unavailable.
Source: V.P. Juneja, *Indo-Pak War: 1971*, New Delhi: New Light Publishers, 1971, pp. 62–105.

particular piece of civilian target was deliberate. Leaflets were dropped asking Niazi to surrender. He surrendered the next day, i.e. 16 December and the war came to an end on 17 December.[60]

Till the mid-1960s, the air to ground mission of NATO air forces emphasized AI which included interdicting rear areas and attacking the follow on forces of the Soviets. The AI was supposed to occur to an operational depth of about 200 km. from the political borders.[61] The IAF's air operations during the 1971 War to an extent conformed to the pattern as charted out by the NATO air doctrine. It could be argued that because the IAF had been able to gain air superiority it was able to shift effectively from CAS to AI especially in East Pakistan. From the technological point of view, both the IAF and the PAF were much behind the aerial technologies at the disposal of the superpowers. The USAF had displayed the potential of smart bombs in Vietnam. Between 1965 and 1968, USA lost 97 aircraft in its attempts to bomb the bridge at Thanh Hoa. The bridge was still standing when in 1972 it was destroyed by a 2000-lb. smart bomb. This smart bomb used laser beam for guidance and television to gain pinpoint accuracy.[62] The PAF and the IAF lacked such bombs. In scale and scope, India–Pakistan air war was a little affair compared to the Vietnam War. The PAF had begun the war with 290 machines and the IAF with 600 aircraft, of which the PAF claimed to have destroyed fifty-seven Indian aircraft.[63] During the twelve years of Vietnam War, the USA lost over 7,000 aircraft.[64] The point to be noted is that from the financial point of view, the Vietnam War was mostly an air war on part of the USA. More than half of the funds spent by US government for this war were absorbed by the USAF, the United States Army's aviation and the United States Navy's air operations.[65]

OPERATION WHITE SEA: 1999

Just after the defeat of Pakistan in 1971, Indian military officers gave serious thought about how to conduct a war in the future. Major Yogi Saksena rightly asserted that in future all the wars were going to be of short duration. So, Douhet's theory of strategic bombing has become useless.[66] In other words, Saksena was apprehensive that if the IAF followed the doctrine of strategic bombing in future then its CAS role for the ground forces might suffer. Saksena's concept of using air power is a microcosm of what the Indian Army officers thought of airpower in war. In fact, the thinking pattern of the Indian Army officers was somewhat similar to the Soviet air doctrine. The Soviet air doctrine emphasized CAS. Till 1991, the Soviet doctrine regarding air support of the ground forces consisted of four stages. In the first stage the attack was to be carried on deep targets far removed from the battlefields. The targets included enemy aircraft on the enemy airfields located far behind the Forward Edge of the Battle Area. In the second stage, the air force would concentrate on a particular frontage along which the friendly ground offensive would roll. And in the third stage the air force would attack those enemy targets which were beyond the range of artillery. In the fourth and final stage, the air force would continue to support the forward moving ground forces which had outstripped its artillery support because of rapid advance.[67] The Indian Army in its anxiety about air power support demanded rotary wing craft. Till 1986, all the helicopters were controlled by the IAF, but after 1986, the Army Aviation Corps acquired helicopters.[68]

From the 1980s, the IAF lagged far behind the NATO air forces when the latter started acquiring precision guided munitions, such as laser-guided bombs, electro-optical guided bombs, precision television guided and infra red guided anti-tank missiles and runway attack munitions.[69] The IAF was incapable of waging the type of integrated operation conducted by the Israeli Air Force in 1982 over Bekaa Valley. The Israeli Air Force used free fall bombs and anti-radiation missiles against the Syrian surface to air missiles (SAM) sites. The Unmanned Aerial Vehicles, E-2Cs and RC-135 aircraft provided a complete picture of the battle space. The Israelis knew the exact time of movement of the Syrian ground formations as well as the launch and flight of Syrian aircraft. Hence, the Israeli Air Force was able to establish effective kill ratios. In 1982, over Bekka Valley, within three days the Israeli Air Force destroyed eighty-six Syrian MIGs at the cost of one Israeli aircraft.[70]

In 1980, the IAF comprised 1,400 aircraft (including training and support types) and about 1,00,000 personnel. In that decade, the IAF was in a numerically inferior position against the PLAAF. The PLAAF possessed 5,000 aircraft against IAF's less than 1,000 operational combat machines. Over the next decade modernization of the IAF was very slow. In 1997, the Indian government opted for induction of multi-role rather than single role dedicated aircraft. The SU-30, a twin seat multi-role fighter aircraft was chosen for deep penetration. However, even these aircraft lacked the operational range to threaten the heartland of China. Then, the strike aircraft of the IAF also lacked night attack capability.[71] And for this defect the IAF would come to grief in the next round of confrontation with Pakistan.

On 21 May 1999, a Canberra which was surveying the border in Kashmir was shot at and though its engines were damaged, it was able to land at Srinagar airport.[72] When the intruders in Kargil were located, the IAF launched operation Safed Sagar. Though the IAF carried out numerous sorties, they were not quite effective because of the mountainous terrain and the shoulder fired SAMs used by the infiltrators.[73] A MIG-21 and a MI-17 were lost to the SAMs, and a MIG-27 was lost due to engine failure. According to the Ministry of Defence report, during the 50 days when Safed Sagar was in full swing, the fighters conducted 1,730 sorties, the transport aircraft carried out 3,427 sorties and the helicopters carried out 2,474 sorties.[74] In comparison, during the First Gulf War (1991), the Coalition air forces launched between 2,000–3,000 sorties per day throughout the conflict.[75]

The IAF remained alert after the Kargil fiasco. A Pakistani Atlantic maritime reconnaissance and anti-submarine aircraft violated Indian airspace over the Rann of Kutch on 10 August 1999 between 1100 and 1117 hours. Immediately two MIG-21s scrambled and one of the MIG-21 shot it down at 1118 hours.[76] In the aftermath of the Kargil Operation, the IAF was entrusted with mammoth task of airlifting supplies to the troops deployed in the Kargil Sector—an inhospitable and unfriendly terrain—guarding against future incursions. With a deadline of 30 October 1999, the IAF had to airlift 6,323 tons by transport aircraft and 4,445 tons by helicopter. Ilyushin-76, Antonov-32 transport aircraft, MI-17 and Cheetah helicopters were used ferry the supplies.[77]

The PAF remained inactive during the Kargil crisis. Despite Ian Anthony's claims that the IAF was losing its qualitative and quantitative edge over the PAF,[78] the IAF remained in a superior position vis-à-vis the PAF throughout the 1990s. The last planes purchased by the PAF were in 1992. However, from the 1990s, there was increasing collaboration between the PAF and China whose aeronautical stores at least till 2000 were not of world class standard. China had provided Pakistan with a F-6—fighter (a modified version of the MIG-19). Islamabad had transferred to Beijing one F-16 (given to Pakistan by America) which the aeronautical engineers of China wanted to study. A grateful China in exchange provided the PAF with the PLAAF's latest F-7 MP interceptors. The F-7 MPs (Chinese version of the MIG-21) are yet to prove their performance in air combat. The PAF in 2000 had around 350 aircraft with its strength lying with the thirty-two F-16s.[79] There constitute the real danger for the IAF but the IAF's Mirages can counter them.

There are some similarities despite more obvious dissimilarities between the experience of USAF in Vietnam, the SAF in Afghanistan and the IAF in Kargil. All the three air forces fought mainly land-based elusive guerrillas. Between 1962 and 1973, the USA lost 3,719 fixed wing aircraft in Vietnam. Of them about

2,300 were lost to the SAMs and small arms. The MIGs of North Vietnam accounted for only 79 US aircraft. Within the same period, the USA lost 4,869 helicopters in Vietnam. And the firearms and missiles of the guerrillas accounted for about 2,600 of these helicopters.[80] The USAF found laser-guided bombs effective against the insurgents in Vietnam. The IAF following this experience used laser-guided bombs against the enemy supply dumps at Muntho Dalo in the Kargil Sector with great effect but its arsenal did not possess adequate number of laser-guided bombs.[81] The IAF should have also learnt from the experience of the SAF which had fought in Afghanistan. The terrain of Afghanistan and the Drass–Kargil Sector of Kashmir are more or less similar. And the equipment of the IAF and the SAF in Afghanistan were also similar. The MI-24 nicknamed Flying Tank entered the Soviet service in 1973. Each MI-24 carried a four-barrelled 12.7-mm. rotary cannon and eight soldiers. With fighter cover, the MI-24 gunships spearheaded the search and destroy mission between 1980 and 1984 in Afghanistan.[82] The SAF used MIG-21, 23 and 27s against the *mujahideens*. But, only with the introduction of the PGMs did the aircraft score hits against the insurgents perched on the mountaintops. However, the advent of shoulder fired Stinger missiles after 1986 forced the Soviet aircraft to fly much higher and their bombing became inaccurate.[83] The Stinger, manufactured by General Dynamics of USA, is an optically aimed infra-red heat seeking guided missile. It could be launched in the general direction of the aircraft and the missile automatically picked its way to the target and had a range of three miles. Since it weighed only 30 lbs., it was easily transportable. In 1986, the Soviet helicopters operating in Afghanistan were equipped with infra-red decoy flares and jamming equipment. These more or less neutralized the SAM-7s. The Soviet helicopters circled a Soviet plane when it was taking off or landing, and ejected flares at two second intervals in order to divert the Stingers from their targets.[84] The IAF should have used these tactics in Kargil to ward off the threat of the Stingers. To sum up, in unconventional operations, the principal threat came from the small arms and SAMs launched by the guerrillas rather than the hostile aerial weapon systems.

Nevertheless, even after the harrowing Vietnam and Afghanistan experiences, the western air forces remain enamoured of the concept of fighting conventional armies under a nuclear umbrella. The theorists of NATO and the RAF emphasize counter-air campaign rather than CAS against the insurgents. Andrew G.B. Vallance expresses the air doctrine in the 1990s in the following words: 'the capability to deter, contain and ultimately defeat the enemy air forces must be the top air power priority. For without control of the air no other type of operations can be sustained.'[85] However, CAS is not totally neglected. It remains secondary within the western air war doctrines. For ground attacks, in the 1970s, the USAF came up with A-10 ground attack aircraft. The Americans also designed Air-Mobile Forces, which are part of the Rapid Reaction Corps. After the Vietnam War, the American Army invested heavily in helicopters to provide its units mobility, flexibility and airborne firepower. The aviation brigades that have been constituted to raise the 'speed of battle' include Apache Attack Helicopters armed with missiles and PGMs. The Apaches can attack targets to a depth of 100 miles beyond the frontlines and are guided to their targets by Scout Helicopters, which are designed for reconnaissance. And behind the Apaches fly the Blackhawk Transport Helicopters ferrying troops to the target zones.[86]

Besides doctrinal and organizational issues, the IAF also faced problems as regards training. Every year about 776 pilots are recruited by the IAF. For over a decade, the IAF had been losing on an average two aircraft per month. Between 1993 and 2003, the IAF lost about 208 aircraft mostly in peacetime manoeuvres. The IAF had lost 85 pilots and 185 aircraft in accidents between 1991 and 1995. The lack of a proper trainer aircraft was at the root of the trouble. The absence of a trans-sonic aircraft, i.e. an advanced jet trainer (AJT) in the IAF means that the pilots moved straight from the subsonic Kirans/Iskaras to the supersonic MIG-21s. The take-off speed of the Kirans was 180 km. per hour and while flying, its maximum speed was 720 km. per hour; the take off speed of the MIG-21 was 350 km. per hour and its cruising speed was 780 km. per hour. In addition, the landing speed of Kiran was 170 km. per hour and that of a MIG-21 was double of that. In 1999, India's All Party Standing Committee on Defence had criticized the government's laxity in acquiring AJTs to bring down the number of accidents. In the same year, Rs. 60 billion (US $1.38 billion) had been sanctioned for buying AJTs. Lakshya, the indigenous pilotless aircraft developed by the Defence Research and Development Organization has been inducted into the IAF. This is a surface launched reusable aircraft flying below the speed of sound and will provide training to aircrew in air to air firing, thus eliminating the need for costly target towing aircraft.[87]

THE IAF IN THE NEW MILLENNIUM: A STRATEGIC DETERRENT OR A TACTICAL AIR FORCE?

The post-Gulf War era is witnessing a surge of enthusiasm among the airpower theorists all over the world. After the Gulf War, the Russians concluded that in order to win, the enemy forces must be attacked not only from the air, but also should be provided air protection to the friendly ground forces so that they can manoeuvre. The latter requires air superiority.[88] Experience from the Russian front during World War II shows that the principal effect of the air strikes lies not in the losses they cause to the enemy ground forces but rather in the disruption and insecurity they create which adversely affects deployment, regrouping and advance of the hostile land forces.[89]

The Indian airpower theorists are also aggressively putting forward their views. Group Captain R.K. Jolly asserts: 'Air power, with its capabilities of parallel warfare, has now begun to dominate the battlefield, quite often dictating the outcome of war.'[90] Air Vice-Marshal S.C. Rastogi asserts that the recent conflicts (Gulf War, Kosovo and Afghanistan) are all won exclusively by the use of air power.[91] This interpretation could be challenged. The post-action analysis shows that the damage done to the military targets, thanks to the excellent camouflage and concealment on part of the Serb Army, was minimal.[92] In contrast to Rastogi, Air Vice-Marshal Kapil Kak writes that in the future air power by itself will not be able to win any war. However, without the successful application of air power, no war can be won.[93]

The IAF theorists are concerned about the nature of war that they might have to fight in the near future. Air Commodore Jasjit Singh asserts that all future conflicts would be limited wars. To deter the enemy from engaging in 'salami slicing' India's territories, New Delhi requires to deter the potentially hostile neighbours with the aid of the IAF. The IAF should be prepared for launching long-range accurate 'surgical' strikes and also controlled punitive strikes as part of coercive diplomacy.[94] Again the use of the IAF in such limited conflicts would prevent rapid escalation. Jasjit Singh assumes that a calibrated air strike against the enemy would result in an airpower response only.[95] Kapil Kak who accepts the concept of limited war wrote in 1999, that limited conventional war in the subcontinent is still a possibility. In his view, such warfare would be characterized by high-intensity, fast manoeuvre operations. Such military operations restricted in space, scope, and political aims would last between five to seven days. For sustaining this sort of warfare, Kak recommends reduction in infantry battalions to maintain mechanized formations, with sophisticated multirole aircraft providing CAS.[96] However Jasjit Singh marks a departure from the traditional CAS concept and picks up Boyd and Warden's theory. He writes that the IAF will target the leadership of the enemy country during a limited war to ensure strategic paralysis of the enemy.[97] It is not clear whether this objective could be achieved by the IAF with the hardware at its disposal and within the short time span that would be available in case of a limited war. And the use of IAF in any border skirmish might escalate the conflict into a full scale war.

The concept of air–land battle has finally reached the subcontinent from the USA. The US Army's concept of air–land battle as developed in 1986 emphasizes a series of simultaneous combined arms mechanized assaults using integrated airpower. The aim was to inflict operational shock on the enemy resulting in disruption of enemy defence.[98] The Indian Army officers in order to prevent the IAF from jettisoning its traditional CAS role use the concept of air–land battle. Brigadier Gurmeet Kanwal had written in 2003: 'The concept of Air–Land Battle would become more relevant on the Indian subcontinent as RMA technologies come to the forefront. Synchronized, simultaneous ground and air strikes would be needed to achieve the greatest impact.'[99] Even the IAF officers are unwilling to turn away completely from the ground support role. Group Captain Jolly writes: 'It is our doctrinal belief that should enemy air threat to friendly surface operations exist, counter-air operations must be the first priority. While limiting the enemy's use of its airpower, it would create conditions for the success of our own forces. In a limited conflict, the classical counter-air campaign may not be feasible. In such a case, the IAF would have to use its assets to inflict punishment and attrition on the adversary.'[100] Gurmeet Kanwal agrees that electronic warfare and information operations for conducting land battles can only be done by the IAF.[101]

Nevertheless, the IAF argues that the threat to CAS has increased due to the advanced generation of air defence weaponry and heavy cost of the fixed wing aircraft. So, BAI offers a more promising solution.[102] According to Air Commodore R.V. Phadke, pure CAS should be left to the attack helicopters of the army, while the fighters of the IAF should provide combat air patrol at higher altitudes.[103] In 2001, a major-general asserted that manoeuvre war would require airborne troops. So, the IAF requires to increase its helilift capacity which would extend the army's reach and enable it to conduct 'deep battle'.[104] Whether the IAF would accept such a policy remains to be seen. In accordance with the Soviet Air doctrine, helicopters provide fire support and logistical support to the ground units operating deep in the enemy's territory. In the Soviet frame of air warfare, the heavy-lift aircraft constitutes an air bridge to the ground units operating deep inside enemy territory and escort/attack helicopters have to guard these aircraft.[105] The IAF uses MI-35 as attack helicopters and MI-17s as transport helicopters—the latter provide logistical aid to the military units deployed in Siachen and Arunachal Pradesh.[106]

As the last decade of the twentieth century came to an end, the IAF and theorists of Indian air warfare begin to take note of the Revolution in Military Affairs (RMA) that has started in the West. One feature of the ongoing RMA is the advent of sensor technologies and PGMs. Jasjit Singh comments that these two technologies make it possible to attack targets across long distances which earlier would have required close combat.[107] Other theorists of the IAF continue to harp on the PGMs which make air bombing very lethal and precise and because of the use of knowledge intensive weapons, reduce the Circular Error of Probability.[108] Air Vice-Marshal Viney Kapila writes that the PGMs has made the notion of mass use of conventional weapons obsolete.[109] In 2003, Air Chief Marshal S. Krishnaswamy asserted that, 'The Indian Air Force has very good and highly accurate precision attack capability, of the order of five metres. We believe that is workable. Now we are aiming at about one metre.'[110] In reality the PGMs are not that precise as can be seen from the Kosovo experience. In 1999, the F-16s flying at 15,000 ft. over Kosovo, used laser targets and smart bombs. But, instead of hitting Serb Army vehicles, most of the munitions hit the refugee columns.[111] Nevertheless, the share of PGMs used in air operations is increasing with the passage of time. In the First Gulf War (1991), only 10 per cent of the ordnance used were precision guided. And during the Kosovo air campaign (1999), 30 per cent of the munitions utilized were precision guided. During Operation Enduring Freedom (2004), the USAF used B-1 and B-2 bombers and also the older B-52 bombers. More than 70 per cent of the ordnance used in the aerial campaign were precision guided/smart weapons. One example of such a weapon is the Joint Direct Attack Munitions which were one ton bombs guided by satellites to targets that had been marked by the ground forces.[112]

Besides the PGMs, the spokesmen of the IAF are also demanding acquisition of the latest technologies that has made possible the RMA in the West. Kapil Kak writes that the IAF needs airborne warning and control systems for reconnaissance, surveillance and target acquisition.[113] S.C. Rastogi is for integrating electronic warfare with the force structure of the IAF. Electronic warfare involves eavesdropping and interfering in the enemy's electro-magnetic spectrum through which all communications and radar transmissions occur.[114] Air Commodore A.K. Tiwary argues that it is necessary for the-state-of-the-art aerial platforms to be fitted with electro-optical cameras, infrared including thermal imaging and radar sensors, etc. Tiwary accepts that the emergence of stealth technology represents a revolutionary asset for attrition-free precision strikes.[115]

The US-NATO forces in air operations over Yugoslavia used UAVs, which played a vital role in surveillance and data gathering.[116] Fog and low clouds made aerial reconnaissance ineffective as the aircraft had to fly very low exposing them to risk. As part of the RMA, Air Commodore C.N. Ghosh emphasizes the role of the UAVs. He asserts:

> It is likely that future wars will be decided by some form of air power being the principal driving force before the surface forces are able to

make contact with the enemy in major battles. And it is more or less definite that the unmanned combat aerial vehicles (UCAVs) will play a definitive role towards augmentation of air power. The UCAVs will be smaller, faster and more versatile than their manned counterparts. They will be used to penetrate airspace that may be considered dangerous for manned aircraft can carry out reconnaissance and surveillance, or be used for jamming the enemy air defence system. This has become a necessity in the face of the quick reaction, low looking missiles that have been developed by almost every warring nation.[117]

According to C.N. Ghosh, UCAVs will be crucial for the IAF in the near future because the MIG-21s are ageing and India lacks the money to buy the-state-of-the-art fighters. In Ghosh's paradigm, the role of the interceptors would be taken over by the much cheaper UCAVS.[118] Squadron Leader P.M. Sinha also argues the case of UCAVs. They could be flown by high-speed fibre optic and satellite communication links. The advantage is that the UCAVs are not bound by the limits of human tolerance.[119]

From theory let us shift to reality. In the new millennium, Pakistan is negotiating with China for buying fifty F-7 fighters that would replace two squadrons of ageing F-6 Chinese fighters in service with the PAF. The PAF has also ordered forty Mirages. By 2000, about twenty-four Mirage-III and Mirage-V had been delivered. Because of inadequate funds and latent hostility of the US-led Western world, the IAF continues to depend on the obsolete Russian MIGs. Due to US–India rapprochement initiated by Bush–Manmohan Singh administrations in 2008, India in the near future might acquire some F-16s. In 1998, there was an attempt by India to acquire MIG-21s, 23s and 25s from Kazakhstan.[120] Against the F-16s and Mirages of the PAF, the only counter available to the IAF is the Mirages. The Sukhoi-30 (SU-30) is still an untested commodity. The MIG-21s and 23s of the IAF would be of no use in countering the F-16s, because way back in 1982 over Lebanon, the F-16s of the Israeli Air Force proved superior against the MIG-21s and 23s of Syria.[121] The PLAAF plans to induct two hundred SU-27s and about fifty SU-30s. China possesses some Russian T-52 bombers which are capable of delivering nuclear warheads. However, it is unlikely that these bombers will be able to penetrate the Indian fighter defence.[122]

On 27 September 2002, the first batch of ten Sukhoi-30 was

inducted into the IAF's 20th Lightning Squadron based in Pune. For engaging several targets simultaneously, these aircraft are equipped with both air to air as well as air to surface missiles. The SU-30s can be used for delivering both conventional as well as nuclear weapons. The IAF plans to induct six air to air refuelling aircraft which till enable mid-air refuelling of the Sukhois increasing their operational range from 3,000 km. to 5,200 km. From the last decade of the twentieth century, the IAF started replacing its ageing AN-12 fleet with IL-76 aircraft which are heavy airlift machines. The medium tactical transport units possess AN-32s. The Dornier light transport aircraft has replaced the Devons and the Otters. India is also studying the supersonic KH-31 (AS-17 Krypton) to equip the Tupolev T-22 M Backfire bombers, the lease of which is being negotiated with Russia. The presence of such bombers along with fighters and refuellers would give the IAF a strategic reach. The IAF could then target the hinterland of China but the tankers and refuellers will be at the mercy of Chinese air superiority fighters.[123]

One lobby within the IAF asserts that till the nuclear submarine is operational, the long-range penetration strike elements of the air force constitute the only reliable platform for India's minimum credible nuclear deterrent.[124] Even civilian analysts like Kanti Bajpai views that the aircraft remains the most secure nuclear delivery vehicles as the Prithvi and Agni missiles can be unreliable and inaccurate.[125] According to Jasjit Singh, India's fleet of Sukhoi-30 and Mirage-2000 should constitute the aerial component of India's nuclear deterrent, and about eighty such aircraft should be set aside for this task. Each aircraft carrying a nuclear warhead requires two escorts. So, only about twenty-five aircraft should be designed to carry nuclear warheads and another fifty escorts assigned to guard them.[126] From the late 1990s, Pakistan's F-16s are able to carry nuclear warheads and India's Mirage-2000, MIG-27 and Sukhoi-30 are able to do the same.[127]

The threat of a limited conventional war by an extra-regional power remains. For deterring regional conflicts, the United States, argues William J. Perry (ex-US Secretary of Defence), must maintain strong, ready, forward-deployed conventional armed forces. For raising the operational reach of the US military power, Perry emphasizes airlift and sealift capacities of the US armed

TABLE 6.2: STRENGTH OF IAF IN 2002

Type of Craft	Number Available
MIG-21	450
MIG-23	180
MIG-25	10
MIG-27	200
MIG-29	75
SU-30	8
Jaguar	124
Mirage-2000	42
AN-32	122
IL-76	30
Dornier-228	43
HS-748	40
MI-8	80
MI-17	100
MI-25	20
MI-26	6
MI-35	40

Notes: Total Manpower = 1,30,000
Transport Aircraft (AN-32, IL-76, Dornier 228 and HS-748)
Total Number of Helicopters = 246
Armed Helicopters = 60.

Source: Air-Marshal Viney Kapila, *The Indian Air Force: A Balanced Strategic and Tactical Application*, New Delhi: Ocean Books, 2002, Appendix, pp. 129–32.

forces which would make possible rapid transmission of troops and equipment to distant theatres.[128] In the near future if an American Task Force appears near the coast of India (a repeat of the US Enterprise Incident of 1971) or if the PLAN attempts to conduct an amphibious landing in the Andaman and Nicobar group of islands, then the IAF has to play an active role. The actions of the Argentine Air Force in the Falklands War of 1982 offer the IAF a probable model although the Argentine Air Force failed to defeat the British Armada which was successful in landing troops and recapturing Port Stanley. But, if the Mirages of the Argentine Air Force had more Exocet missiles in its inventory and possessed tankers with air refuelling capacity, the situation might have been different.[129]

In the new millennium the airmen are trying to grab space as their sphere of influence. Aircraft is confined to the troposphere and with a slight extension upto the lower limits of the strato-sphere. This is because the air-breathing engines of the aircraft fail beyond 32 km. of the earth's surface (i.e. the upper limits of the stratosphere), due to lack of air.[130] For operations beyond that limit, space shuttles and satellites are required. An American airpower analyst Major Bruce Deblois asserts that in the near future space weapons will be used for ballistic missile defence. It is assumed that by using space-based BMD, the current stalemate of mutually assured destruction can be broken. Space-based weapons include anti-satellites, Directed Energy Weapons (DEW), air to space airborne laser and kinetic miniature homing vehicles, etc.[131] The whole baggage is termed as Strategic Defence Initiative and popularly known as Star Wars.

The IAF is also aware of the implications of the Star Wars. Air Commodore P.S. Ahluwalia writes: 'Control of space implies ensuring free access of one's own forces to the new high ground, while denying access to potential adversaries.'[132] Air Marshal Vinod Patney asserts that space assets will gain increasing economic, political and military importance in the near future. So space-based assets will need to be defended from enemy action which will require manned space-combat vehicles. The militarization and weaponization of space will result in the introduction of DEWs and lasers.[133] Air Commodore C.N. Ghosh writes that the next RMA will involve the emergence of the DEWs. The generation of high power microwaves weapons that can disable electronic circuitry in computers and communication equipment is a possibility. Ghosh says that the Indian military establishment accepts that it lacks the financial muscle and technological know-how to manufacture DEW systems. But, continues Ghosh, the Indian military should at least consider the rudimentary steps for protection against the various DEWs like lasers, Particle Beam and High Power Micro Wave Weapons. Ghosh goes on to say that in the near future the UAVs should be equipped with DEWs.[134]

CONCLUSION

In total war, even the enemy's home front is treated at par with the frontline resulting in massive casualties amongst civilians. An example is the strategic bombing of German and Japanese cities by the RAF and USAF between 1942 and 1945.[135] Such features are conspicuous by their absence in the limited wars. India and

Pakistan's successive air war could be categorized as a series of limited encounters as regards the objective, scope and the number of aircraft used. There was no plan by either the PAF or the IAF for attacking civilian targets.[136] Of course, at times non-military targets were attacked by mistake such as the IAF attack on a passenger train at Gujranwala during the 1965 War.[137] Both the IAF and the PAF lacked the intention as well as the hardware to conduct strategic bombing in the style conducted by the USAF over North Vietnam[138]—no Indian or Pakistani leader could have uttered a statement like US General Curtis Le May who said that North Vietnam 'should be bombed into the stone age.'[139] The Korean War saw the destruction of 800 MIGs.[140] The Soviet Union had lost over 2,000 aircraft before its withdrawal from Afghanistan.[141] During the whole period of Soviet occupation of Afghanistan, neither the Pakistani nor the Indian air forces possessed 2,000 first line aircraft. Even the medium-sized powers of the Middle East like Israel and Syria fought aerial battles which were bigger compared to those conducted by the PAF and the IAF.

In the midst of rapid technological changes, the debate regarding the use of airpower between the Indian Army officers and the IAF continues. While the IAF want to use their machines beyond the battlefields, the army want to use the IAF as flying artillery. After the display of US airpower during the Gulf War, most of the analysts all over the world have accepted the primary role of airpower in future battles. To conclude, it is essential to note Lieutenant-General Vinay Shankar's comment: 'The temptation to adopt the American template of airpower to our environment should be avoided. Our requirements are quite different and we must find our unique cost-effective solutions.'[142]

NOTES

1. 'Air Marshal Thomas Elmhirst, 15 August 1947 to February 1950, Looking Back: First Person', in Group Captain Ranbir Singh, ed., *Indian Air Force: In the Footsteps of Our Legends*, Noida: Book Mates, 1998, p. 23.
2. S. Gopal and Uma Iyengar, eds., *The Essential Writings of Jawaharlal Nehru*, vol. 2, New Delhi: Oxford University Press, 2003, p. 406.
3. Thomas Elmhirst, in Singh, ed., *In the Footsteps of Our Legends*, p. 20.
4. Ibid.

5. Vijay Seth, *The Flying Machines: Indian Air Force 1933 to 1999*, New Delhi: Seth Communications, 2000, p. 30.
6. Brigadier Ashok Malhotra, *Trishul: Ladakh and Kargil, 1947-1993*, New Delhi: Lancer, 2003, p. 10; Brigadier H.S. Sodhi, *Top Brass: Critical Appraisal of the Indian Military Leadership*, Noida: Trishul Publications, 1993, pp. 6, 14; Air Commodore N.B. Singh, *Air Power in the New Millennium*, New Delhi: Manas, 2000, p. 203; Group Captain S. Das Sarma, 'Meteorology and War', *JUSII*, vol. CII, no. 427 (1972), p. 176; John Keegan, *Intelligence in War: Knowledge of the Energy from Napoleon to Al Qaeda*, New York: Alfred A. Knopf, 2003, pp. 161–83; Seth, *The Flying Machines*, p. 35; Lieutenant-General S.K. Sinha, *Operation Rescue: Military Operations in Jammu and Kashimir 1947-49*, 1977; rpt., New Delhi: Vision Books, 2002, pp. 11–12, 14, 21.
7. Report for the Month of March 1944 for the Dominions, India, Burma and the Colonies and Mandated Territories to the War Cabinet, Report by the Secretary of State for India, Para 79, 27 April 1944, WP(44)229, NMML, New Delhi; Colonel Bhaskar Sarkar, *Outstanding Victories of the Indian Army: 1947-71*, New Delhi: Lancer, pp. 25–6.
8. Seth, *The Flying Machines*, p. 33; Air Marshal Bharat Kumar, 'The Pakistani Air Force: An Asessment', in Vice-Admiral K.K. Nayyar, ed., *Pakistan at the Crossroads*, New Delhi: Rupa, 2003, p. 360; Pushpindar Singh, *The Battle Axes: No. 7 Squadron Indian Air Force, 1942–1992*, New Delhi: The Society for Aerospace Studies, 1993, pp. 31–2, 37–8; George K. Tanham and Marcy Agmon, *The Indian Air Force: Trends and Prospects*, 1995; rpt., New Delhi: Vikas, 1996, p. 41; Francis K. Mason, *Hawker Aircraft Since 1920*, London: Putnam & Company, 1961, pp. 300–2.
9. 'Mehar Baba', in Air Vice-Marshal S.S. Malhotra and Group-Captain Ranbir Singh, eds., *In and Out of Cockpit: Yadein Memories*, New Delhi: Ebouz Classics, 1993, pp. 35, 38.
10. Kumar, 'The Pakistani Air Force', in Nayyar, ed., *Pakistan at the Crossroads*, pp. 351, 360; Singh, *The Battle Axes*, p. 35.
11. 'Thomas Elmhirst', in Singh, ed., *In the Footsteps of Our Legends*, p. 3.
12. Seth, *The Flying Machines*, p. 39; Air-Commodore T.K. Sen, 'Development of the IAF after Independence', in Major-General Afsir Karim, ed., *The Indian Armed Forces: A Basic Guide*, New Delhi: Lancer, 1995, p. 223.
13. 'Thomas Elmhirst', in Singh, ed., *In the Footsteps of Our Legends*, p. 4.
14. Oral Transcript of Interview with Prof. P.M.S. Blackett, pp. 4, 6.
15. Colonel Phillip S. Meilinger, 'Introduction', in idem, ed., *The Paths of Heaven: The Evolution of Airpower Theory*, New Delhi: Lancer, 2000, p. xvii.
16. David Hall, 'Lessons not Learned: The Struggle Between the Royal Air Force and Army for the Tactical Control of Aircraft, and the Post-Mortem on the Defeat of the British Expeditionary Force in France in 1940', in

Gary Sheffield and Geoffrey Till, eds., *The Challenges of High Command: The British Experience*, Basingstoke: Macmillan, 2003, pp. 113–14; Harold R. Winton, 'An Ambivalent Partnership: US Army and Air Force Perspectives on Air-Ground Operations, 1973–90', in Meilinger, ed., *The Paths of Heaven*, p. 408.

17. Azar Gat, *A History of Military Thought: From the Enlightenment to the Cold War*, Oxford: Oxford University Press, 2001, pp. 570, 573, 577–80; John Gooch, 'Clausewitz Disregarded: Italian Military Thought and Doctrine, 1815–1943', *JSS*, vol. 9, nos. 2–3 (1986), p. 316.
18. Ravinder Kumar and H.Y. Sharada Prasad, eds., *Selected Works of Jawaharlal Nehru*, Second Series, vol. 25 (1 February 1954–31 May 1954), New Delhi: Oxford University Press, 1999, p. 296.
19. Ibid., pp. 299–300.
20. Ibid., p. 295.
21. Andrew G.B. Vallance, *The Air Weapon: Doctrines of Air Power Strategy and Operational Art*, London: Macmillan, 1996, p. 17.
22. Kumar and Prasad, eds., *Selected Works of Nehru*, Second Series, vol. 25, pp. 295–6.
23. Wing-Commander M.K. Chopra, 'The Indian Air Force and the Nation', *JUSII*, vol. LXXXVII, no. 366 (1957), p. 13; Seth, *The Flying Machines*, p. 30; Mason, *Hawker Aircraft*, pp. 362, 462; Sen, 'Development of the IAF', in Karim, ed., *The Indian Armed Forces*, p. 225.
24. Bill Sweetman, 'The Modernization of China's Air Force', in Ray Bonds, ed., *The Chinese War Machine: A Technical Analysis of the Strategy and Weapons of the People's Republic of China*, London: Salamander Book, 1979, pp. 130–1; Mark A. O'Neill, 'Air Combat on the Periphery: The Soviet Air Force in Action During the Cold War, 1945–89', in Robin Higham, John T. Greenwood and Von Hardesty, eds., *Russian Aviation and Air Power in the Twentieth* Century, London: Frank Cass, 1998, pp. 212, 214, 219; Edgar O'Ballance, *Korea: 1950-53*, Dehra Dun: Natraj, 1969, p. 70; Chopra, 'The Indian Air Force', p. 18.
25. R. Sukumaran, 'The 1962 India–China War and Kargil 1999: Restrictions on the Use of Air Power', *Strategic Analysis*, vol. 27, no. 3 (2003), pp. 334, 336.
26. *Birth of an Air Force: The Memoirs of Air Vice-Marshal Harjinder Singh*, ed. Air Commodore A.L. Saigal, New Delhi: Palit & Palit, 1977, p. 284.
27. Ibid., p. 285; Malhotra, *Trishul*, pp. 33, 70.
28. *Birth of an Air Force*, pp. 274–6; *The Indian Air Force and Its Aircraft: IAF Golden Jubilee, 1932–82*, London: Ducimus Books, n.d., p. 3.
29. Roopinder Singh, *Arjan Singh: Marshal of the Indian Air Force*, New Delhi: Rupa, 2002, p. 51.
30. Ibid., p. 52.
31. Air Chief-Marshal P.C. Lal, *My Years with the IAF*, 1986; rpt., New Delhi: Lancer, 1987, pp. 127–8; Air Marshal M.S. Chaturvedi, *History of the Indian Air Force*, New Delhi: Vikas, 1978, pp. 139–40; Sen,

'Development of the IAF after Independence', in Karim, ed., *The Indian Armed Forces*, p. 225; Sukumar Biswas, *Three Weeks' War*, Calcutta: M.C. Sarkar & Sons, 1966, pp. 53–4.

32. Lieutenant-Colonel David S. Fadok, 'John Boyd and John Warden: Airpower's Quest for Strategic Paralysis', in Meilinger, ed., *The Paths of Heaven*, p. 363; Christopher Chant, 'Yalu River: 1952', in idem, Richard Holmes and William Koenig, *Two Centuries of Warfare*, London: Octopus Books, 1978, pp. 453–60; Lal, *My Years with the IAF*, p. 128.
33. Ibid., p. 132; Biswas, *Three Weeks' War*, pp. 57–8; Chaturvedi, *History of the Indian Air Force*, p. 139.
34. Lal, *My Years with the IAF*, p. 132; Pushpindar Singh, Ravi Rikhye and Peter Steinemann, *Fiza'ya: Psyche of the Pakistan Air Force*, New Delhi: Society for Aerospace Studies, 1991, pp. 30–1.
35. Chaturvedi, *History of the Indian Air Force*, pp. 141, 143.
36. Seth, *The Flying Machines*, p. 33.
37. P.A. Petersen, '*Perestroyka* and Planning Soviet Air Power', in Air Commodore E.S. Williams, ed., *Soviet Air Power: Prospects for the Future, Perestroyka and the Soviet Air Forces*, London: Tri Service Press, 1990, p. 57.
38. Lieutenant-General Vijay Oberoi, 'Doctrinal Challenges', in Air Commodore Jasjit Singh, ed., *Air Power and Joint Operations*, New Delhi: Knowledge World, 2003, p. 231.
39. Marshal of the Indian Air Force Arjan Singh, 'Presidential Address', in Singh, ed., *Air Power and Joint Operations*, p. 18.
40. Kumar, 'The Pakistani Air Force', in Nayyar, ed., *Pakistan at the Crossroads*, p. 351.
41. Pervaiz Iqbal Cheema, *The Armed Forces of Pakistan*, 2002; rpt., Karachi: Oxford University Press, 2003, p. 117.
42. Biswas, *Three Weeks' War*, p. 116.
43. Chaturvedi, *History of the Indian Air Force*, p. 147.
44. Singh, *Arjan Singh*, p. 65.
45. *Birth of an Air Force*, pp. 278–9.
46. T.K. Sen, 'Operational Role and Evolution: 1962–1977', in Karim, ed., *The Indian Armed Forces*, p. 237.
47. Roopinder Singh, *Arjan Singh*, p. 65.
48. Sen, 'Operational Role and Evolution' in Karim, ed., *The Indian Armed Forces*, p. 238.
49. Killen, *The Luftwaffe*, pp. 177–80.
50. Christopher Chant, 'Six Day War: 1967', in idem, Holmes and Koenig, *Two Centuries of Warfare*, pp. 466–70.
51. Mark McNeilly, *Sun Tzu and the Art of Modern Warfare*, Oxford/New York: Oxford University Press, 2001, p. 51.
52. U.P. Juneja, *Indo-Pak War: 1971*, New Delhi: New Light Publishers, 1971, p. 65.
53. Ibid., p. 63.

54. Major-General D.K. Palit, *The Lightning Campaign: Indo-Pakistani War, 1971*, New Delhi: Thomson Press, pp. 77–8; Juneja, *Indo-Pak War*, p. 64.
55. Kumar, 'The Pakistani Air Force', in Nayyar, ed., *Pakistan at the Crossroads*, p. 362; D.R. Mankekar, *Pakistan Cut to Size*, Delhi: Hind Pocket Books, 1972, p. 97; Palit, *The Lightning Campaign*, p. 78; Mason, *Hawker Aircraft*, p. 464; *The Indian Air Force and Its Aircraft*, p. 5; Juneja, *Indo-Pak War*, pp. 64–5.
56. J.S.B. Arora, *War with Pakistan: 1971*, Delhi: Army Educational Stores, n.d., pp. 54–5; 'Reminiscences: Air Chief Marshal P.C. Lal Recalls The Fourteen Day 1971 War', in Malhotra and Singh, eds., *Yadein*, p. 56; Palit, *The Lightning Campaign*, p. 83; Mankekar, *Pakistan Cut to Size*, p. 95; *The Indian Air Force and Its Aircraft*, pp. 5, 7.
57. *The Indian Air Force and Its Aircraft*, p. 10; Arora, *War with Pakistan*, pp. 46–7, 79; Antia, 'Indo-Pak War 1971, p. 112; 'Reminiscences: P.C. Lal', in Malhotra and Singh, eds., *Yadein*, p. 54.
58. Lieutenant-Colonel David S. Fadok, 'John Boyd and John Warden: Airpower's Quest for Strategic Paralysis', in Meilinger, ed., *The Paths of Heaven*, pp. 357–98.
59. Ranbir Singh, ed., *In the Footsteps of Our Legends*, pp. 102–3.
60. Ibid., p. 103.
61. Colonel Maris 'Buster' McCrabb, 'The Evolution of NATO Air Doctrine', in Meilinger, ed., *The Paths of Heaven*, p. 448.
62. Lieutenant-General Phillip B. Davidson, *Vietnam at War, The History: 1946-75*, 1988; rpt., London: Sidgwick & Jackson, 1989, p. 704.
63. Mankekar, *Pakistan Cut to Size*, pp. 49, 94.
64. Richard A. Pawloski, 'Lanes, Trains and Technology', in Williams, ed., *Soviet Air Power: Prospects for the Future, Perestroyka and the Soviet Air Forces*, p. 129.
65. Vallance, *The Air Weapon*, p. 18.
66. Major Yogi Saksena, 'Indo-Pak War: Some Lessons', *JUSII*, vol. CII, no. 428 (1972), 1971, p. 256.
67. Lieutenant-Colonel Edward J. Felker, 'Soviet Military Doctrine and Air Theory: Change Through the Light of a Storm', in Meilinger, ed., *The Paths of Heaven*, p. 497.
68. Brigadier Vijay K. Nair, 'Employment of Military Helicopters, Part II: The Indian Experience and Compulsions', *IDR* (January 1992), p. 107.
69. McCrabb, 'NATO Air Doctrine', in Meilinger, ed., *The Paths of Heaven*, p. 450.
70. Air Vice-Marshal J.E. 'Johnnie' Johnson, *The Story of Air Fighting*, London: Hutchinson, 1985, p. 292; Commander Sanjay J. Singh, 'The Past Five Decades', in Singh, ed., *Air Power and Joint Operations*, p. 57.
71. *MODAR: 1997–98*, p. 34; Georg Mander, 'Sukhoi Pushes Again for Reform', *JDW*, vol. 33, no. 21, 24 May 2000, p. 18; Jaswant Singh, ed.,

Indian Armed Forces Yearbook: 1981–82, Bombay, n.d., p. 608; *The Indian Air Force: 1932–82*, p. 5.

72. Vinod Anand, 'India's Military Response to the Kargil Aggression', *Strategic Analysis*, vol. 23, no. 7 (1999), p. 1054.
73. Satish K. Jain, 'Operation Safed Sagar', *IDR*, vol. 16, no. 3 (2001), p. 17.
74. *MODAR: 1999–2000*, p. 34.
75. Vallance, *The Air Weapon*, p. 20.
76. *MODAR: 1999–2000*, p. 34; Ranjit Bhushan, 'Peace, Anyone?', *Outlook*, 23 August 1999, p. 21.
77. *MODAR: 1999–2000*, p. 36.
78. Ian Anthony, 'Arms Exports to Southern Asia: Policies of Technology Transfer and Denial in the Supplier Countries', in Eric Arnett, ed., *Military Capacity and the Risk of War: China, India, Pakistan and Iran*, Oxford: Oxford University Press, 1997, p. 295.
79. Cheema, *Armed Forces of Pakistan*, pp. 104–5; *The Story of the Pakistan Air Force: 1988–1998: A Battle Against Odds*, Islamabad: Shaheen Foundation in collaboration with Oxford University Press, 2000, p. 27; Arvind R. Deo, 'Pakistan's Unending Search for a Viable Foreign Policy', in Nayyar, ed., *Pakistan at the Crossroads*, p. 14.
80. Rene J. Francillon, *Vietnam Air Wars*, London: Temple Press, 1987, Appendix A, Table 1, p. 208.
81. Vinod Anand, *Joint Vision for the Indian Armed Forces*, New Delhi: IDSA, 2001, p. 76.
82. O'Ballance, *Afghan Wars*, p. 102.
83. Philip Towle, 'Air Power in Afghanistan', in Williams, ed., *Soviet Air Power*, pp. 184, 190, 193.
84. O'Ballance, *Afghan Wars*, pp. 155, 163–4.
85. Vallance, *The Air Weapon*, p. 22.
86. Tim Ripley, *The New Illustrated Guide to Modern US Army*, London: Salamander, 1992, pp. 17–18; Meilinger, 'Introduction', and Winton, 'An Ambivalent Partnership', in Meilinger, ed., *The Paths of Heaven*, pp. xxx, 427.
87. *MODAR: 1999–2000*, pp. 39–40; Rahul Bedi, 'India Urged to Stop Stalling over ATJ Delays', *JDW*, vol. 31, no. 11, 1 March 1999, p. 16; Rahul Bedi, 'India Opts for Hawk Jet Trainers', *JDW*, vol. 32, no. 13, 29 September 1999, p. 16; Ravi Sharma, 'An Ailing Fleet', *Frontline*, 15 August 2003, pp. 54–6.
88. Felker, 'Soviet Military Doctrine', in Meilinger, ed., *The Paths of Heaven*, p. 504.
89. Niklas Zetterling and Anders Frankson, *Kursk 1943: A Statistical Analysis*, London: Frank Cass, 2000, p. 133.
90. Group-Captain R.K. Jolly, 'Joint Operations: The Way Forward', in Singh, ed., *Air Power and Joint Operations*, p. 202.
91. Air Vice-Marshal S.C. Rastogi, 'Indian Air Force and Technology', *JUSII*, vol. CXXXIII, no. 552 (2003), p. 246.

92. Major-General Julian Thompson, 'The World in 1945: The View from May 2001', in idem., ed., *The Imperial War Museum Book of Modern Warfare: British and Commonwealth Forces at War, 1945–2000*, 2002; rpt., London: Pan Macmillan, 2004, pp. 3–4.
93. Kapil Kak, 'A Century of Air Power: Lessons and Pointers', *Strategic Analysis*, vol. 24, no. 12 (2001), p. 2111.
94. Jasjit Singh, 'Dynamics of Limited War', *Strategic Analysis*, vol. 24, no. 7 (2000), pp. 1209, 1219.
95. Jasjit Singh, 'Our Wars in Future', in idem., ed., *Air Power and Joint Operations*, p. 140.
96. Kapil Kak, 'India's Conventional Defence: Problems and Prospects', *Strategic Analysis*, vol. 22, no. 11 (1999), pp. 1644, 1653, 1660.
97. Jasjit Singh, 'Our Wars in Future', in Singh, ed., *Air Power and Joint Operations*, p. 139.
98. Shimon Naveh, *In Pursuit of Military Excellence: The Evolution of Operational Theory*, 1997; rpt., London: Frank Cass, 2000, p. 310.
99. Brigadier Gurmeet Kanwal, 'Airland Operations', in Singh, ed., *Air Power and Joint Operations*, p. 171.
100. Jolly, 'Joint Operations', in Singh, ed., *Air Power and Joint Operations*, p. 206.
101. Kanwal, 'Airland Operations', in Singh, ed., *Air Power and Joint Operations*, p. 165.
102. Oberoi, 'Doctrinal Challenges', in Singh, ed., *Air Power and Joint Operations*, pp. 231–2.
103. R.V. Phadke, 'Response Options: Future of Indian Air Power Vision 2020', *Strategic Analysis*, vol. 24, no. 10 (2001), p. 1802.
104. V.K. Shrivastava, 'Indian Air Force in the Years Ahead: An Army View', *Strategic Analysis*, vol. 25, no. 8 (2001), p. 944.
105. Major James F. Holcomb, 'Developments in Soviet Helicopter Tactics', in Williams, ed., *Soviet Air Power*, pp. 164, 173.
106. *MODAR: 1989–90*, p. 22.
107. Jasjit Singh, 'Strategic Framework for Defence Planners: Air Power in the 21st Century', *Strategic Analysis*, vol. 22, no. 12 (1999), p. 1814.
108. G.D. Bakshi, 'Yugoslavia: Air Strikes Test of the Air War Doctrine', *Strategic Analyis*, vol. 23, no. 5 (1999), pp. 792, 794.
109. Kapila, *The Indian Air Force*, p. 23.
110. Air Chief Marshal S. Krishnaswamy, 'Air Power Today and Tomorrow', in Singh, ed., *Air Power and Joint Operations*, p. 30.
111. 'Casualties of War', *Newsweek*, 26 April 1999, pp. 10–11.
112. Tom Lansford, *Terrorism, NATO and the United States*, Aldershot: Ashgate, 2002, pp. 109, 127.
113. Kak, 'A Century of Air Power', p. 2115.
114. Rastogi, 'Indian Air Force and Technology', p. 247.
115. Air Commodore A.K. Tiwary, 'Future of Fighter Operations', *IDR*, vol. 15, no. 1 (2000), pp. 54–5, 59.

116. G. Jacobs, 'Unmanned Aerial Vehicles: Some Programmes', *IDR*, vol. 14, no. 2 (1999), pp. 101, 106.
117. C.N. Ghosh, 'Unmanned Combat Aerial Vehicles in Future Battles of the Subcontinent', *Strategic Analysis*, vol. 25, no. 4 (2001), pp. 599–600.
118. Ibid., pp. 601, 605.
119. Squadron-Leader P.M. Sinha, 'Air Power in the Information Age', *IDR*, vol, 16, no. 3 (2001), p. 48.
120. *Vayu 2000 Aerospace Review*, 3 (1998), p. 50; Umer Farooq, 'Pakistan in Talks to Buy Chinese F-7 MGs', and 'Pakistan Receives More Mirages', *JDW*, vol. 33, no. 22, 31 May 2000, pp. 14, 15.
121. Johnson, *Air Fighting*, p. 290. Certainly one could argue that the training of the Arab pilots was not of high quality and this was one of the reasons behind the aerial victory of the Israelis.
122. M.V. Rappai, 'China's Nuclear Arsenal and Missile Defence', *Strategic Analysis*, vol. 26, no. 1 (2002), p. 69; 'China Assembled SU-27s Make Their First Flight', *JDW*, vol. 31, no. 8, 24 February 1999, p. 16.
123. Sandeep Unnithan, 'Indian Navy Looks to Russia for Successor to Sea Eagle', *JDW*, vol. 33, no. 22, 31 May 2000, p. 15; *MODAR: 1989–90*, pp. 21–2; Lieutenant-General R.K. Jasbir Singh, ed., *Indian Defence Yearbook: 2003*, Dehra Dun: Natraj, 2003, pp. 226, 228.
124. Phadke, 'Future of Indian Air Power', p. 1798.
125. Kanti Bajpai, 'India's Future Wars', in Singh, ed., *Air Power and Joint Operations*, p. 129.
126. Jasjit Singh, 'Nuclear Command and Control', *Strategic Analysis*, vol. 25, no. 2 (2001), pp. 156–7.
127. K. Subrahmanyam, 'After Pokhran II', in J. Baranwal, ed., *SP's Military Yearbook: 1998–99*, New Delhi: Guide Publications, 1998, p. 9.
128. William J. Perry, 'Defence in an Age of Hope', *Foreign Affairs*, November–December (1996), pp. 73, 75.
129. Johnson, *Air Fighting*, p. 293.
130. Air Commodore P.S. Ahluwalia, 'Application of Military Power in Space', *IDR*, vol. 16, no. 3 (2001), p. 50.
131. Major Bruce M. DeBlois, 'Ascendant Realms: Characteristics of Airpower and Space Power', in Meilinger, ed., *The Paths of Heaven*, pp. 534, 536.
132. Ahluwalia, 'Application of Military Power in Space', p. 51.
133. Air Marshal Vinod Patney, 'Manned Combat Space Flights', *Defence Watch*, vol. 3, no. 5 (2004), pp. 12–13.
134. C.N. Ghosh, 'EMP Weapons', *Strategic Analysis*, vol. 24, no. 7 (2000), pp. 1333, 1343; C.N. Ghosh, 'Directed Energy Weapons', *Strategic Analysis*, vol. 24, no. 11 (2001), pp. 2055, 2064.
135. John Pimlott, 'The Theory and Practice of Strategic Bombing', in Colin McInnes and G.D. Sheffield, eds., *Warfare in the Twentieth Century: Theory and Practice*, London: Unwin Hyman, 1988, pp. 113, 129–33.

136. Kapila, *The Indian Air Force*, p. 17.
137. Mohammad Ashgar Khan, *The First Round: Indo-Pakistan War*, 1965, Ghaziabad: Vikas, 1979, p. 15.
138. McNeilly, *Sun Tzu*, p. 21.
139. Quoted in Vallance, *The Air Weapon*, p. 16.
140. Johnson, *Air Fighting*, p. 229.
141. Pawloski, 'Lanes, Trains and Technology', in Williams, ed., *Soviet Air Power*, p. 130.
142. Lieutenant-General Vinay Shankar, 'Dynamics of Future Wars', in Singh, ed., *Air Power and Joint Operations*, p. 119.

SEVEN

Evolution of the Indian Navy

The Indian Ocean is the third largest ocean in the world and covers an area of 74 million sq. km. which comprises some 20 per cent of the total oceanic area of the world. It touches the shores of three continents—Asia, Africa and Australia, and its southern end reaches up to Antarctica. The Indian Ocean from Sumatra in the Indonesian archipelago to the east coast of Africa extends over 3,000 miles. Till 1947, the Royal Navy tackled the duty of guarding the Indian Ocean. The Royal Indian Navy (RIN) emerged as a small coastal force; a mere adjunct of the Royal Navy during World War II. After Independence, the Indian Navy faced challenges from the Pakistan Navy and had to guard about 7,516 km. long coastline plus 1,200 islands.[1] The presence of the United States Navy's (USN) detachments stationed in the Indian Ocean also posed occasional threat to the Indian Navy. In the new millennium two additional threats emerged: the Peoples Liberation Army's Navy (PLAN) of China and the non-state threats. Let us trace the origin and growth of independent India's navy to evaluate how far had it been able to tackle maritime security of the nation.

THE MEANING OF MARITIME SECURITY

For some time past, in fact long time past, I felt how people in this great bulk of the north of India are, what might be said, land minded. They are not so conscious of the sea, naturally they are not as the people on the sea coast and the south of India. There is a definite thing which I have felt repeatedly. How will you think in terms of defence? You think in terms of army in the north. In terms of defence in the south, one would of course think of an army but more immediately of the sea you think about, whatever it is, trade, etc. There is the land consciousness in the north and the sea consciousness in the south, and we have to be equally conscious of both land and sea apart from the air, which is common to both.

JAWAHARLAL NEHRU[2]

Maritime security or sea power, depends on maritime resources which is the sum total of naval power and non-naval maritime components of a nation. Ensuring power over the sea requires proper strategies. Colin S. Gray and Roger W. Barnett write: 'Maritime strategy refers to the purposeful exercise of the sea-using national assets of all kinds for the political goals set by government. Naval strategy refers more narrowly to the purposeful exercise of naval forces, again for the political goals set by government.'[3] This could be accepted as a working definition. The influence of British naval theorist Julian S. Corbett (who operated in the first decade of the twentieth century) on Barnett and Gray is clear. Corbett differentiates between maritime and naval strategy in the following words: 'By maritime strategy we mean the principles which govern a war in which the sea is a substantial factor. Naval strategy is but that part of it which determines the movements of the fleet when maritime strategy has determined what part the fleet must play in relation to the action of the land forces.'[4] So, for Corbett naval strategy is about movement of the fleet while maritime strategy links naval power with non-naval military assets of a maritime nation.

Some Indian naval theorists concentrate on force structure while chalking out the framework of naval strategy. They demand a blue water navy for ensuring maritime security of India. John W. Garver defines blue water naval capability as the ability to sustain intense combat operations on the high seas, hundreds of miles away from a country's coast.[5] One feature of blue water navy is possession of one or more aircraft carriers. In 2000, Lieutenant-Commander Dean Mathew had written that aircraft carriers are essential for India's sea control strategy. India needs aircraft carriers to contain the Pakistani naval threat to India's Sea Lines of Communications (SLOC) in the Arabian Sea as well as to check the PLAN in the Indian Ocean.[6] Long before Mathew, in 1957 Lieutenant-Commander Narapati Datta also emphasized the case for an aircraft carrier. He asserted that an aircraft carrier could enhance India's maritime reach as regards reconnaissance during maritime operations, and provide fighter defence to the navy's detached task force. In addition, aircraft from the carrier could conduct strikes against the enemy's surface ships as well as against the hostile submarines. Finally the carrier's aircraft could destroy the enemy's shore-based facilities.[7] The last is an instance of power projection and this is what exactly happened in East Pakistan in 1971.

Power projection is also possible without actually engaging in a naval war. Vice-Admiral S.K. Chand in the last decade of the twentieth century analysed sea power in the following words:

> The term sea power in its generic sense indicates a nation's ability to exploit the oceans to its advantage and to protect this ability from interference by others. At the same time, sea power also provides the nation with an ability to influence others in their decision-making process. To simply put it, it acts as an instrument of enforcing a national policy. The inherent flexibility of Naval Forces and their ready availability for deployment in the pursuance of foreign policy objectives is another aspect of maritime power which needs to be borne in mind when subjecting sea power to the realms of evaluation.[8]

Thus for Chand naval power and maritime capability are more or less coterminous. P.K. Ghosh following in Chand's footsteps wrote in 2001, that naval power could be an instrument of coercive state policy and also a deterrent to the actual breaking out of war. He argues that coercive gunboat diplomacy is muscle-flexing diplomacy and is the alternative to war. A successful gunboat diplomacy should never escalate to war. Ghosh continues that with the increasing reach of the missiles, gunboat diplomacy instead of merely shaping the coastal enclaves will also influence the hinterlands hundreds of miles beyond.[9] Both Chand and Ghosh are deviating from Mathew and Datta's Mahanite concept of using naval power for fighting a decisive battle with the enemy in order to sweep the latter's war ships from the ocean.

Datta and Mathew are speaking of naval power projection by pursuing a sea control strategy. The concept of sea control can be traced back to Captain Alfred Thayer Mahan (1840–1914) of the USN. He was the first proponent of modern sea power. Mahan emphasizes that principles of naval strategy remain constant with time. As far as naval warfare is concerned, Mahan points out that every great power requires to establish command over the sea and this could be achieved only by destruction of the enemy's battle fleet through a decisive encounter at the high sea.[10]

The Indian naval officers are aware of a strategy which is the polar opposite of sea control concept. Captain A.H. Chitnis and Commodore C.S. Patham divide maritime strategy into sea control and sea denial categories. According to them, 'Sea control is said to have been achieved by a force when the sea area, the air space above it and the underwater area are in complete control of own forces enabling unhindered movement of war ships and merchant ships in that area. Sea denial is the ability to deny the use of a certain sea area to the enemy.'[11] The concept of sea denial can be traced back to the late nineteenth century.

In opposition to Mahan's sea control concept, the *Jeune Ecole* School of thought (which evolved within the French Navy in the 1880s) put forward the policy of *guerre de course*. The objective is not destruction of the enemy's battle fleet but its mercantile marine.[12] The point to be noted is that in the first decade of the twentieth century, the British naval theorist Corbett, like the *Jeune Ecole* School, argued that there are ways to challenge enemy's naval supremacy rather than engaging its battle fleet directly in a decisive encounter at sea.[13]

Credit is due to the Japanese for inventing the theory of amphibious operation, which emphasizes avoiding encounters with the enemy's battle fleet. In the Japanese view, the navy is to be used for transporting troops. The principal task of the naval forces is not to search and destroy the enemy's main battle fleet but to escort the troop carriers. The troop carriers are supposed to land the troops along the flanks of the enemy. In the Japanese conceptual framework there is no necessity for establishing first command of the sea in accordance with the Mahanite version through a decisive sea battle, as this would delay transportation of troops and enable the enemy land forces to strengthen their positions. Further, every ship detached for searching the enemy battle fleet means a ship less for the primary mission of protecting troop transportation.[14]

The Soviet Navy evolved the 'sea denial strategy' in the 1950s. It is actually an adaptation of the French School's commerce raiding strategy. The Soviet Navy's policy was to severe the Western power's SLOCs. Sea denial is the weapon of the weak.

The objective of sea denial is not to use the sea for itself but to prevent the enemy from doing so and thus could be categorized as a defensive strategy.[15]

Admiral S.M. Nanda who was the Chief of Naval Staff (CNS) of India during the 1971 India–Pakistan War is a bit over-ambitious as he wants India to pursue both sea control and sea denial strategies simultaneously. In his memoirs published in 2004, he writes:

> Our capability for sea control will depend a great deal on the number of aircraft carriers and long range maritime patrol aircraft that we can field, while our sea denial capability will depend largely on the effectiveness and strengths of our submarine forces. We must create independent carrier battle groups with escorts, amphibious assault ships, fleet submarines and attendant air early warning aircraft to safeguard our maritime interests.[16]

Long back in 1954 Commander V.A. Kamath wrote that India ought to pursue sea control strategy in order to ensure freedom of ocean and sea to India's merchant marine. In addition India should also pursue sea denial strategy to prevent the enemy country from getting any supplies. This is necessary, says Kamath, because future wars are going to be protracted.[17]

In contrast in 2002, Commander Sanjay J. Singh had written: 'Overall, Navies can also expect an increasing constabulary role, and there may be greater cooperation between the maritime forces of different nations to provide collective security at sea.'[18] Sanjay Singh seems to offer a passive coastal strategy devoid of any warfighting component, relying mostly on deterrence and diplomacy. This strategy is designed to deter the enemy not by power projection as Chand and Ghosh argue but by building friendly coalitions. Singh's strategy could at best deny certain portions of coastal water to the enemy's use. Before Singh, a civilian analyst K.M. Panikkar was the greatest advocate of sea denial strategy because he believed that submarines were the weapons for future sea warfare. In Panikkar's paradigm, India needs merely to protect its coast from the neighbouring local powers. Panikkar continues that in case of any intervention by the big powers, India should not fight alone but resort to maritime diplomacy to pursue a sea denial strategy.[19]

A middle path between sea control and sea denial strategies

seems to be the concept of littoral warfare which involves fusion of maritime and continental strategies. Rear Admiral Raja Menon is the spokesman of this school. Menon's perspective is narrow and is concerned with warfighting and jointery. Corbett first introduced the concept of 'jointery' as part of his critique of Mahan's theory of sea power. Mahan's biggest lacuna is his failure to realize that with the onset of the industrialization, the age of sea borne commercial capitalism is over. Henceforth, a nation's wealth would no more depend mostly on sea borne commerce as Mahan assumes. Domestic industrial capacity would now be more important than commercial mercantilism and colonies of the sailing era. In 1904, the British geo-politician Halford Mackinder argued that steam locomotion is opening up the vast heartland of Asia and America, which were hitherto sparsely populated, and the railways and the internal combustion engine would result in cheap and quick land transportation. So, the vast continental states would be able to utilize their inland resources which would allow a continental state to check the incursions by an enemy maritime power at its periphery. In other words the battle fleet of a maritime nation would be quite useless against a continental state practicing *autarkic* economic policy.[20] Hence, emerged the necessity of fusing land power with sea power, and the modern name for this fusion is jointery.

Unlike Mahan who conceptualizes the high sea fleet sailing in the deep ocean and shaping the course of war, Corbett says that problems can hardly be solved on naval considerations alone. This is because naval strategy is only a part of maritime strategy. Corbett defines maritime strategy as 'the higher learning which teaches us that for a maritime State to make successful war and to realize her special strength, army and navy must be used and thought of as instruments no less intimately connected than are the three arms ashore.'[21] Corbett emphasizes jointery (fusion of sea and land operations) and writes:

> We are accustomed, partly for convenience and partly from lack of a scientific habit of thought, to speak of naval strategy and military strategy as though they were distinct branches of knowledge which had no common ground. It is the theory of war which brings out their intimate relation. It reveals that embracing them both is a larger strategy which regards the fleet and army as one weapon, which coordinates their action, and indicates the lines on which each must move to realize the

full power of both. It will direct us to assign to each its proper function in a plan of war; it will enable each service to realize the better the limitations and the possibilities of the function with which it is charged, and how and when its own necessities must give way to a higher or more pressing need of the other.[22]

At present a maritime strategy emphasizing jointness is also propounded by the Australian military. Lieutenant-General Frank Hickling says:

A Maritime Strategy recognizes that Australia's maritime approaches are an environment in which the battlefield effects of sea, land and air power are interwoven. It demands that the three services must be able to operate as a team; and it demands that the Army must be able to operate offshore of the continental land mass as well as onshore in Australian territory. A Maritime Strategy also recognizes the need to defend Australia's interests in the region and globally, as well as direct defence of our sovereignty. And it also means that we must be able to exert influence by adding weight to diplomacy, particularly in our region.[23]

If instead of 'Australia', the word 'India' is inserted then Hickling's definition of maritime strategy emphasizing jointness would suit New Delhi's purpose also.

Following Corbett, Menon in 1998 had written that sea control by itself is not an end. Rather the navy should try to influence events ashore by creating a maritime flank and pursuing a joint strategy with the other branches of the armed forces. Menon asserts that the navies of the continental powers like India should be geared for defending its own littoral and attacking the enemy's littoral. Because of the ongoing Revolution in Naval Affairs (RNA), Menon continues, the naval fleet could attack the enemy's shore defence and also conduct amphibious operations as well as blockade the enemy's coast. Such operations, continues Menon, would make the navy an equal partner of the ground and air forces in conducting a continental war.[24]

The rise in naval combat effectiveness is due to the RNA which Menon believes is the product of the rise of revolutionary hardware. The use of new technology raises the speed of battle and also enables domination of the battlespace. The integration of the latest hardware of course demands, writes Menon, changes in organization and doctrinal aspects in managing the navy. He warns, however, that the new technologies to which he is referring are lethal weapons and do not constitute instruments merely for conducting information warfare. Menon rightly adds in a sarcastic tone that information warfare regarding which at present there is so much enthusiasm among the supporters of Revolution in Military Affairs is probably not even true warfare.[25]

Some naval theorists of India tend to take a broader view of sea power. They bring into focus the economic and scientific dimension of maritime power. K.M. Panikkar in 1960 had pointed out that naval defence or for that matter any sort of defence depends on a nation's scientific progress and industrial development.[26] According to Rahul Roy-Chaudhury India's maritime security includes political, economic, technological, military, scientific and environmental components.[27] Vice-Admiral P.S. Das points out the economic aspects inherent in India's maritime strategy in the following words:

> It is hardly necessary to emphasize the maritime, not continental, nature of India's interests. Our history has been inextricably linked to the rise and decline in our sea-going traditions. . . . If India became enslaved and remained so for three centuries, it was largely because we could not protect our sea frontiers. Today, almost our entire incoming and outgoing commerce moves across the seas. India's dependence on this trade and on our ports and harbours is critical to economic growth. Our large Exclusive Economic Zone, which may soon exceed three million sq. km. is host to a cache of undisclosed and unexploited wealth. Once technology to harness these resources becomes available, and this may happen sooner rather than later, one can expect conflict of interest at sea.[28]

S.N. Kohli an ex-CNS argues like P.S. Das that with the passage of time, the oceans' importance will continue to grow as they increasingly provide food and minerals to the people at large.[29] Kohli explains his concept of sea power in the following words:

> There is, however, a mistaken belief that 'sea power' is a synonym for a strong Navy. This is a myopic view which limits the meaning of the term. Sea power really connotes the power to use the seas, during peace and war, to the best advantage. The well-known American writer, Alfred Thayer Mahan, put it more expansively when he wrote: "Sea power embraces all that tends to make a people great upon the sea, or by the sea." Though he wrote about the days of sailing ships, he wisely related history to the steam-powered times in which he lived so that some of his prognoses could attain an enduring validity.[30]

To be fair to Mahan, he was aware of the economic, political and geographical dimensions of sea power. Mahan says that naval supremacy has given a country like Britain a monopoly over trade and colonies. The Royal Navy's denial of these assets to Britain's enemies and the policy of subsidizing her continental allies with the money generated from sea-borne commerce enabled Britain to remain a world power in the nineteenth century.[31] For Mahan, terrain, climate, population, commerce, industry and people's attitude towards the sea, etc., constitute sea power of a nation. Kohli accepts Mahan's ingredients of sea power.[32] A nation's peacetime commerce, says Mahan, is an index of its staying power during a naval war. There must be a large reserve among the population with those skills that are essential to the maintenance of ships both in times of peace and war. A nation with the tradition of trading and shipbuilding possesses the human and technical resources which are essential for success during naval war.[33] Kohli continues that Mahan rightly harps on attitudes and infrastructure for assessing sea power of a nation in question. Kohli points out that banking infrastructure, insurance and ports are also vital in the growth of sea power.[34] Two American analysts rightly say: 'Justly it has been said that naval power without merchant shipping is like a locomotive without freight or passenger cars.'[35]

Mahan was influenced by Social Darwinism. He asserts that virile states compete with each other and this in turn breeds struggle and suffering. He argues that while war is wasteful, in the long run, war creates conditions for future greatness.[36] Before Mahan, the German philosopher Immanuel Kant in a somewhat similar tone had written: 'Conflict and selfishness, even avarice and lust for power, do, after all, awaken our human powers and stir us out of complacency. Selfishness drives us to accomplish things, competition sharpens our abilities. We are driven toward the fullest development of our power and faculties.'[37] Kant's observation about human nature remains relevant even now. Towards the end of the twentieth century, the relevance of naval warfare for the future was put forward by the American theorist James John Tritten. Nations, says Tritten, will continue to need navies because oceans cover 70 per cent of the earth.[38] And oceans remain not only the medium of commerce but also the source of new minerals.

FROM A COASTAL FORCE TOWARDS A SEA GOING FLEET: 1947–1971

In September 1944, Claude Auchinleck (commander-in-chief of India) set up the reorganization committee to deliberate upon the nature of post-war India's force structure. The navy and the air force's share of the defence budget amounted to Rs. 35 crore each. And the army's share was fixed at Rs. 42 crore. Towards the end of the war, India was able to acquire a significant amount of sterling balance vis-à-vis Britain. During October 1944, A. Wavell, the Viceroy informed Leo Amery, the Secretary of State for India, that with the sterling balance India could buy modern ships from Britain after the war. And this would benefit both the colony and the mother country.[39]

Before Partition, the RIN possessed forty-eight vessels. Pakistan's share included two sloops, two frigates, four minesweepers, two motor minesweepers, two trawlers and four harbour defence launches. Independent India's principal warships included four anti-aircraft frigates, two anti-submarine frigates, one corvette, twelve fleet minesweepers and one survey ship with 30,478 naval personnel. As most of the principal shore establishments like HMIS *Himalaya*, the Gunnery School, HMIS *Chamak*, the Radar School, HMIS *Dilawar* and *Bahadur*, the Boys' Training Establishments, were located at Karachi, they went to Pakistan and the Indian Navy faced great difficulties in training its officers and sailors. In the RIN while most of the officers were Hindu, the lower deck personnel were predominantly Muslims. The Pakistan Navy had an advantage because most of the sailors were Muslims from west Punjab (now a portion of Pakistan) and had fought in World War II. But, the problem of the Pakistan Navy lay with the officer corps. Out of 200 officers, only nine had regular commissions and only three out of them were in the executive branch. The senior-most Pakistani officer was a captain.[40]

The first action of the Indian Navy involved transporting the Indian Army's units to Jafarabad on 17 October 1947 in an action against the princely state of Junagadh. It was a crude joint operation launched by a brigadier and under him R.D. Katari functioned as the Naval Force Commander. His force included the frigates *Krishna*, *Cauvery* and *Sutlej* and a Landing Ship Transport (LST). The duty of the navy was to land a few infantry companies and a couple of tanks at Porbandar, Mangrol and Veraval. Moreover, the navy had to supply the Indian forces going ashore. The Junagadh forces shortly surrendered and the nawab with his beautiful begums fled to Pakistan in his private aircraft.[41] However, the Indian Navy was yet to learn the skill of conducting amphibious operations on shores defended by the enemy.

In the aftermath of Independence, Admiral Edward Perry tried to sell four fleet carriers to India. P.M.S. Blackett, Nehru's Defence Advisor obstructed the scheme. Blackett's reasoning was that to protect the four carriers, India would need twenty destroyers and the Indian Navy at that time possessed only three destroyers. Blackett at that time visualized the Indian Navy as a coastal force. Just after 1947, Britain wanted the Indian Navy to protect the Western bloc's SLOCs in the Indian Ocean against the Soviet submarines in case of a Third World War. So, London wanted the Indian Navy to focus exclusively on anti-submarine warfare and pressurized India to purchase frigates and minesweepers from Britain. Through the British officers who continued in the Indian Navy, London influenced the doctrine and force structure of the Indian Navy. Complete Indianization of the navy occurred only in 1958 when Vice-Admiral R.D. Katari succeeded Vice-Admiral Stephen Carlill as the CNS.[42]

India's political elite were not critical of the close connection between the Indian Navy and the Royal Navy. On 25 March 1954, Jawaharlal Nehru paid tribute to the Royal Navy and the British officers in the Parliament, when he said:

> Our Navy is small, and we really had to build up from scratch. It is not a question of passing an examination merely, but it requires experience of training in all kinds of conditions, technical and other; and one cannot produce a trained person suddenly out of a hat. We have been connected more specially in regard to the Navy, with the British Royal Navy, in the sense of receiving training, sometimes joining their manoeuvres, and receiving our equipment and the rest. Without some such experience, we would have been far more backward than we are. We could only have that experience from some other navy, and the obvious thing was for us to have it from the British Navy, because our pattern of development of the Navy. . . . was the British pattern. We had developed that way, and unless we reject that pattern and adopt some entirely different pattern, it is obviously desirable for us to improve

along the pattern that we had followed thus far—and it is a good pattern.[43]

After Independence, the Indian Navy went for a slow but steady expansion. In fact, the Indian Navy was lucky that it was not stagnating, for Nehru's concept of maritime defence was indeed myopic. Nehru influenced by Blackett wanted a coastal navy. On 12 May 1954, Nehru wrote to the Defence Secretary:

> So far as we are concerned, we have no intention of sending expeditionary forces to other countries or to have any operations, military, naval or air, far from our country. The whole conception of our defence forces is one for defence, that is round about our frontiers. That means that our Navy will at no time be charged with protecting the sea routes for us or to bring in food supplies etc. Indeed, this is completely beyond our capacity. Our Navy has to perform the smaller but very important task of protecting our ports and making it hot for any enemy ships which seek to attack us. This also leads to the conclusion that we do not normally require big ships (except for training and like purposes). What is far more important is a number of small but swiftly moving ships, well armed. Normally speaking again, long distance submarines are not a necessity for us. This applies to an aircraft carrier also, which is really needed foı attacking distant places. It is very expensive and we could utilize the money much more effectively by having more land based aircraft.[44]

In Nehru's paradigm, only a country like Britain which is dependent on import of food and essential materials and possesses dependencies and colonies needs to ensure sea control through a blue water navy.[45] Besides India's maritime strategy, Nehru had something to say on the required force structure of the Indian Navy. In a note addressed to the Ministry of Defence dated 25 May 1955, Nehru authoritatively commented: 'I think it is perfectly clear now that battleships, cruisers and other big ships are completely out of date from the point of view of waı. Their chief importance lies in ceremonial purposes and possibly in training. Even in training, I rather doubt if their utility is now appreciable.'[46] Unlike Nehru, Indonesia's leader Dr. Soekarno had faith in the big ships. In the 1950s, the Indian Navy had no answer to Indonesia's Sverdlov class Russian built cruiser. Each such cruiser with a crew of 1,000 had a range of 8,700 miles. These ships were armed with missiles, guns and torpedoes.[47]

If Nehru lacked a vision of maritime security, the same charge could be applied to Mao Tse-Tung. In 1949, the PLAN came into existence and pursued a coastal defence strategy under the 'People's War' doctrine.[48] Between 1950 and 1959, the Soviets upgraded the PLAN's combat effectiveness as a coastal force. Over 2,500 Soviet naval advisers were sent to China. In the mid 1950s, the PLAN received four destroyers, thirteen submarines, twelve large patrol boats, two minesweepers and fifty torpedo boats from USSR. Most of these were of World War II vintage.[49] Nehru and Mao had a Western counterpart in Lord Trenchard of the Royal Air Force who asserted in 1953 that the land-based bomber would be able to penetrate the enemy's anti-air naval defence and sink the aircraft carriers. In other words, Trenchard was implying that the Royal Navy's carriers could as well be scrapped.[50]

Overall, the induction of ships remained *ad hoc* and opportunistic. It is not clear who were actually calling the shots in the North and South Blocks. Subsequent to the Junagadh operation, a decision was taken by the government, despite Nehru's admonition, to acquire a cruiser named H.M.S. *Achilles* of the Leander class from the Royal Navy. This cruiser had fought in the Battle of River Plate against the German pocket battleship *Graf Spee*. In 1948, it was commissioned as *Delhi*. In 1950, India acquired three R-class destroyers from the UK. These destroyers were named *Rajput*, *Ranjit* and *Rana* and formed the 11th Destroyer Squadron. In 1951, the Indian Navy obtained three Hunt Class destroyers on loan from the Royal Navy. Later these ships were purchased and named *Godavari*, *Ganga* and *Gomati*. In 1954, they constituted the 22nd Destroyer Squadron. In 1955, Nehru had a brainwave that because of the development of sophisticated mines, steel hulled ocean class minesweepers had become outdated. On 25 May, the same year, the Defence Ministry was informed that India probably did not require these sorts of vessels. So, in 1956, four coastal minesweepers were acquired and constituted the 149th Minesweeping Squadron. These minesweepers were designed for clearing mines laid down by the enemy along the coast and near the mouths of the harbours.[51]

Lord Mountbatten convinced Nehru that a small aircraft carrier would be of some aid to the Indian Navy and the CNS of India was also eager to acquire an aircraft carrier. Nehru referred the issue to Blackett who had already told Nehru that India did not require even cruisers and that destroyers were adequate for India's need. Despite Blackett's advice, India acquired a second cruiser from Britain in 1954. It joined the Indian Navy in 1957 and was named *Mysore.* Ultimately Mountbatten was able to persuade the Indian Prime Minister that an aircraft carrier would be more useful than a cruiser. In Mountbatten's view the aircraft carrier would not result in India having a blue water navy but would be a component of her coastal navy useful in protecting India's long coastline. As construction of airfields all along the coast would be very costly, the aircraft carrier could function as a mobile airfield. In addition, the helicopters in the aircraft carrier would be of much use in anti-submarine warfare. On 30 January 1955, Nehru informed the Defence Ministry that instead of another cruiser India should acquire an aircraft carrier. However, as India could not afford a new aircraft carrier an old British carrier would suit her well.[52] A similar line was pursued by the Indian civilian analyst Panikkar who claimed that India needs aircraft carriers not for projecting power but for guarding the coast.[53] Nehru was clearly torn between two policies. In a note addressed to the Defence Secretary dated 8 January 1955, Nehru had written that aircraft carriers were only necessary in case of global warfare involving thermonuclear weapons. And India had no wish to join any side in such a conflict. Nehru continued that it would have been better if the money was spent on industrial development instead of an aircraft carrier.[54] But, finally in 1957, India bought the British aircraft carrier named *Hercules* which was modernized, refitted and commissioned as *Vikrant* in 1961.[55]

One day in 1962, Admiral Katari was summoned to the office of the Defence Minister. A very agitated Krishna Menon said: 'Admiral I want the Navy to take over Anjidev Island.' Anjidev, a small island off Goa, was at that time under Portugal. The Portuguese troops in the island used to shoot at Indian coastal steamers.[56] When one of the Indian sailors was hit, there was an uproar in the Parliament. So, Menon decided to act. Katari told Menon that it would be simpler and less expensive in men and materials to attack Goa from the landward side. And with the fall of Goa, Anjidev would also capitulate. Assaulting a defended island from the sea, said Katari, might prove to be an expensive operation. Probably, at that time, the Indian Navy lacked adequate training for conducting amphibious operations against enemy-defended shores. Hence, Katari's uneasiness at Menon's proposal. Menon replied: 'Oh, yes, I see. I shall consult the Prime Minister, but in the meantime make your plans for taking Anjidev.'[57]

In 1962, the navy got only 4.7 per cent of the defence budget.[58] Katari explains the step-motherly attitude of the government towards the navy in the following words:

> To revert to our own situation, Pakistan's hostile noises meant that at least the Army and the Air Force should be maintained at adequate strength to meet any threat from that quarter. This inevitably meant that the Navy received a very tiny piece of the already small financial cake that was set apart for defence. While one could not legitimately quarrel with this, I sometimes wished that my colleagues in the other services could have tempered their understandable desire to take advantage of the situation with the exercise of their undoubted awareness of the need to take a balanced, global view of our national interest. The result of this enforced circumscribed outlook was that we in the Navy were not in a position to hold out any tangible prospects of developing the service into a worthwhile force in the foreseeable future.[59]

Nearer home the situation eased a bit when Britain vacated its naval and air facilities in Sri Lanka in 1956.[60] The Pakistan Navy was also going through troubled times. In 1959, Vice-Admiral Choudri resigned after the government's refusal to expand the Pakistan Navy. In 1953, Britain turned down Pakistan Navy's request for submarines. The Pakistan Navy like the Indian Navy also emerged as an anti-submarine force because USA conceived it as an auxiliary instrument to combat Soviet submarines in case of a general war. However, the situation changed in the next decade. Thanks to the USA, the Pakistan Navy acquired its first submarine in 1963.[61] The Indian Navy demanded submarines not only for attacking enemy ships but also for acting as targets for anti-submarine ships conducting exercises. In 1959, the Indian Navy sent a proposal to the government demanding three submarines. In 1964, an inter-service defence delegation headed by Defence Minister Y.V. Chavan went to the USA where the Johnson administration poured cold water over the demands of the Indian Navy. The Americans told the Indian team that since Britain was the traditional supplier of naval materials to India,

New Delhi should approach London, but Britain was not eager to supply India with submarines. By mid-1965, the USSR was keen to supply India with warships. Relations between Moscow and Beijing was at that time at an all time low, and the Soviets concluded that a powerful Indian Navy could contain a hostile China.[62]

During the 1965 India–Pakistan War, the government forced upon the Indian Navy a passive strategy. The navy's role was to protect the tankers bringing oil to India from the Persian Gulf. Only if enemy ships approached within 12 nautical miles of India's coast, was the navy to go into action. The Indian Navy was not allowed to proceed north of Porbandar. Permission to capture Pakistani naval vessels and naval blockade of Karachi harbour were not allowed. The strategic mission of the Pakistan Navy was to defend the port of Karachi and to protect the coastline of Pakistan against amphibious assaults. In addition the Pakistan Navy was also tasked with keeping Islamabad's SLOCs open. In September 1965, in the Arabian Sea the Indian Navy deployed one cruiser, four destroyers and seven frigates under Fleet Commander Rear Admiral B.A. Samson. Pakistan responded by concentrating one cruiser, five destroyers, one frigate, one submarine and one tanker under Commodore S.M. Anwar. Unfortunately, the Indian aircraft carrier *Vikrant* and the cruiser *Delhi* were undergoing refit at Bombay. Otherwise, the Indian Navy would have been in a far superior position vis-à-vis its antagonist in the Arabian Sea. In fact, out of the twenty-three ships (one aircraft carrier, two cruisers, nineteen destroyers and frigates and one tanker) of the Indian Navy, ten ships were undergoing refits. The Pakistan Navy sent a naval surface raiding force against the Indian ports and installations on the western seacoast. The Indian Navy was literally caught with its pants down and the Indian surface crafts failed to intercept the raiders. The Pakistani bombardment of Dwarka in Gujarat on the night of 7 September did not result in any great material loss for India but certainly resulted in a psychological setback.[63]

In addition to the government's passive doctrine there was another reason behind the Indian Navy's lack of aggressive stance. The Indian Navy also perceived a threat from Indonesia in the Bay of Bengal. The Indonesian Navy had a nebulous plan of capturing the Andaman and Nicobar Islands while the

Pakistanis engaged the Indians. Indonesia transferred two Whisky class submarines and four Komar class missile boats to Pakistan which it had received earlier from the USSR. Each Komar class small missile boat with a displacement of 70 tons and powered by four diesel engine was armed with surface to surface N-2A missiles. A Whisky class submarine powered by diesel-electric motors and armed with eighteen torpedoes or forty mines and a crew of sixty could move at fifteen knots under water. Overall the 1965 War in the sea was limited with neither Pakistani submarines nor the Indian ships attacking the merchant shipping.[64] But, this would change in the next round of war.

After 1965, India embarked on a sea control strategy against its local adversary Pakistan while the latter with the aid of its submarines pursued sea denial strategy. Between 1962–71, the Indian Navy was given lower priority within the defence budget compared to the navy's share chalked out by Auchinleck's re-organization committee just before Independence. The navy got only 4 per cent of the defence budget between 1962–71. Nevertheless, the Indian Navy acquired ships from the Soviet arsenal before the 1971 War. Between 1968 and 1970, four Foxtrot class submarines were given to India. Each of these diesel submarines, with a crew of seventy and armed with twenty torpedoes,[65] could cruise at fifteen knots while submerged.

The Indian Navy's 'finest hour' came during the 1971 India–Pakistan war. The Pakistan Navy comprised of six destroyers, three frigates, four submarines, one tanker, one salvage tug and eight minesweepers. The Indian prime minister encouraged the CNS Admiral S.M. Nanda to take an aggressive posture. The Indian Naval Fleet was divided into the Western Fleet and the Eastern Fleet. The Western Fleet under Rear Admiral E.C. Kuruvilla comprised of one cruiser, one destroyer, eight frigates, two submarines, three patrol vessels and eight Osa class missile boats imported from the USSR. The Osa boats were bigger than Komar class boats and each had a displacement of 165 tons and was only 128 ft. long armed with surface to surface missiles, and capable of all weather operations. The Eastern Fleet under Rear Admiral S.H. Sarma comprised of one aircraft carrier, one destroyer, two frigates, one submarine, three LSTs and seven patrol vessels.[66]

During the 1971 War, the Indian Navy achieved the twin

objectives of gaining supremacy in the Arabian Sea and the Bay of Bengal. India carried out missile attacks with Osa boats on Karachi and its aircraft carrier blockaded East Pakistan. The Osa class boats turned Karachi into a burning hell—the Indian Navy was taking revenge for the sneaking raid of Pakistani raiders in Gujarat during the last war.[67] Small missile boats were emerging as a lethal weapon system in naval warfare even outside South Asia. In 1965, the Israeli destroyer *Eilat* was destroyed by a missile-armed fast-attack craft (Komar class) of the Egyptian Navy,[68] and the Royal Navy concluded that helicopters armed with anti-tank missiles could function as potent killers of the fast moving missile-armed crafts.[69]

The Indian Air Force established complete air superiority over East Pakistan and the Bay of Bengal. In a situation which would have pleased the Japanese Admiral Chuichi Nagumo, the Indian aircraft carrier *Vikrant* carried out continuous air strikes against Pakistani military assets. As the *Vikrant* sailed towards East Pakistan, the Commander-in-Chief of the Eastern Fleet, Vice-Admiral S.H. Sarma, evaluated the threat that could be posed by the Pakistan Air Force and the Pakistan Navy against his armada. The Pakistani submarine *Ghazi* (a gift from the USA) under Commander Zafar Muhammad Khan could pose a deadly threat to the *Vikrant*, and the *Ghazi* was operating somewhere in the Bay of Bengal. Luckily for the Indian Navy, *Ghazi* was sunk accidentally on the night of 3/4 December. Human errors on part of the *Ghazi's* crew were probably responsible for its unworthy demise. Sarma and India's naval command feared that the Sabre jets stationed at Dacca might resort to a 'do or die mission' against the old *Vikrant*. Hence, Admiral Sarma was ordered to keep his carrier about a 100 miles away from the coast of East Pakistan. But, this meant that he could not carry out air strikes against the coastal targets in East Pakistan because the range of the Sea Hawks that could be launched from the *Vikrant* was around a 100 miles. Sarma took a courageous decision to move his carrier within 55 to 60 miles of East Pakistan's coastline. Round the clock strikes from the *Vikrant* by the Alizes and Sea Hawks equipped with rockets and 500 lb. bombs between 6 to 14 December 1971, was a classic case of a aircraft carrier projecting power in the littoral. The Alizes struck targets even during the night. Aircraft from the *Vikrant* bombed Chittagong cantonment area, airfield, ordnance factory, Cox Bazar and merchant shipping along the various rivers and creeks. The air strikes from the *Vikrant* were able to sink more than 57,000 tons of Pakistani merchant shipping.[70]

December 10 was an anxious day for Sarma. He received intelligence report that elements of the US Seventh Fleet under the American Admiral Elmo Zumwalt was entering the Bay of Bengal through the Malacca Straits.[71] This was done at the behest of Henry Kissinger who wanted to provide support to Pakistan, a US client. At that time the USA was fighting in Vietnam and these ships were detached from the American naval armada concentrated at the Gulf of Tonkin Bay. The American armada included the *Enterprise*, a nuclear-powered aircraft carrier equipped with hundred aircraft and four guided missile destroyers. The *Enterprise* weighed 75,000 tons, while the *Vikrant* weighed only 16,000 tons and was leaking badly.[72] Sarma did not get much help from Admiral Nanda. The latter's instructions to Sarma was that if he encountered American ships then the American crew were to be invited aboard for a drink.[73] But, Sarma was well aware that the nuclear powered American aircraft carrier with the guided missile destroyers had not sailed all the way from South-East Asia just for having a drink with the Indian sailors. Sarma was worried about what would happen if the American nuclear aircraft carrier launched an attack against him. Meanwhile, the submarine *Khanderi* which was patrolling southwards reported that the US armada was only two hours flight time away from the Indian carrier group. Sarma though anxious was ready to fight in the event the worse-possible-scenario became a reality. He writes in his memoirs:

> Anything was possible in war. I had not personally talked to our pilots, but some thoughts like sending in all our strike aircraft at one go, with the *Enterprise* as their prime target were playing in my mind. Not all of them would have got through, but the few that did might do a repeat of the *kamikaze* attacks on the British ships *Prince of Wales* and *Repulse*. No ship was unsinkable; I was full of confidence that if we asked for volunteers from our pilots, it would have been difficult to keep anyone back. These and other thoughts constantly recurred to me.[74]

The point to be remembered is that the two British ships *Prince of Wales* and *Repulse* were not sunk by *kamikaze* attacks. The land-based Japanese torpedo planes and bombers sank these two ships near the coast of Malaysia in 1941. Both British capital ships were operating without any air cover at a time when the Japanese war machine had gained complete air supremacy along South-East Asia.[75] Sarma's forlorn hope was that of a desperate man. Even if a handful of pilots from the *Vikrant* attempted an attack on the *Enterprise*, they would not have got anywhere near the US carrier and would have been knocked out of the air as the Japanese aircraft were in the Mariana campaign.[76] The *Enterprise's* combat air patrol would have picked up each and every Indian aircraft and sent the whole of India's Eastern Fleet to the bottom of the sea within a single afternoon.

Thanks to the bilateral treaty between India and USSR, the entry of the Soviet ships on the scene saved the situation for the Indians. The Soviet warships had first entered the Indian Ocean in 1968. On 6/7 December 1971, a Task Group including a missile cruiser and its escorts sailed from Vladivostok. These ships were part of the Soviet Pacific Fleet. On 9 December, the Soviet Task Force was discovered by an American reconnaissance aircraft at the Tsushima Strait near Japan. Of course the Soviet Task Force was there by design, to be observed by the US and prevent anything nasty happening in the Bay of Bengal. On 16 December a Second Soviet Task Group comprising of a missile cruiser and its escort ships sailed from Vladivostok. On 17 December the First Soviet Task Group entered the Indian Ocean through the Malacca Straits in full view. On 18 December the Soviet Task Group shadowed the American Task Force in the Bay of Bengal. Probably a Soviet hunter-killer submarine was also tracking the American ships. In early January both the Soviet Task Groups and the American Task Force withdrew from the Indian Ocean.[77] The two superpowers neutralized each other—and the field was left to the two regional powers to settle their conflict.

One of the principal factors behind Pakistan's failure to defend East Pakistan was an inadequate navy. This was because Pakistan's army dominated polity had no appreciation of maritime strategy and East Pakistan was 2,600 nautical miles from West Pakistan.[78] About Ayub Khan, Asghar Khan had written: 'Born and brought up in the feudal environment of the north, the sea and maritime strategy found no place in his thinking.'[79] But the Indian Navy did not escape totally unscathed either. The 1,200 ton anti-submarine Indian frigate, *Khukri* was sunk by a

Pakistani submarine on the chilly night of 9 December in the Arabian Sea. Three torpedoes hit the frigate and it sunk in a few minutes taking down with her the eighteen officers present plus 176 shipmates and her captain, Mahendra Nath Mulla.[80] The Indian Navy had yet to learn effective anti-submarine warfare.

INDIA, 1972–2004: A REGIONAL NAVAL POWER

By 1971, the Royal Navy withdrew from the east of the Suez and the Indian Navy emerged as the strongest regional naval power in the Indian Ocean. The defence allocation to the Indian Navy increased from 3.7 per cent in 1964–5 to 7.8 per cent in 1970–1 and then to 9.7 per cent in 1973–4.[81] In the 1970s the Indian Navy's strength exceeded 55,000 personnel.[82] Geoffrey Till writes that in the 1990s, the Indian Navy emerged as a well-balanced force capable of offshore power projection. The Indian Navy represented a synergistic mix of platforms and weapon systems, which offered New Delhi a variety of responsive options.[83] However, Till's assertion the about Indian Navy's capability of projecting power even against its neighhouring countries is questionable.

During the 1970s and the 1980s, the Pakistan Navy was allocated about 7 per cent of the total defence budget. Towards the end of 1980s, the Pakistan Navy with a manpower of 16,000 maintained six submarines, eight destroyers and twenty-nine patrol and coast defence vessels but remained committed to the sea denial strategy. At present nine submarines constitute the core of the Pakistan Navy. Though the Pakistan Navy is much smaller than the Indian Navy, the former poses still a credible threat to the latter. Pakistan's three French Agosta 90-B diesel electric submarines are reserved for a sea denial role. The American PC-3 Orion maritime strike aircraft enables the Pakistani naval air force to cover the entire west coast of India.[84] According to K.R. Singh, the Pakistan Navy in case of a future war with India would follow a defensive strategy based upon a series of concentric rings. The PC-3 Orion and the Atlantic maritime aircraft armed with anti-ship missiles has a range of 200 miles from the coast. These aircraft escorted by Mirages would constitute the outermost rings about 200 miles from the coast. The next ring would comprise of midget submarines armed with guided torpedoes and Sea King helicopters armed with Exocet anti-ship missiles. The third ring would comprise surface combatants armed with anti-ship missiles. And finally the land-based anti-ship missile would constitute the last ring.[85] In all probability a 1971-style raid on Karachi will not be possible for the Indian Navy in the near future given the recent Pakistani development. And the Falklands War of 1982 provides a pointer in this regard.

The Sea King anti-submarine warfare helicopters of the Royal Navy maintained a continuous anti-submarine defensive screen some dozen miles ahead of the British Task Force at Falklands. Yet the Task Force suffered enormous losses from the land-based Mirages of the Argentine Air Force in the Falklands War. The Falklands War proved that no naval force could operate near hostile offshore waters with any degree of security, even with a protective air umbrella.[86] Ashley Tellis rightly says that the Indian carrier group cannot operate with any security in Pakistan's coastal water as Pakistan during the late 1980s acquired three Atlantics, eight F-27s, four E-2C and six Orions. These aircraft provide sophisticated airborne early warning and battle management systems with lethal Exocet/Harpoon cruise missiles variety.[87]

The Indian Navy's strategic role has become crucial in the twenty-first century for it has to protect the SLOCs, maritime assets plus the island territories. It is the task of the Indian Navy to protect 2 million sq. km. of India's exclusive economic zone (EEZ) with all its under-water economic resources. In the near future, India could exploit nickel, copper, cobalt and manganese from its EEZ. India has made a huge investment in its offshore oil fields which account for over 60 per cent of natural gas for India. Another task of the Indian Navy is to protect India's vulnerable SLOCs with West Asia from Pakistani submarines and missile boats. Saudi Arabia, United Arab Emirates and Iran were the largest source of crude oil for India, accounting for 54 per cent of the country's oil imports in 1996. India's industrial infrastructure as well as the war machine are dependent on oil imported from West Asia in tankers. In 1998, out of 57 million tons of oil imported, about 54 million was imported through the Arabian Sea. In the coming decade, energy dependence on sea will increase from 80 to 90 per cent. And India cannot allow Pakistan to threaten its oil route as well as its oceanic trade. Sea-

In 1987, the Indian Navy acquired a Charlie-II type nuclear submarine from the USSR. The Charlie class submarine entered Soviet service way back in 1968. It was about 305 ft. long and with 100 crews was able to cruise at about thirty knots while submerged. In 1991, due to financial problems, the V.P. Singh government cancelled the programme of retaining the Russian nuclear submarine on lease.[99] The Indian *Ministry of Defence Report for 2000–2001* accepts that the Chinese nuclear submarines equipped with submarine launched ballistic missiles pose a threat to India[100] and some Indian analysts believe that the PLAN is equipped with tactical nuclear weapons. As early as 1974, the PLAN deployed a Han-class nuclear submarine. The Indian nuclear doctrine emphasizing limited deterrence with second strike capability requires sea-based nuclear deterrent.[101] Commander Vijay Sakhuja demands that India should maintain nuclear-powered submarines armed with nuclear warheads.[102] The indigenously manufactured nuclear submarine named Advanced Technology Vessel (ATV) is yet to be inducted for operational service. The ATV is based on Russia's Severodvinsk Class Type-885 attack submarine and the Russians are reportedly helping the Indians with fabrication of the hull.[103]

Since the Indian Navy unlike the pre-1942 Royal Navy lacks the power to implement sea control around the Malacca Straits, India is using the navy for maritime cooperation. The first bilateral exercise with the Republic of Singapore Navy ships was

TABLE 7.2: INDIA–PAKISTAN–CHINA NAVAL BALANCE 2003

Ships	Number of Ships of the Various Navies		
	Indian Navy	Pakistan Navy	Peoples Liberation Army's Navy
Submarines	16	7	66
Aircraft Carriers	1	0	0
Destroyers	8	3	22
Frigates	12	8	36
Corvettes	25	0	20
Fast Attack Missile Craft	2	9	83
Fleet Tanker	3	1	6
Maritime Patrol Aircraft	33	4	12

Source: Sandeep Unnithan, 'Ship Shape', *India Today*, 12 May 2003, pp. 50–1.

conducted in March 1999.[104] Both India and Singapore have a common security purpose to check the entry of PLAN ships through the Malacca Straits. India seems to be following Sanjay Singh and Panikkar's strategy of building naval–diplomatic alliances for deterring extraregional maritime presence.

The Indian Navy is also charged with the responsibility of conducting hydrographic survey and publication of hydrographic documents. The Indian Naval Hydrographic Office is located at Dehra Dun where the Survey of India is located as well. In collaboration with the Survey of India, the navy carries out the task of charting the seas, coastline and islands. Information about ocean depth at various points as well as the publications of nautical charts aids even the merchant marine. In 1993–4, the Indian Navy undertook hydrographic work for Oman.[105] Such activities not only lead to the establishment of better diplomatic relations with foreign countries but also increase the Indian Navy's knowledge regarding the water bodies surrounding foreign countries, a knowledge that would aid the Indian Navy in case of operational activities along such areas in the future.

LOW-INTENSITY THREATS

Gold and silver smuggling between the Persian Gulf and the western coast of India is a traditional activity. From the 1990s, transhipment of explosives and small arms also started. And over the years narco-terrorism through this route has increased. The vast unguarded coastline of India and innumerable small islands are an ideal place for the activities of Inter Service Intelligence sponsored terrorists. On 25 June 1999, a North Korean vessel was impounded by the Indian custom authorities at Kandla Port for carrying equipment for the production of tactical surface-to-surface missiles. The cargo comprised of 148 boxes that included parts of the guidance system, blueprints, drawings and instruction manuals. The owners of the vessel admitted that the consignment was to be off-loaded at Karachi. From 2000 onwards, the Indian Navy is also engaged in deterring contraband goods landing along the coast of Gujarat. For countering low-intensity maritime threats, a retired vice-admiral writes that what is required is greater jointness between the Indian Navy and the Coast Guards for conducting maritime policing. In October 2003, the Indian Naval Fast Attack Craft (INFAC) T-82 with a speed of forty knots was inducted into the Western Naval Command. This boat is used in counter-insurgency and anti-smuggling operations.[106]

The LTTE gun trade route passes through the Bay of Bengal and the Andaman Sea. The arms originating from Cambodia are loaded on small fishing trawlers in the port of Ranong in southern Thailand. The LTTE also uses merchant shipping for smuggling narcotics. The Sea Tigers (the maritime wing of the LTTE) have the capacity to destroy the small attack boats of the Sri Lankan Navy. In 1994, the Sri Lankan Navy lost a Jayasagar-class patrol vessel armed with two heavy machine-guns and two 25-mm. cannons to the LTTE. On 5 June 2000, the Sri Lankan Navy lost two Super Dovra-class fast attack craft with twenty-one personnel. In 1999, the Indian Navy's Eastern Fleet conducted exercise in the Bay of Bengal with the objective of increasing surveillance on the coastline of northern Sri Lanka. In 2000, Sri Lanka's Foreign Minister Lakshman Kadirgamar said that Delhi and Colombo were discussing the possibility of jointly patrolling the sea in the north and east of Sri Lanka to prevent the LTTE from bringing in arms by sea. In 2000–1, the Indian Navy patrolled the coast of Tamil Nadu to counter the activities of the terrorist groups engaged in gun running and narcotics smuggling. The insurgents of north-east India also import arms through the Bay of Bengal. During a forty-eight hour combined operation, code named Leech involving the army, the air force and the navy, a consignment of light weapons was seized in February 1998, in the Andaman Sea. The arms consignment was coming from South-East Asia and was heading for north-east India.[107]

Surveillance is also required to check illegal fishing by sailors of foreign nations. For surveillance of the region around Andaman and Nicobar Islands against poachers, the Indian Navy is stationing fast attack craft. Incidents of piracy have also increased in the eastern part of the Indian Ocean. Between 1991 and 1999, the number of armed robbery at sea had gone up from 107 to 285. And out of the 285 incidents in 1999, 173 cases occurred in South-East Asia and 43 in South Asia. About 210 incidents occurred in the area that concerns India. As India's east coast is becoming vulnerable to piracy, Indian naval ships and aircraft are now engaged in patrolling the coast of Orissa. In April 2004, a T-84 fast attack craft was inducted in the Indian Navy. Vice-Admiral O.P. Bansal, commanding the Eastern Naval Command said that this craft would undertake rapid reaction patrol along the eastern coast of India.[108]

CONCLUSION

To conclude, it must be said that maritime security was of marginal concern to the Indian strategic elites till recent times. Most of the theorization by the Indian naval analysts has been influenced by western theories. Because of Jawaharlal Nehru's Non-Aligned movement, India refused to get drawn under the protective umbrella of either the USA or the USSR. Hence, after Independence, India followed a strategy that may be termed as 'regional approach' to security affairs. By the 1970s, India's navy became a regional force through, it lacked the capacity to control the eastern entry points (Malacca Strait and Singapore) and the western entry points (Aden and Cape of Good Hope) of the Indian Ocean. The Indian Navy is now a battle-hardened force, and by post-World War II standards, the 1971 sea war between India and Pakistan was quite significant.

In recent times, the economic dimension of sea power is becoming important. And the navy in the coming decades also has to protect India's increasing non-military maritime commitments. At present, sea control beyond the subcontinent's coastal water is beyond the capacity of the Indian Navy. Further, India is helpless to prevent deployment of superpower naval units in the Indian Ocean. The Indian Navy also has to take up the probability of fighting an extra-regional naval power. This is because in the twenty-first century, India is faced with the threat of PLAN's entry into the Indian Ocean. Again there is a contradiction between the threats posed by the navies of hostile states and the maritime threats posed by non-state actors. While the response to the former threat requires heavy ships, the latter threat necessitates fast attack crafts. Whether India will be able to meet the Sino–Pakistani maritime challenge on the one hand and the burgeoning low-intensity maritime threats on the other is yet to be seen.

NOTES

1. Anil Kumar Singh, 'India's Maritime Security: Challenges Ahead', *Contemporary India*, vol. 2, no. 2 (2003), p. 45; John Keegan, *Intelligence in War: Knowledge of the Enemy from Napoleon to Al-Qaeda*, New York: Alfred A. Knopf, 2003, p. 130; Anand Mathur, 'Growing Importance of the Indian Ocean in the Post-Cold War Era and Its Implication for India', *Strategic Analysis*, vol. 26, no. 4 (2002), p. 550.
2. 'The Conception of Sea Power', Speech at a Congress Parliamentary Party Meeting, 5 April 1955, in Ravinder Kumar and H.Y. Sharada Prasad, eds., *Selected Works of Nehru*, Second Series, vol. 28, New Delhi: Jawaharlal Nehru Memorial Fund, 2001, pp. 525–6.
3. Colin S. Gray and Roger W. Barnett, 'Reflections', in Gray and Barnett, eds., *Seapower and Strategy*, London: Tri Service Press, 1989, p. 378.
4. Julian S. Corbett, *Some Principles of Maritime Strategy* (With an Introduction and Notes by Eric J. Grove), 1911; rpt., London: Brassey's, 1988, p. 15.
5. John W. Garver, *Protracted Contest: Sino-Indian Rivalry in the Twentieth Century*, New Delhi: Oxford University Press, 2001, p. 278.
6. Dean Mathew, 'Aircraft Carriers: An Indian Introspection', *Strategic Analysis*, vol. 23, no. 12 (2000), pp. 2154–7.
7. Lieutenant-Commander Narapati Datta, 'An Aircraft Carrier for the Indian Navy', *JUSII*, vol. LXXXVII, no. 367 (1957), p. 112.
8. Vice Admiral S.K. Chand, 'The Naval Traditions of India', in K.K.N. Kurup, ed., *India's Naval Traditions*, New Delhi: Northern Book Centre, 1997, p. 2.
9. P.K. Ghosh, 'Revisiting Gunboat Diplomacy: An Instrument of Threat or Use of Limited Naval Force', *Strategic Analysis*, vol. 24, no. 11 (2001), pp. 2006–7, 2015.
10. Margaret Tuttle Sprout, 'Mahan: Evangelist of Sea Power', in Edward Mead Earle, ed., *Makers of Modern Strategy: Military Thought from Machiavelli to Hitler*, 1943; rpt., Princeton: Princeton University Press, 1971, p. 418; Azar Gat, *The Development of Military Thought: The Nineteenth Century*, 1992; rpt., Oxford: Clarendon Press, 1999, p. 190.
11. Captain A.H. Chitnis and Commodore C.S. Patham, 'The Naval Tactics of Kunhali Marakkars and Their Relevance to Modern Warfare at Sea', in Kurup, ed., *India's Naval Traditions*, p. 35.
12. Theodore Ropp, 'Continental Doctrines of Sea Power', in Earle, ed., *Makers of Modern Strategy*, pp. 446, 448.
13. Gat, *Military Thought*, pp. 220–1.
14. Alexander Kiralfy, 'Japanese Naval Strategy', in Earle, ed., *Makers of Modern Strategy*, pp. 463, 466.

15. Mathew, 'Aircraft Carriers', pp. 2138–9.
16. Admiral S.M. Nanda, *The Man who Bombed Karachi: A Memoir*, New Delhi: HarperCollins, 2004, p. 291.
17. Commander V.A. Kamath, 'India and Sea Power', *JUSII*, vol. LXXXIV, no. 354 (1954), p. 82.
18. Commander Sanjay J. Singh, 'The Nature of War in the 21st Century', in Singh, ed., *Air Power and Joint Operations*, p. 100.
19. K.M. Panikkar, *Problems of Indian Defence*, Bombay: Asia Publishing House, 1960, pp. 106, 108, 110.
20. Gat, *Military Thought*, pp. 184–5.
21. Corbett, *Principles of Maritime Strategy*, p. 11. The three arms that Corbett has in mind are infantry, cavalry and artillery. Instead of cavalry, we could now add airforce.
22. Ibid., pp. 10–11.
23. Peter Lewis Young, 'Australia's Military Doctrine: Outlines of Changes', *IDR*, vol. 14, no. 2 (1999), p. 99.
24. Rear Admiral Raja Menon, *Maritime Strategy and Continental Wars*, London: Frank Cass, 1998, pp. 184, 187, 190.
25. Ibid., pp. 185, 191.
26. Panikkar, *Indian Defence*, p. 114.
27. Rahul Roy-Chaudhury, *India's Maritime Security*, New Delhi: Knowledge World, 2000, p. 186.
28. Vice-Admiral P.S. Das, 'A View from the Sea', in Singh, ed., *Air Power and Joint Operations*, pp. 240–1.
29. S.N. Kohli, *Sea Power and the Indian Ocean*, New Delhi: Tata McGraw-Hill, 1978, p. 23.
30. Kohli, *Sea Power*, pp. 23–4.
31. Gat, *Military Thought*, p. 184.
32. Kohli, *Sea Power*, p. 24.
33. Sprout, 'Mahan', in Earle, ed., *Makers of Modern Strategy*, pp. 419–20.
34. Kohli, *Sea Power*, p. 25.
35. Gray and Barnett, 'Reflections', in Gray and Barnett, eds., *Seapower and Strategy*, p. 380.
36. Gat, *Military Thought*, pp. 188–90.
37. Philip K. Kain, 'Kant's Political Theory and Philosophy of History', *CLIO*, vol. 18, no. 4 (1989), p. 328.
38. James John Tritten, 'Is Naval Warfare Unique?', *JSS*, vol. 12, no. 4 (1989), pp. 494, 496.
39. Field Marshal Wavell to Leo Amery, Telegram, no. 69, 26 October 1944, in Nicholas Mansergh, Editor in Chief and Penderel Moon, Asst. Editor, *The Transfer of Power*, vol. 5, *The Simla Conference: Background and Proceedings*, 1 September 1944–28 July 1945, London: Her Majesty's Stationery Office, 1974, p. 140; *Reorganization of the Army and Air*

Forces in India, Report of a Committee set up by the Commander-in-Chief in India, vol. 1, pp. xxiv, xxix, 355.00954, IND, 31893, Institute of Defence Studies and Analyses Library, New Delhi.

40. Vice-Admiral R.B. Suri, 'The Pakistani Navy: Force Level Options', in Vice-Admiral K.K. Nayyar, ed., *Pakistan at the Crossroads*, New Delhi: Rupa, 2003, pp. 324–5; Commander K. Sridharan, *A Maritime History of India*, Delhi: Publications Division, 1965, pp. 111–12; Kenneth McPherson and Peter Reeves, 'The Foundations of Indian Naval Power', in Robert H. Bruce, ed., *The Modern Indian Navy and the Indian Ocean*, Perth: Centre for Indian Ocean Regional Studies, Curtin University of Technology, 1989, p. 88; Vijay Sakhuja, 'Pakistan's Naval Strategy: Past and Future', *Strategic Analysis*, vol. 26, no. 4 (2002), p. 494; Pervaiz Iqbal Cheema, *The Armed Forces of Pakistan*, 2002; rpt., Karachi: Oxford University Press, 2003, pp. 90–1.
41. N.M.L. Saksena, 'Major Naval Operations since Independence', in Major-General Afsir Karim, ed., *The Indian Armed Forces: A Basic Guide*, New Delhi: Lancer, 1995, p. 150; Admiral R.D. Katari, *A Sailor Remembers*, New Delhi: Vikas, 1982, p. 52.
42. Rahul Roy-Chaudhury, *Sea Power and Indian Security*, London/ Washington: Brassey's, 1995, pp. 40–1, 44–7; A.K. Chatterji, *Indian Navy's Submarine Arm*, New Delhi: Birla Institute of Scientific Research, 1982, p. 36; Lorne J. Kavic, *India's Quest for Security: Defence Policies, 1947-65*, Berkeley/Los Angelas: University of California Press, 1967, p. 117; Oral Transcript of Interview with Prof. P.M.S. Blackett, p. 3.
43. Ravinder Kumar and H.Y. Sharada Prasad (eds.), *Selected Works of Jawaharlal Nehru*, Second Series, vol. 25 (1 February 1954–31 May 1954), New Delhi: Jawaharlal Nehru Memorial Fund, 1999, 302–3.
44. Ibid., p. 307.
45. Ibid.
46. Kumar and Prasad, eds., *Selected Works of Nehru*, Second Series, vol. 28, p. 530.
47. Captain John E. Moore, *The Soviet Navy Today*, London: Macdonald and Jane's, 1975, p. 104; Kavic, *India's Quest for Security*, pp. 124–5.
48. Vijay Sakhuja, 'Dragon's Dragonfly: The Chinese Aircraft Carrier', *Strategic Analysis*, vol. 24, no. 7 (2000), p. 1377.
49. Hugh Lyon, 'China's Navy for Coastal Defence Only', in Ray Bonds, ed., *The Chinese War Machine: A Technical Analysis of the Strategy and Weapons of the People's Republic of China*, London: Salamander, 1979, p. 150.
50. Desmond Wettern, 'The Navy Post-1945', in Antony Preston, ed., *History of the Royal Navy in the 20th Century*, London: Bison, 1987, pp. 168–9.
51. Commodore N.M.L. Saksena, 'Role of the Indian Navy', in Karim, ed., *Indian Armed Forces*, p. 115; Sridharan, *Maritime History of India*,

pp. 113–14; Kumar and Prasad, eds., *Selected Works of Nehru*, Second Series, vol. 28, p. 530; Katari, *A Sailor Remembers*, p. 53.

52. Ravinder Kumar and H.Y. Sharada Prasad, eds., *Selected Works of Jawaharlal Nehru*, Second Series, vol. 27 (1 October 1954–31 January 1955), New Delhi: Jawaharlal Nehru Memorial Fund, 2000, pp. 495–6.
53. Panikkar, *Indian Defence*, p. 113.
54. Kumar and Prasad, eds., *Selected Works of Nehru*, vol. 27, p. 495.
55. Joel Larus, 'India and Its Ocean: The Atypical Relationship Ends', in Bruce, ed., *Modern Indian Navy and the Indian Ocean*, p. 65.
56. Katari, *A Sailor Remembers*, p. 111.
57. Ibid., p. 112.
58. Larus, 'India and Its Ocean', in Bruce, ed., *Modern Indian Navy and the Indian Ocean*, p. 65.
59. Katari, *A Sailor Remembers*, p. 96.
60. McPherson and Reeves, 'Foundations of Indian Naval Power', in Bruce, ed., *Modern Indian Navy and the Indian Ocean*, p. 86.
61. Suri, 'The Pakistani Navy', in Nayyar, ed., *Pakistan at the Crossroads*, pp. 325–6.
62. Chatterji, *Indian Navy's Submarine Arm*, pp. 37–8, 41–2, 44.
63. Vice-Admiral Mihir K. Roy, *War in the Indian Ocean*, New Delhi: Lancer, 1995, pp. 82–4; Ashley J. Tellis, 'Securing the Barrack: The Logic, Structure and Objectives of India's Naval Expansion', in Bruce, ed., *Modern Indian Navy and the Indian Ocean*, p. 13; Vice-Admiral G.M. Hiranandani, *Transition to Triumph: History of the Indian Navy, 1965–1975*, New Delhi: Lancer, 2000, pp. 31, 33; Saksena, 'Major Naval Operations', in Karim, ed., *Indian Armed Forces*, p. 157.
64. Hiranandani, *Transition to Triumph*, p. 54; Moore, *Soviet Navy*, pp. 90–2; Bill Gunston, 'Soviet Warships', in Ray Bonds, ed., *The Soviet War Machine: An Encyclopedia of Russian Military Equipment and Strategy*, 1976; rpt., London: Salamander, 1977, p. 146; Roy, *War in the Indian Ocean*, p. 82.
65. Moore, *Soviet Navy*, p. 88; Jasjit Singh, 'Foreword', in Roy-Chaudhury, *Maritime Security*, p. xvi; K.R. Singh, *Navies of South Asia*, New Delhi: Rupa, 2002, pp. 2–3.
66. Nanda, *A Memoir*, pp. 207, 306–9; Gunston, 'Soviet Warships', in Bonds, ed., *Soviet War Machine*, p. 145.
67. Tellis, 'Securing the Barrack', in Bruce, ed., *Modern Indian Navy and the Indian Ocean*, p. 13; Eric Groove, 'Maritime Forces and Stability in South Asia', in Eric Arnett, ed., *Military Capacity and the Risk of War: China, India, Pakistan and Iran*, Oxford: Oxford University Press, 1997, p. 299.
68. Wettern, 'The Navy Post-1945', in Preston, ed., *History of the Royal Navy*, p. 182; Gunston, 'Soviet Warships', in Bonds, ed., *Soviet War Machine*, p. 146.

69. Wettern, 'The Navy Post-1945', in Preston, ed., *History of the Royal Navy*, p. 182.
70. Hiranandani, *Transition to Triumph*, pp. 131–40; Sarma, *My Years at Sea*, pp. 172, 177–8; Nanda, *A Memoir*, p. 309.
71. Vice-Admiral S.H. Sarma, *My Years at Sea*, New Delhi: Lancer, 2001, p. 178. Nanda in his memoir gives the date as 12/13 December. Nanda, *A Memoir*, pp. 242–3
72. Sarma, *My Years at Sea*, p. 178.
73. Roy, *War in the Indian Ocean*, p. 213.
74. Sarma, *My Years at Sea*, pp. 179–80.
75. Major-General, J.F.C. Fuller, *The Second World War: 1939-45, A Strategical and Tactical History*, 1954; rpt., New York: Da Capo, 1993, pp. 140–1.
76. In June 1944 the United States Navy's carrier borne aviation wiped out the qualitatively and quantitatively inferior Japanese carrier borne aviation in the Mariana campaign. This aerial campaign constituted the opening gambit of the Battle of Philippine Sea. According to one American pilot, the Japanese aircraft 'fell like leaves'. Christy Campbell, *Air War Pacific: The Fight for Supremacy in the Far East, 1937–45*, London: Hamlyn, 1991, pp. 101–5.
77. Hiranandani, *Transition to Triumph*, pp. 162–5; Moore, *Soviet Navy*, p. 39; Bharat Karnad, *Nuclear Weapons and Indian Security: The Realist Foundations of Strategy*, Delhi: Macmillan, 2002, p. 303; Garver, *Protracted Contest*, p. 277.
78. Sakhuja, 'Pakistan's Naval Strategy', p. 493.
79. Mohammad Ashgar Khan, *The First Round: Indo-Pakistan War, 1965*, Ghaziabad: Vikas, 1979, p. 4.
80. J.S.B. Arora, *War with Pakistan: 1971*, Delhi: Army Educational Stores, n.d., p. 76.
81. Raju G.C. Thomas, 'The Sources of Indian Naval Expansion', in Bruce, ed., *Modern Indian Navy and the Indian Ocean*, pp. 97–8.
82. Cheema, *Armed Forces of Pakistan*, p. 88.
83. Geoffrey Till, 'Maritime Strategy and the Twenty-First Century', *JSS*, vol. 17, no. 1 (1994), p. 185.
84. Rahul Roy-Chaudhury, 'India's Maritime Challenges in the Early 21st Century', *IDR*, vol. 14, no. 2 (1999), p. 93; Cheema, *Armed Forces of Pakistan*, p. 87; Rasul B. Rais, 'Indian Naval Developments: Implications for Pakistan', in Bruce, ed., *Modern Indian Navy and the Indian Ocean*, pp. 122, 127.
85. K.R. Singh, *Navies of South Asia*, pp. 424–5.
86. Anthony J. Watts, 'The Falklands War', in Preston, ed., *History of the Royal Navy*, pp. 197, 203.
87. Ashley J. Tellis, 'Aircraft Carriers and the Indian Navy: Assessing the Present, Discerning the Future', *JSS*, vol. 10, no. 2 (1987), p. 144.

88. *MODAR: 1989-90*, p. 17; *MODAR: 1993–94*, p. 13; Lieutenant-General R.K. Jasbir Singh, ed., *Indian Defence Yearbook 2003*, Dehra Dun: Natraj, 2003, p. 144; Roy-Chaudhury, *Maritime Security*, pp. 1, 4, 13; Rahul Roy-Chaudhury, 'The Limits to Naval Expansion', in Kanti P. Bajpai and Amitabh Mattoo, eds., *Securing India: Strategic Thought and Practice*, New Delhi: Manohar, 1996, p. 192; Vice Admiral K.K. Nayyar et al., *National Security: Military Aspects*, New Delhi: Rupa, 2003, p. 181; Raja Menon, 'Maritime Strategy for India', in J. Baranwal, ed., *SP's Military Yearbook: 1998–99*, New Delhi: Guide Publications, 1998, p. 20; Anil Kumar Singh, 'India's Maritime Security', p. 46.
89. Vice-Admiral Arun Prakash, 'Evolution of the Joint Andaman and Nicobar Command (ANC) and Defence of Our Island Territories', Part II, *JUSII*, vol. CXXXIII, no. 551 (2003), pp. 25, 31; Sandy Gordon, 'Conclusion', in idem and Ross Babbage, eds., *India's Strategic Future: Regional State or Global Power?* Delhi: ford University Press, 1992, pp. 173–4; Roy-Chaudhury, 'India's Maritime Challenges', p. 93; Mathur, 'Growing Importance of the Indian Ocean', p. 552.
90. Donald L. Berlin, 'The Indian Ocean and the Second Nuclear Age', *Orbis*, vol. 48, no. 1 (2004), p. 58; K.R. Singh, *Navies of South Asia*, p. 414; Bill Gunston, 'Soviet Aircraft' in Bonds, ed., *Soviet War Machine*, p. 90; Sandeep Unnithan, 'Indian Navy Looks to Russia for Successor to Sea Eagle', *JDW*, vol. 33, no. 22, 31 May 2000, p. 15.
91. Ross Babbage, 'India's Strategic Development: Issues for the Western Powers', in idem and Gordon, eds., *India's Strategic Future*, pp. 155–6; Jaswant Singh, ed., *Indian Armed Forces Year book: 1981–82*, Bombay, n.d., p. 613; *MODAR: 1974–75*, pp. 2–3; Nayyar et al., *National Security*, p. 13; Garver, *Protracted Contest*, p. 277; Steven Ekovich, 'Iran and New Threats in the Persian Gulf and Middle East', *Orbis*, vol. 48, no. 1 (2004), p. 72.
92. Vijay Sakhuja, 'Maritime Power of People's Republic of China: The Economic Dimension', *Strategic Analysis*, vol. 24, no. 11 (2001), pp. 2023, 2028; Garver *Protracted Contest*, p. 276; R.K. Jasbir Singh, ed., *Indian Defence Yearbook 2003*, p. 145.
93. Srikanth Kondapalli, 'China's Naval Strategy', *Strategic Analysis*, vol. 23, no. 12 (2000), pp. 2040, 2055; Srikanth Kondapalli, 'China's Naval Equipment Acquisition', *Strategic Analysis*, vol. 23, no. 9 (1999), pp. 1516–17; Srikanth Kondapalli, 'China's Naval Structure and Dynamics', *Strategic Analysis*, vol. 23, no. 7 (1999), pp. 1097, 1099; Srikanth Kondapalli, 'China's Naval Training Programme', *Strategic Analysis*, vol. 23, no. 8 (1999), pp. 1343, 1345, 1347, 1349; Srikanth Kondapalli, 'Chinese Navy's Political Work and Personnel', *Strategic Analysis*, vol. 23, no. 10 (2000), p. 1757; Yiong Zhang, 'Beijing Develops New Radar-absorbing Materials', *JDW*, vol. 31, no. 8, 24 February 1999, p. 3; Vice-Admiral Mihir Kumar Roy, 'Asymmetry of India's Defence

Forces: A Sailor's view', *Strategic Analysis*, vol. 13, no. 3 (1990), p. 241; Swaran Singh, 'Continuity and Change in China's Maritime Strategy', *Strategic Analysis*, vol. 23, no. 9 (1999), p. 1498; Ken Gause, 'Profile of the PLA Navy', in Baranwal, ed., *SP's Military Yearbook: 1998–99*, pp. 29–30.

94. K.R. Singh, *Navies of South Asia*, p. 417.
95. Jaswant Singh, *Defending India*, Chennai: Macmillan, 1999, p. 251.
96. Nayyar et al., *National Security*, p. 191.
97. Tellis, 'Aircraft Carriers and the Indian Navy', p. 143.
98. Sakhuja, 'Dragon's Dragonfly', p. 1370; Swaran Singh, China's Maritime Strategy', p. 1502; Ross Babbage, 'Introduction', in Babbage and Gordon, eds., *India's Strategic Future*, p. 1; John Pay, 'Full Circle: The US Navy and its Carriers, 1974–93', *JSS*, vol. 17, no. 1 (1994), pp. 124–5; Sandeep Unnithan, 'Ship Shape', *India Today*, 12 May 2003, p. 50; Rahul Bedi, 'India's Air-defence Ship Plan', *JDW*, vol. 31, no. 11, 17 March 1999, p. 15; Garver, *Protracted Contest*, p. 278.
99. Bharat Verma, 'MAKS 2001 International Air Show: Moscow on the Comeback Trail', *IDR*, vol. 16, no. 3 (2001), p. 60; Moore, *Soviet Navy*, pp. 78–9; Manoj K. Joshi, 'Directions in India's Defence and Security Policies', in Babbage and Gordon, eds., *India's Strategic Future*, p. 69.
100. *MODAR: 2000–2001*, p. 3.
101. Matin Zuberi, 'The Proposed Indian Nuclear Doctrine', *Contemporary India*, vol. 1, no. 1 (2002), pp. 61–2; Colonel William V. Kennedy, 'The Defence of China's Homeland', in Bonds, ed., *The Chinese War Machine*, p. 119; M.S. Mamik, 'Tactical Nuclear Weapons at Sea: Are They Essential?', *Strategic Analysis*, vol. 13, no. 3 (1990), p. 246.
102. Vijay Sakhuja, 'Sea Based Deterrence and Indian Security', *Strategic Analysis*, vol. 25, no. 1 (2001), p. 22.
103. Berlin, 'The Indian Ocean', p. 59; Karnad, *Nuclear Weapons and Indian Security*, p. 656.
104. *MODAR: 1999–2000*, p. 29.
105. *MODAR: 1993–94*, p. 13; Saksena, 'Role of the Indian Navy', pp. 118, 140 and N.H.L. Saksena, 'Manpower Recruitment and Branches of the Navy', in Karim, ed., *Indian Armed Forces*, pp. 118, 140.
106. 'Israel Ship Boost to Fleet', *Telegraph*, 10 October 2003, p. 8; Das, 'A View from the Sea', in Singh, ed., *Air Power and Joint Operations*, p. 243; *MODAR: 2000–2001*, pp. 28–9; Vijay Sakhuja, 'Indian Ocean and the Safety of Sea Lines of Communication', *Strategic Analysis*, vol. 25, no. 5 (2001), p. 700; Nayyar et al., *National Security*, p. 182.
107. G.V.C. Naidu, 'India and South-East Asia: An Activist Role for Indian Navy', *Journal of Indian Ocean Studies*, vol. 11, no. 2 (2003), p. 199; *MODAR: 2000–2001*, p. 28; Rahul Bedi, 'Indo-Sri Lanka Naval Exercise', *JDW*, vol. 33, no. 21, 24 May 2000, p. 3 and 'Carnage in Sri Lanka', *JDW*, vol. 33, no. 24, 14 June 2000, p. 12; K.R. Singh, *Navies of*

South Asia, p. 438; Rohan Gunaratna, 'Transnational Terrorism: Support Networks & Trends', in K.P.S. Gill and Ajai Shani, eds., *FAULTLINES: Writings on Conflict & Resolution*, New Delhi: Bulwark Books and the Institute for Conflict Management, 2000, p. 9; Sakhuja, 'Sea Lines of Communication', p. 694.

108. 'Fast Attack Craft Joins fleet', *Telegraph*, 20 April 2004, p. 5; *MODAR: 1997–98*, p. 25; K.R. Singh, 'Regional Cooperation in the Bay of Bengal: Non-Conventional Threats: Maritime Dimension', *Strategic Analysis*, vol. 24, no. 12 (2001), pp. 2200–1; *MODAR: 2000–2001*, p. 30.

EIGHT

Defence Industries of India

Effective integration of innovative technology with the military organization is not merely a matter of hardware. In contrast to the assertion of the technological and economic determinists, geo-economics, geo-strategy, societal fissures and the nature of polity also shape a country's ability to eliminate the technology gap. Weapons production implies collaboration between the military, scientists and the industry. The working of this alliance can be categorized as the scientific–military–industrial cartel.[1] The military–industrial complex first emerged in late-nineteenth century Europe and reached its apogee in the age of globalization. Globalization may be referred to as the worldwide intensification of relations. The Communications Revolution in the last two decades of the twentieth century has further accentuated globalization. The strengthening of the inter-state affairs due to the globalizing process has increased the reach of innovative technology. Thanks to rapid advancement in the information and transportation technologies, power projection by a polity with the aid of 'wonder weapons' has become more potent.

In the post-colonial age, technological innovations remain the only way to provide a high standard of living to the populace. Michael Adas' assertion that due to differing local conditions appropriate technologies which could function well in the context of the Third World countries might be different from those pursued by the West is questionable.[2] And Itty Abraham's underhand criticism that independent India's pursuit of nuclear technology is a failed race on part of the post-colonial polity for acquiring post-colonial modernity,[3] is part of the post-modernist critique against modernization. However, there is no way out except modernization and concomitant industrialization. To a great extent, the necessity of Western technology is universal.

This chapter attempts to show how far post-colonial India till the new millennia has been able to emanicipate itself from the technology trap. The focus is on military hardware whose import constitutes the biggest burden on a developing polity's exchequer.

MONEY AND THE POLITICS OF TECHNOLOGY TRANSFER

War is a matter not so much of arms as of money, which makes arms of use.

THUCYDIDES[4]

Along with the availability of cash, international politics determines the quantity and quality of weapons flowing to the non-western countries. Importing arms in order to maintain an aggressive defence posture has resulted in economic bankruptcy of certain Third World countries. The case of Pakistan illustrates this point. Defence expenditure in Pakistan consumes 55 per cent of the country's total revenue.[5] The Western firms have to be paid in foreign currency which remains scarce in a developing society. Buying foreign arms resulted in a huge external debt for Pakistan. During the 1950s, Pakistan spent an average of $40 million in foreign exchange annually for buying military equipment.[6] As a result, by 1987–8, its external debt as the share of GNP was 30.1 per cent. For the same period, debt service payment amounted to 2.8 per cent of the GNP. The only way out for Pakistan was to acquire funds from USA and in turn agreeing to become a client state of Washington. In 1987, USA agreed to provide US $4.02 billion as loan in return for Islamabad agreeing to become an ally in the USA's fight against the USSR in Afghanistan.[7]

For high technology military goods, too, the countries in the subcontinent remain dependent on the policies of the big powers. In 1951, Britain was not eager to supply Goblin engines for the Vampire aircraft bought by India from the former as it was trying to force India to reduce tension with Pakistan.[8] In Pakistan, the American arms embargo during 1965 forced Ayub Khan to emphasize indigenous production.[9] Complete dependence on the USA even for the simplest spares, especially in case of the Pakistan Air Force (PAF), prevented Pakistan from conducting a protracted war.[10] Pakistan's importance in American strategy declined after the Soviet withdrawal from Afghanistan and the US Senate passed the Pressler Amendment. After the Pressler Amendment in 1990, the US government refused to provide spare parts as well as new F-16s to Pakistan. For spare parts, the PAF had to identify alternative sources and initiated the process of indigenization through research and development and reverse engineering.[11]

Rivalry between the two superpowers in the Cold War era occasionally enabled India to acquire sophisticated technology from the USSR. With the emergence of Beijing-Washington axis from the mid-1960s, India became important to the USSR for maintaining strategic stability in Southern Asia. So India was able to acquire MIG-21 and frigates from Moscow on easy rupee terms. In the same period, when Britain refused to supply submarines to India, Moscow provided these vessels.[12]

Technology transfer, including the latest Russian aircraft design, in the new millennium is possible due to the break up of the USSR and Russia's dire need for cash. The Hindustan Aeronautics Limited (HAL) at Bangalore will manufacture 140 Russian designed SU-30 which will require an investment of $650 million. The first manufactured SU-30s were handed over to the Indian Air Force in 2004. For obtaining 100 per cent technology transfer, HAL paid $286 million to Russia.[13] Russia and India are also funding joint development programmes for producing the Ilyushin II-24 Military Transport Aircraft. One production line will be in Russia and another in India and these aircraft will enter service in 2007.[14]

With the collapse of the USSR, the US control over the flow of military technology to non-Western countries has become more stringent. As a consequence of this policy, Washington has introduced certain technology denial agreements. In 1996, the US Congress ceased delivery of the F-16s to Pakistan because of Washington's suspicion that Islamabad was developing nuclear weapons.[15] In 1997, the US Congress discussed the introduction of a code of conduct aimed to limit military technology supplies to countries with poor human rights and limited transparency concerning arms import. By introducing the Wassenaar Arrangement, the USA could prevent technology transfer by identifying the recipient state as a rogue state.[16]

Compared to conventional military technology, technology denial is more rigorous regarding the know-how of atomic weapons. This is because nuclear-tipped missiles in the hands of the Asian powers have made the United Sates' forward military bases in Afro–Asia vulnerable.[17] In order to ensure custodial safety of the nukes in the nuclearized countries, a centralized command and control system is necessary. Communication tools for such a concentrated top down command infrastructure is not only costly but has to be imported by India. But, USA is not willing to transfer these tools to India and Pakistan. The Comprehensive Nuclear Test Ban Treaty (CTBT) is an instrument to prevent the non-nuclear countries from acquiring nuclear technology. The CTBT was adopted in the United Nations General Assembly under American initiative in 1996. However, this treaty is yet to win ratification from forty-four states in order to become a law.[18] Then, the Fissile Material Cut Off Treaty (FMCT) is another technique to restrict the quantitative and qualitative aspects of the nuclear arsenals of recently emerging nuclear powers like India. Once a country signs this treaty then its nuclear stockpile remains limited.[19] To deliver nuclear warheads to the targets, missiles are required. To prevent the developing countries from having access to missile technology, the Western countries are trying to push the Missile Technology Control Regime down the throats of the Third World nations. India till this date has resisted pressure to become a signatory to this discriminatory regime. The United States government pressurized Russia not to provide India with the cryogenic engines necessary for the Indian Geo Stationary Launch Vehicle (GSLV). USA is continuously pressurizing the Afro–Asian countries to sign CTBT and FMCT. The American strategy regarding South Asia is to cap the atomic programmes of both Pakistan and India and to bring their nuclear energy programmes under the International Atomic Energy Agency's safeguards.[20] Washington by pulling the strings of the World Bank and the International Monetary Fund is trying to bring the 'recalcitrant' nations in line. Again, the USA remains India's guarantor for secure supply of oil from the turbulent Gulf region and so India is not in a position to annoy the USA.[21] The only way out for a Third World Country like India is to go for indigenization of weapons production.

LIMITATIONS OF INDIGENOUS MILITARY MODERNIZATION IN INDIA

One thing I should like to say, and that is that the growth of defence industries in this country has been particularly satisfactory. All of them are not functioning; some of them are in the process of being built up. I think that our record in that respect is very satisfactory, because, ultimately the defence depends on the growth of industry generally and more especially the defence industry. More and more the art of warfare becomes mechanized; technical improvements come in daily. It becomes necessary to rely more and more on these technical improvements.

JAWAHARLAL NEHRU[22]

The Nehru government was aware that in order to pursue an independent foreign policy, self-reliance was essential. That Jawaharlal Nehru was hell bent on military industrialization of India comes from a note to the Ministry of Defence dated 25 May 1955 where he wrote: 'Obtaining everything in India, even if it is a little costlier, is better because it provides employment and training, and foreign exchange is saved.'[23] Hence, under the industrial policy resolution of 30 April 1956, the government took the decision that production and development of arms, ammunition and aircraft will be the exclusive responsibility of the state.[24]

The industrial base inherited from the colonialism by independent India was defective. The British deliberately destroyed whatever little indigenous industry existed in pre-colonial times. In 1947, there were sixteen ordnance factories in the subcontinent. From 1951 onwards, while government expenditure on the pay and allowances of the staff of the ordnance factories were increasing, the value of production was falling, an indication that the productivity of labour in the factories was declining. Presence of surplus staff and labour characterized the ordnance factories of India. The problem was exaggerated by the fact that only in wartime, the ordnance factories went into full production. During peacetime, the ordnance factories experienced cut in production, but had to retain all the staff on its payroll. This was one of the factors which made the ordnance factories loss-making enterprises.[25]

In response, the government set up several committees to study the problem. In January 1954 under the Chairmanship of

Baldev Singh, a committee was set up to study the problems and its recommendations submitted. But, the government instead of acting on the recommendations appointed a Parliamentary Estimates Committee in 1956.[26] This committee like its predecessor hit out against the inefficient bureaucratization of the ordnance factories. Its curt recommendations are worth quoting. The report says:

> The Committee is of the firm view that all industries in the Public Sector, whether defence or civil should be run as industries are intended to be run anywhere in the world, i.e. not under the departmental system of management but under the Company System of management. . . . The Committee considers it necessary to associate a few prominent industrialists with the Defence Production Board. . . . The Committee does not consider the appointment of the Minister for Defence Production as the Chairman of the Defence Production Board as a satisfactory arrangement.[27]

The Estimates Committee also suggested that decentralization of authority was necessary in order to secure businesslike and efficient working of the ordnance factories. Further, the paperwork needed to be drastically cut down.[2]

In 1957, the Defence Production Advisory Committee was established to ensure effective liaison with civil industry and consisted of representatives from various ministries and private industry, but, bureaucratic expansion continued unchecked. The Defence Production and Supply Committee under the chairmanship of the Controller General of Defence Production was in charge of exploring possibilities for establishing indigenous production of stores imported from abroad.[29] Another Parliamentary Estimates Committee was set up in 1958. While submitting its report next year, this committee admitted that surplus labour existed, and the problem was how to get rid of them. Gradual retrenchment of the surplus labour was considered virtually impossible. After 1953, surplus labour was got rid off and other departments of the government reabsorbed some 8,000 among them. The factories' staff, however, sympathized with those retrenched labour who were not compensated and the net result was strikes during September 1956. Both the MoD and the Estimates Committee in 1958 wished to avoid the recurrence of such a problem. While on the one hand, the ordnance factories were facing problems due to the presence of excess labour and their low productivity, on the other hand, shortage of skilled senior officers further reduced the economic viability of the factories. The absence of pension scheme and slow rate of promotion compared to the other departments were the principal factors behind the unattractiveness of service in the factories. The MoD in 1958 decided to improve the pay scales and other terms of service for retaining and attracting people with technical abilities. The Parliamentary Estimates Committee in 1958 advocated the necessity of attracting experienced persons from private industry, but nothing was done in this regard.[30]

Domestic politics occasionally encouraged indigenization of advanced technology in India. Between 1957 and 1962, when Krishna Menon was the Defence Minister, he stressed on establishing a military-industrial base in India to woo voters.[31] Worsening international relations also encouraged military industrialization. The Department of Defence Production was set up in 1962, in the wake of Chinese attack.[32] From the 1960s, the government of India began setting up several new factories. At present India possesses thirty-nine ordnance factories.[33]

From the 1950s, the ordnance factories started manufacturing 30-mm. Browning belt ammunition and 4.2-inch mortar bomb, and the next year saw the production of 5.5-inch and 7.2-inch explosive shells. In 1975, the ordnance factories were producing picrite for use in flashless propellants (necessary for night fighting), nitrocellulose powder for the ammunition of 7.62-mm. semi-automatic rifle, 9-mm. sten gun, 30-mm. anti-aircraft gun and 40-mm. anti-aircraft gun. By 1980, India was self-reliant in the field of small arms and ammunition. As far as field guns were considered, the indigenously prepared 105-mm. guns were replacing the 25-pounder guns of Second World War vintage. The air defence capability was represented by the L-70 guns manufactured in India.[34]

Nevertheless, the products of the ordnance factories manifested certain defects. Even in 2000, the Indian ordnance factories have failed to establish proper facilities to overhaul BMP-2 infantry fighting vehicles, despite the fact that these machines were in service for twenty-two years. The engines and 373 spare parts are still being imported from Russia. To make matters worse, the ordnance factories are finding it difficult to produce rifles. In 2001, Ordnance Factory, Medak, which was already infamous for supplying defective infantry combat vehicles, faced another problem. Medak manufactured and supplied defective investment castings that are used in manufacturing 5.56-mm. rifles. Because of blow holes, cracks and dimensional differences, 44,000 out of 1.46 lakh investment castings had to be rejected, resulting in a loss of Rs.78.16 lakh.[35]

The ordnance factories manufacture low-technology goods like jonga jeeps, Shaktiman trucks, etc. Three ordnance factories are also associated with manufacturing armoured vehicles, but they are not able to manufacture state-of-the-art tanks. The indigenous Leyland L-60 engine which was fitted in to the Vijayanta tanks (whose development started during the 1960s with inputs from the British firm Vickers-Armstrong) proved troublesome. So, in the 1980s these tanks were fitted with Rolls Royce engine. During the 1990s, these factories were merely capable of repowering obsolete Vijayanta tanks with the engines of imported T-72 tanks. The Main Battle Tank (MBT) named Arjun, which India started manufacturing in the 1990s developed serious defects. Being sixty tons, it was too heavy and the power unit imported from France was not adequately powerful. In 2000, the Indian Army had to withdraw 770 faulty 125-mm. barrels built for the T-72 MBTs. These barrels worth Rs. 450.70 million ($10.2 million) were built by the Field Gun Factory at Kanpur using a defective tempering process. The net result was that between 1992 and 1999, eleven barrels burst during practice firing. Another thirty-five accidents also occurred within this period. The T-72s were built under licence at the Heavy Vehicles Factory, Avadi since the 1980s, yet about 90 per cent of the parts have to be imported from Russia. Since Pakistan acquired T-80s from Ukraine, the Indian Army is demanding T-90s because the army regards Arjun MBT as under-powered. The cost of estimated 300 T-90s which the Indian Army wants, is about Rs.50 billion ($1.1 billion). The T-90s are operationally little better than the T-72s. But, the Russians capitalizing on the Indian Army's eagerness to buy the T-90s increased the unit price from $2.1 million to $2.8 million. And if the T-90s are inducted, then for spares and ammunition, the Indian Army will remain dependent on Russia for another twenty-five to thirty years.[36]

Unlike India, Israel's record as regards indigenous production of military hardware is much more impressive. In contrast to the Arjun, Israel's Merkava MBT is a successful project.[37] Tank production in Pakistan seems to be much better than it is in India. In 1988, Pakistan announced the plan for production of a new MBT, and in 1999, work on Pakistan's Al-Khalid MBT-2000 started. This tank has a 125-mm. smoothbore gun and a maximum speed of 70 km. per hour with an operational range of 400 km. Due to the automatic loader feeding the gun, a crew of three is adequate. The automatic loader feeding the gun is an advancement from the World War II *panzer* whose gunner had to load the gun manually. Khalid is the product of an alliance between Pakistan's Heavy Industries Taxila and China North Industries Corporation. About 45 per cent of Khalid's components are imported from China. From 1996 onwards, Ukraine's Malyshev Plant is also assisting the Khalid project.[38]

At present India has eight Defence Public Sector Undertakings (DPSUs)—HAL, Bharat Electronics Limited, Bharat Earth Movers Limited, Mazagon Docks Limited (MDL), Garden Reach Shipbuilders and Engineers Limited, Goa Shipyard Limited, Bharat Dynamics Limited, and Mishra Dhatu Nigam Limited. Of these eight undertakings, HAL is perhaps the most famous. As regards the development of missiles and their delivery systems, the Defence Research Development Organization (DRDO), DPSUs and the ordnance factories are significantly successful. In 1983, plans for the Integrated Guided Missile Development Programme were announced. A civilian rocket specialist named Abdul Kalam (later President of India) headed the team. Previously he was with the Indian Space Research Organization (ISRO) where he developed the Satellite Launch Vehicle. The Indian Army was not over-enthusiastic about the missile programme initiated by the DRDO, and its attitude was similar to the pose taken by the United States Navy on guided missile development. The Tomahawk missiles were pressed on the USN by their civilian masters. The Prithvi short-range missiles are developed autonomously by the DRDO and now Bharat Dynamics Limited is manufacturing the Prithvi. Around 1993, the Ordnance Factory Medak was able to manufacture Sarath vehicles with missile launchers.[39]

Unlike missiles, indigenous productions of aircraft were not that successful. In a note to the Defence Minister dated 11 April

1955, Nehru emphasized: 'It is quite essential that we should think of some aircraft which we can manufacture in India. Our whole approach in defence must be production in India even if the weapon or aircraft so produced is not up to the high standard of some new weapon or aircraft elsewhere. This should apply to almost every kind of equipment for our Army, Navy or Air Force.'[40] Nehru was eager that India should manufacture Gnat light fighters because they could be more easily produced than other complex aircraft. Nehru's policy was indigenous military industrialization even at the cost of qualitative superiority of the hardware imported from abroad. In 1955, the Indian government got in touch with the Follands in the UK, the company, that manufactured the Gnats. Though the Gnat was rejected by the North Atlantic Treaty Organization, Louis Mountbatten pointed out to Nehru the positive features of that aircraft. If India was ready to buy between 50–100 Gnats then Follands was willing to manufacture the aircraft in collaboration with the HAL at Bangalore. The IAF was against the acquisition of Gnats and preferred the French Ouregon. In the 1960s though HAL was able to manufacture the engines of the Gnats, for the vital parts, the production process remained dependent on inputs from the United Kingdom. After the limited success of the Marut programme, designing of aircraft took a back seat in India. Rather the aircraft industry in India focussed on licensed production of MIG-21 aircraft. Due to subsequent stress on acquisition of production technology rather than design technology, India has a license to produce rather than the capability to design. India's aviation industry is still backward especially in the field of engine manufacture. India's indigenous Unmanned Aerial Vehicle named Nishant (with a British engine) underwent army trials in 1999. The 4.6 m long Nishant with a 6.5-m wingspan weighed about 350 kg. The Central Vehicle Research and Development Establishment in Pune is still trying to develop an indigenous engine.[41] But India is not alone is being dependent on foreign technology. Even China, to a limited extent, is dependent on foreign technology. The Shenyang Aircraft Industry Group of China with the aid of about 100 Russian engineers is trying to manufacture SU-27s.[42]

Among the three services of the Indian military, the navy is foremost as regards indigenous production, though it is not adequate. In May 1955, Nehru enquired about the prospect of India trying to build ocean-class minesweepers. In the 1970s, India was moderately successful in manufacturing war ships. The first Leander Class frigate Nilgiri of British design was launched from MDL in 1972. In 2003, the MDL in Mumbai started building six destroyers which are larger, modified version of the 6,700 ton Delhi-class destroyers. The MDL's East Yard has started manufacturing six Scorpene-class French submarines. But, the three Krivak-III class frigates demanded by the Indian Navy are under construction at the Baltisky Shipyard in St Petersburg. Each such frigate will carry eight 3M–54E anti-ship missile with a range of 300 km. These missiles are also produced in Russia.[43]

The gap between the Third World countries and the First World becomes clear from the fact that while the USA is investing heavily in manufacturing laser rays designed to destroy enemy combat vehicles and personnel,[44] India is still unable to manufacture spare parts for the Bofors guns. In 1989, the Bharat Earth Movers failed to manufacture 105-mm. Indian Field Gun MK-II,[45] and the Indian Army continues to rely on Bofors field guns. Even for spare parts for field artillery, India remains dependent on foreign manufacturers. During the 1999 Kargil War, the Bofors guns served the Indian Army well. India placed an order in September 1999 about spares and ammunition for the 410 Bofors guns bought by India in 1986. In 2000, the Indian government awarded Bofors another contract worth $12.3 million for spare parts of the Bofors FH-77 B towed 155-mm. howitzer.[46]

The DRDO and the Indian ordnance factories' deficiency in producing high-grade sensor equipment is most noticeable. In 1989, the Indian Army demanded Weapon Locating Radars (WLR) which the DRDO is yet to deliver. In order to guard the Line of Control in Kashmir, in October 2002 India had agreed to buy 1,000 mobile radars from Israel at a cost of Rs. 350 crore. They are to be used in tracking down the militants while they are crossing the LOC. Again, India is negotiating with Israel for 300 thermal imaging systems which will provide night vision to the T-72 tank crews. In 1999, India signed a deal with Elta Electronics Industries of Israel for 56 EL/M-2129 battlefield surveillance radars and 200 manportable radars under a contract worth about Rs. 800 million ($14.28 million). This deal includes transfer of technology with the Bharat Electronics Limited.[47] Even Pakistan with its flagging economy and an inferior scientific-industrial base has been able to overtake India in some spheres. The Pakistan Institute of Optronics has produced night vision sight for heavy machine-guns and recoilless rifles. The range is 2,000 metres in moonlight.[48] Such hardware could have given India an edge in Kargil during 1999. In 1995, DRDO, the DPSUs and the ordnance factories only accounted for 30 per cent of India's defence needs. The government has planned to increase the share of indigenized production of military hardware from 30 to 70 per cent by 2010.[49] Going by the past records of DRDO, the DPSUs and the ordnance factories, this sort of transition will be impossible.

To sum up, the Indian military industry through indigenization has been able to achieve some sort of import substitution[50] but failed to come up with new weapon systems. Pakistan's case is also somewhat similar. In 1985, General Zia ordered that all arms transfer should be accompanied by at least some technology transfer. The Pakistani ordnance factories are capable of limited reverse engineering.[51] During 1990, the aim of India's MoD as regards defence production was articulated in the following words: 'While it is neither feasible nor necessary to achieve hundred per cent self-reliance in all items required by services, it is essential to aim at production of crucial systems indigenously, even if doing so is not economically viable due to the low volume of production.'[52] This is yet to be achieved. While the Indian economy does not allow procurement of sophisticated arms from western countries on a commercial basis, India is not in a position to manufacture the requisite arms by itself.

OPTIONS FOR THE FUTURE

The defence *versus* development debate is at times a sterile one. The relationship between military spending and economic health of the nation is at best non-linear. As in the Soviet case, after 1975 the military apparatus became a burden on economic development. The share of the arms industry in total industrial output for the USSR in 1980 was 9 per cent. The USSR devoted 10–15 per cent of its GNP on defence.[53] But at times, defence expenditure accelerates economic growth indirectly.[54] Dual-use technology ensures that a strong economic-cum-industrial base is a necessary prerequisite for military prowess. Heavy military expenditure leading to spread of dual use technology with their spin off effects, is not always wasteful despite the assertion of the Marxists. As regards advanced electronics requirements, computers, lasers, communications and special alloys, etc., there are considerable commonalities in technologies in both civil and military spheres.[55] For instance in the late twentieth century, the United States Air Force sponsored automated machine tools development aided the American civilian sector in the long run.[56]

The Indian ordnance factories and the DPSUs have a spin off effect on the civil sector also. In 1957, the ordnance factories of Ambarnath and Katni manufactured mild-and spring-steel billets and non-ferrous products for the civilian industries. In addition, Aruvankadu Cordite Factory sold acetone to the industries in the civil sector. Factories producing tanks could also manufacture tractors. And tractors are a necessity for an agricultural country like India. In the 1960s, the Gun and Shell Factory at Calcutta assembled Japanese Komatsu tractors. And the tracks of the tractors were manufactured at the Ordnance Factory Kanpur. From 1978 onwards, the ordnance factories manufactured explosive devices for the ONGC, coal industry, Tear Smoke Ammunition and small arms for the police, paramilitary forces and also non-lethal goods like microscopes, binoculars, etc. During 1993, the Metal and Steel Factory Ishapore developed Cold Rolling Mill Rolls required by Bokaro Steel Plant, which were earlier imported. The defence production units procure large amount of raw materials and assemblies from the civil sector for the main equipment manufactured by them. Between 1986 and 1989, the defence industry bought goods worth more than 2,000 crore from the civil sector. Thus the defence production units aid in utilizing the full capacity lying dormant in the civil sector. In 1976, the ordnance factories employed 91,483 workers. In 1977, the total strength of employees in the defence public sector undertakings was 92,000. And in 1978, defence production employed 1,50,000 skilled and semi-skilled workers. Between 1972 and 1986, the number of workers in the Indian defence industry had increased from 1,93,000 to 2,92,000. In 1990, only the ordnance factories employed about 1,77,000 workers.[57] So, an expanding military-industrial base would go a long way in reducing unemployment among the skilled labour.

The Third World countries like Iran, Pakistan and India till the 1990s developed indigenous technology in order to assemble and maintain imported equipment. Indigenous production activities based on licenced production of the existing weapon systems are imitative modernization. However, research and development (R&D) is necessary for innovative modernization. This is because each technology follows a cycle of invention, adaptation, proliferation, mass production and obsolescence. The cost of error involved in designing a new weapon system has increased several times. The R&D of new weapon systems are not only costly but also time consuming.[58]

Production of an innovative type of technology requires a long lead-time and enormous expenditure over a long-term period. One of the chief constituents of the Revolution in Military Affairs is integration of the latest communication technology with the military machine. Acquisition and organization of detailed information in order to reduce the reaction time to a threat requires computers. The present superiority of the US in the field of computer technology is because of long-term expenditure of resources in this sphere. Even as early as 1964, the US Defence Department spent $6 billion annually (i.e. 10 per cent of the defence budget) for the introduction, operation and maintenance of the computer-aided communication system.[59] Before 1947, no R&D was done in India. All R&D was conducted in the UK. The ordnance factories of colonial India received design and drawing for the products from Britain.[60] In 1958 the DRDO was established and R&D gradually shifted from the factories to the laboratories. In the long run, the net result has been total lack of any innovative activities at the factory level. And the laboratories continue with research unrelated with the requirements of the armed forces. The DRDO runs fifty laboratories of which 20 per cent need to be merged or closed down. What is required is greater integration between the researchers and end user of the products (i.e. the armed forces) and concentration on a few relevant projects. At present, defence R&D constitutes only 5 per cent of the Indian defence budget. This is very low compared to the defence R&D in countries with developed defence industries. The USA spends 10 per cent of its defence budget on defence related R&D. France during the 1980s spent 30 per cent of the research budget on defence matters.[61] In the new millennium, China spends about 28 per cent of the military budget on R&D. As regards R&D in military technology, China is far ahead of India. In 1995, while China spent US $1,000 million for research on new weapons, India could only provide US $510 million for the military R&D.[62]

How can India generate resource for greater investment in R&D? Between 1964 and 1977, India's defence expenditure amounted to 3.6 per cent of the Gross Domestic Product.[63] It may not be possible to further increase the share of GNP that goes to the MOD but the GNP can be raised. Opening up of the economy is the only possible path for economic growth. Before Deng Xiaoping's introduction of his 'Modernization Programme', the Chinese military-industrial base was three decades behind the advanced western countries.[64] In the next two decades, thanks to Deng's policy, China had rapidly modernized. In the era of economic reform, a mild expansion of independent R&D on civilian technologies is a positive step by India to attempt to get out of the technology trap. Massive spending on education, research and development, including the replacement of generalist bureaucratic managerial elite with technocratic elite, has paved the way for systematic innovation in the Western world. Capitalism and its concomitant free market economy aids innovation. This is because individual choices about work, consumption and leisure activities, etc., are available within the capitalist framework. Again, the history of the Cold War shows that dimnished allocation of resources through direct governmental action and the enhanced scope for fluctuating prices as regulators of economic behaviour improved the productive efficiency of the West vis-à-vis the Communist world. After all, the corporate managers of the West has to sell goods plus services to those who are not bound to buy these because of government orders. All these resulted in a managerial revolution in the leading industrialist countries of the West.[65] Gigantic private combines manufacture all the state-of-the-art military products. For instance General Dynamics produces M-1 main battle tanks.[66] Competition is the mother of all innovations. In its heyday, the Soviets developed through their several design bureaus different aircraft prototypes. For instance the MIG-15 faced competition from Lavotchkin and Yakovlev. Later, the quality of hardware declined because the Soviet state was under obligation to buy whatever the factories manufactured.[67]

The Indian ordnance factories and the Public Sector Undertakings must compete with the private sectors.[68] One could argue that excess government subsidies and the lack of 'hire and fire' principle in the Indian public sectors have made many such undertakings inefficient. The subsidy regime must end. In 1996, loss-making state enterprises absorbed $459 million. Between 1996 and 1997, the subsidy for providing power to the farmers rose from $5.3 billion to $6 billion.[69] What India needs to do is cut the red tape by reducing the number and power of the civil bureaucrats and to end unproductive subsidies to the rich segments of the rural society. Instead, the rich and middle peasantry ought to be taxed.

Collaboration with the private entrepreneurs is a must, if the Indian defence industry has to survive. It is worth noting that South Africa has been able to develop a sustainable defence-industrial complex by encouraging participation of the civilian industries in military production. About 80 per cent of all military production are done by the private industries. The South African defence industry is not only surviving but also making profit by merging with civilian firms. Further, the South African firms are also thinking about joint ventures and strategic alliance with foreign companies.[70]

In 1954, the Baldev Singh Committee recommended that at least two private industrialists be appointed on the Board of Management of the Ordnance Factories.[71] But the left leaning politicians and the bureaucrats were able to shoot down this proposal. In 1957, the Estimates Committee in the report submitted to the government pointed out the necessity of integrating the industrialists with the defence production sector. In the Committee's report, it was stated: 'The Committee regrets to observe that the Defence Production Advisory Committee which was set up on 2 January 1956 has so far held only one meeting to which also no private industrialist was invited.'[72] In 1965, after the India–Pakistan War, a Defence Supplies Department was established for mobilizing the capabilities in the civil sector industry to supplement the efforts of the defence industrial units. This department and the Defence Production Department were merged in 1984 to constitute the Defence Production and Supplies Department. This department failed to deliver state-of-the-art hardware in large numbers because of the dominance of the bureaucrats at the secretarial level.[73]

Using the civil sector for critical assemblies and sub-assemblies by the defence industry reduces the cost of production of certain weapon systems. Of the 186 assemblies and sub-assemblies required for the Sarath vehicles, the defence industry contracted about 176 of the assemblies and sub-assemblies to the civilian sector in 1992. The Avadi Ordnance Factory from 1961 onwards manufactures tents, overalls, uniforms, etc. The ordnance factories should stop manufacturing low-technology commodities like clothing, cables, shoes, rations, fuels, jeeps, etc., because private companies can manufacture them better and at a cheaper price.[74] Again, greater coordination between the civilian and the military agencies like ISRO and the DRDO is necessary for rapid development and effective use of space technologies. For example ISRO is quite successful in launching GSLVs. In the near future, the ISRO will come out with a replacement for the Russian cryogenic rocket motors. The GSLV launch with their new motor would cost the ISRO $300 million.[75] The GSLVs could be used to manufacture weather satellites as well as for nuclear equipped Intermediate Range Ballistic Missiles and Inter Continental Ballistic Missiles. Both MoD and the ISRO could share the cost and pool their talents together since both these organizations would be the end user of this particular piece of technology.

Arms production programmes that are geared purely for military considerations rather than economic and commercial ones are bound to collapse. The combined arms-sales of the top arms producing countries amounted to $156 billion in 1996. In other words, the military-industrial base must be oriented for selling arms abroad in order to cover the high cost of developing and deploying new technologies. In 2001, the global military expenditure accounted for 2.6 per cent of the world's GNP. The European and the American defence industries account for 92 per cent of the total arms sales across the globe.[76] In general, the productive capacity of the Indian ordnance factories producing military goods remains unutilized because they go for full production only in wartime. In peacetime because labour and capital are under utilized, the ordnance factories run at a loss.

One way to get over this trap is to make full use of the productive potential of the factories and to sell goods to foreign customers (both military and civilian), and another technique could be to sell their products to the domestic civilian customers. As early as 1952, in China the Central Ordnance Commission in a report to the party's Central Committee concluded that every defence factory should also produce civilian goods. In 1959, the value of civilian output produced by the Chinese defence industries reached 52 per cent of their total production value.[77] Between 1987 and 1991, Beijing acquired $8 billion from arms sales.[78] In recent times, the Pakistan Aeronautical Complex is involved selling its services to friendly countries.[79] The defence firms of Israel make a lot of profit by upgrading the weapon systems of various countries. For a fraction of the cost of new weapons platform, the Israeli defence industry would take any ageing system and give it more reach by fitting them with improved digital electronics. The new technologically advanced components give the older systems enhanced survivability and lethality.[80]

At least some beginning in India has been made in this direction. The ordnance factories of India are diversifying their production mainly for the paramilitary formations.[81] India should enter into the lucrative arms trade and use the profit for modernizing its defence industry. In 1983, zero export performance was reported. In 1989, India earned Rs. 675 million from arms sales abroad but it was much less than the profit made by Pakistan. The then Defence Minister Chintamani Panigrahi said that India plans to fund 40 per cent of the cost regarding military modernization from arms sales to West Africa and South-East Asia.[82] Between 1991 to 1995, the value of sales by the ordnance factories to non-defence customers has increased from Rs. 174 crore to Rs. 500 crore. For the same period, the value of sales to the non-defence customers on part of the DPSUs had gone up from Rs. 1,373 crore to 4,386 crore. However, exports remain minimal and between 1992 and 1994, exports have only increased from Rs. 64 crore to 125 crore. Towards the end of last decade of twentieth century, there has been some improvement. Between 1998–99 and 1999–2000, total sales of the defence sector (ordnance factories and DPSUs combined) have gone up from Rs. 9,548.1 crore to 10,914 crore.[83]

CONCLUSION

There is no denying that certain key aspects of industrial technology emerge from military settings. The only way for survival is to go for a technocratic state[84] where finance, technology and political power interact with each other in a dynamic manner for generating synergy. India has experienced relative success in shipbuilding and has passed the stage of licenced production of vessels, but is yet to achieve indigenous design for construction. As regards arms production in general, in China, the transition is from self-sufficiency to self-reliance. The former requires autarky hence unattainable in the present globalized world. However, self-reliance means production of selected commodities in order to prevent any disruption of operation of the weapon systems during war, especially if foreign countries cut the supply of critical spare parts.[85] However, India is yet to achieve this level. For the sorry state of the Indian ordnance factories, the responsibility lies on lack of cooperation between the scientists and the military and also on the IAS officers. The latter lack managerial capacities and accountability but administer the defence production infrastructure. The armed forces continue to resist the introduction of homemade second grade weapon systems and force the government to go for import from foreign countries. For the immediate strategic purpose, the armed forces, policy makes sense but in the long run hampers indigenous production of high technology goods.

The disintegration of the bipolar world has made it difficult for the less developed countries to get out of the technology trap. The elimination of Soviet Union from the great power politics enables the USA to put more pressure on the less developed countries to fall in with the discriminatory technology denial regimes. Simultaneously, the globalizing process by strengthening the flow of information across the nations and accelerating the genesis of an international market has also opened a window of opportunity in front of the less developed nations. In the new international environment, commercialization of the defence industry is a necessity. For generating funds required for investment, weeding out of inefficient state enterprises and ending the subsidy regime are necessary. But, the irony is that any democratic government attempting to take such steps would face an electoral collapse. It now depends on the political managers of India, as to how well they will play the game.

NOTES

1. Jacques Sapir, *The Soviet Military System*,1987, tr. David Macey, Cambridge: Polity Press, 1991, p. 269.
2. Michael Adas, *Machines as the Measure of Men: Science, Technology, and Ideologies of Western Dominance*, 1989; rpt., New Delhi: Oxford University Press, 1990, pp. 15, 417.
3. Itty Abraham, *The Making of the Indian Atomic Bomb: Science, Secrecy and the Postcolonial State*, 1998; rpt., New Delhi: Orient Longman, 1999.
4. Quoted in Michael I. Handel, 'Clausewitz in the Age of Technology', *JSS*, vol. 9, nos. 2–3 (1986), p. 84.
5. Veena Kukreja, *Civil-Military Relations in South Asia: Pakistan, Bangladesh and India*, New Delhi: Sage, 1991, p. 48.
6. Sultana Afroz, 'The Cold War and United States Military Aid to Pakistan 1947–60: A Reassessment', *South Asia*, vol. 17, no. 1 (1994), p. 63.
7. Kukreja, *Civil-Military Relations*, pp. 104, 122.
8. George K. Tanham and Marcy Agmon, *The Indian Air Force: Trends and Prospects*, 1995; rpt. New Delhi: Vikas, 1996, p. 42.
9. Ayesha Siddiqa-Agha, *Pakistan's Arms Procurement and Military Buildup, 1979-1999: In Search of a Policy*, Basingstoke: Palgrave, 2001, p. 110.
10. Mohammad Ashgar Khan, *The First Round: Indo-Pakistan War, 1965*, Ghaziabad: Vikas, 1979, p. 10.
11. *The Story of the Pakistan Air Force, 1988-1998: A Battle Against Odd*, Islamabad: Shaheen Foundation in Collaboration with OUP, 2000, p. 139.
12. Rahul Roy-Chaudhury, *Sea Power and Indian Security*, London/ Washington: Brassey's, 1995, p. 149; Lorne J. Kavic, *India's Quest for Secuurity: Defence Policies, 1947-65*, Berkeley/Los Angeles: University of California Press, 1967, pp. 200-1.
13. Lieutenant-General R.K. Jasbir Singh, ed., *Indian Defence Yearbook: 2003*, Dehra Dun: Natraj, 2003, p. 228.
14. D.D. Nair, 'Russian Defence Industry in Asia', *IDR*, vol. 18, no. 4, 2003, p. 88.
15. Eric Arnett, 'Beyond Threat Perception Assessing Military Capacity and Reducing the Risk of War in Southern Asia', in Arnett, ed., *Military Capacity and the Risk of War: China, India, Pakistan and Iran*, Oxford: Oxford University Press, 1997, p. 21.

16. Ian Anthony, 'Arms Exports to Southern Asia: Policies of Technology Transfer and Denial in Supplier Countries', in Arnett, ed., *Military Capacity*, p. 285; Siemon T. Wezeman and Pieter D. Wezeman, 'Transfer of Major Conventional Weapons', in *SIPRI Yearbook 1998: Armaments, Disarmament and International Security*, Oxford: Oxford University Press, 1998, p. 295.
17. Paul Bracken, *Fire in the East: The Rise of Asian Military Power and the Second Nuclear Age*, New Delhi and San Francisco: HarperCollins, 1999, p. 163.
18. Shannon Kile, 'Nuclear Arms Control', in *SIPRI Yearbook 1998*, pp. 403, 406.
19. Ashley J. Tellis, *India's Emerging Nuclear Posture: Between Recessed Deterrent and Ready Arsenal*, New delhi: Oxford University Press, 2001, p. 412.
20. J.N. Dixit, 'Technically Bilateral', *The Telegraph*, 10 December 2002, p. 12; Squadron-Leader B.G. Prakash, 'Indigenous Cryogenic Engine', *IDR*, vol. 14, no. 2 ,1999, p. 107; Pranay Sharma, 'Delhi Rejects Space Regime', *The Telegraph*, 17 November 2002, p. 7.
21. Sandy Gordon, 'Indian Security Policy and the Rise of the Hindu Right', *South Asia*, vol. 17, Special Issue, 1994, pp. 193, 204.
22. Ravinder Kumar and H.Y. Sharada Prasad, eds., *Selected Works of Nehru*, Second Series, vol. 25 (1 February 1954–31 May 1954), New Delhi: Jawaharlal Nehru Memorial Fund, 1999, p. 298.
23. Ibid., vol. 28, p. 530.
24. A.L. Venkateswaran, *Defence Organization in India: A Study of Major Developments in Organization and Administration since Independence*, Delhi: Publications Division, 1967, p. 290.
25. *Report of the Indian Parliament Estimates Committee: 1958–59*, New Delhi: Ministry of Defence, 1959, p. 12.
26. *54th Report of the India Estimates Committee: 1956–57*, New Delhi: Lok Sabha Secretariat, 1957, p. 49.
27. Ibid., pp. 61–2.
28. Ibid., p. 64.
29. Venkateswaran, *Defence Organization*, p. 299.
30. *Parliament Estimates Committee*, pp. 1–2, 8, 10.
31. Kavic, *India's Quest for Security*, p. 139.
32. J. Banarwal, ed., *SP's Military Yearbook: 1992–93*, New Delhi: Guide Publications, 1992, p. 612.
33. *MODAR: 1994–95*, pp. 23, 26.
34. Major-General Pratap Narain, *Indian Arms Bazaar*, Delhi: Shipra Publications, 1994, p. 62; *MODAR: 1974–75*, p. 50; *MODAR: 1979–80*, p. 14.
35. Sushil Rao, 'Ordnance Unit Sends Defective Castings to Rifle Factory', *Times of India*, 2 November 2001, p. 4; Rahul Bedi, 'Indian Army

Withdraws Faulty MBT Barrels', *JDW*, vol. 33, no. 24, 14 June 2000, p. 38.

36. Rahul Bedi, 'India Delays Army MBTs', *JDW*, vol. 33, no. 21, 24 May 2000, p. 13; Chris Smith, *India's Ad Hoc Arsenal: Direction or Drift in Defence Policy*, Oxford: Oxford University Press, 1994, pp. 148, 151; Rahul Bedi, 'Indian Army Withdraws Faulty MBT Barrels', *JDW*, p. 38; Narain, *Indian Arms Bazaar*, pp. 77–9; *MODAR: 1994–95*, pp. 26–8.
37. Farah Naaz, 'Israel's Arms Industry', *Strategic Analysis*, vol. 23, no. 12, 2000, p. 2080.
38. Umer Farooq, 'Production under Way on Pakistan's Al-Khalid MBT', *JDW*, vol. 32, no. 13, 29 September 1999, p. 15.
39. Baranwal, ed., *SP's Military Yearbook: 1992–93*, p. 615; *MODAR: 1989–90*, p. 13; *MODAR: 1993-94*, p. 20; Arnett, 'Beyond Threat Perception: Assessing Military Capacity and Reducing the Risk of War in Southern Asia', in Arnett, ed., *Military Capacity*, pp. 10, 21–2; Itty Abraham, 'India's "Strategic Enclave": Civilian Scientists and Military Technologies', *Armed Forces and Society*, vol. 18, no. 2, 1992, pp. 246–7.
40. Kumar and Prasad, eds., *Selected Works of Nehru*, Second Series, vol. 28, New Delhi: Jawaharlal Nehru Memorial Fund, 2001, p. 527.
41. Rahul Bedi, 'India Set to Put Nishant UAV to the Real Test', *JDW*, vol. 31, no. 7, 17 February 1999, p. 29; Jasjit Singh, *India's Defence Spending: Assessing Future Needs*, New Delhi: Knowledge World, 2000, pp. 146, 148; B.N. Mullik, *My Years with Nehru: 1948-64*, Bombay: Allied, 1972, pp. 125, 127, 130; Kumar and Prasad, eds., *Selected Works of Nehru*, Second Series, vol. 28, p. 528.
42. 'China-assembled SU-27s make their first flights', *JDW*, vol. 31, no. 8, 24 February 1999, p. 16.
43. Nikolai Novichkov, 'Russian Anti-Ship Missile to Arm Indian Kilo Submarines', *JDW*, vol. 32, no. 13, 29 September 1999, p. 15; Sandeep Unnithan, 'Ship Shape', *India Today*, 12 May 2003, p. 50; Vice-Admiral G.M. Hiranandani, *Transition to Triumph: History of the Indian Navy, 1965-75*, New Delhi: Lancer, 2000, pp. 78–82; Kumar and Prasad, eds., *Selected Works of Nehru*, Second Series, vol. 28, p. 530.
44. *IDR*, vol. 22, no. 2 (1989), pp. 132–3.
45. Smith, *India's Ad Hoc Arsenal*, p. 152.
46. 'India Awards Bofors Howitzer Contract', *JDW*, vol. 33, no. 8, 23 February 2000, p. 16.
47. Rahul Bedi, 'India Buys Radars from Israel', *JDW*, vol. 31, no. 8, 24 February 1999, p. 5; *Strategic Digest*, vol. 32, no. 11, 2002, p. 1346; *Tenth Report, Ministry of Defence*, pp. 4–5, 20.
48. *Defence and Armament*, no. 86 (July–August 1989), p. 80.
49. Chandra Shekhar, 'Indigenization of the Defence Industry', *JUSII*, vol. 131, no. 544 (2001), pp. 176–7.

50. *MODAR: 1994–95*, p. 23.
51. Siddiqa-Agha, *Pakistan's Arms Procurement*, pp. 110–11.
52. *MODAR: 1989–90*, p. 28.
53. Sapir, *Soviet Military System*, pp. 258, 262.
54. Alex Mintz and Randolph T. Stevenson, 'Defence Expenditures, Economic Growth, and the "Peace Dividend"', *Journal of Conflict Resolution*, vol. 39, no. 2, 1995, p. 299.
55. *MODAR: 1977–78*, p. 6.
56. Barton C. Hacker, 'Engineering a New Order: Military Institutions, Technical Education, and the Rise of the Industrial State', *Technology and Culture*, vol. 34, no. 1 (1993), p. 18.
57. *MODAR: 1975–76*, p. 29; *MODAR: 1977–78*, pp. 33, 36, 52; *MODAR: 1989–90*, pp. 28, 31; R.G. Matthews, 'The Development of India's Defence-Industrial Base', *JSS*, vol. 12, no. 4, 1989, p. 411; J. Baranwal, ed., *SP's Military Yearbook: 1993–94*, New Delhi: Guide Publications, 1993, p. 264; Narain, *Indian Arms Bazaar*, pp. 72–3; *Annual Accounts of the Ordnance and Clothing Factories and Agency Factories for the Year 1958–59*, Calcutta: Govt. of India Press, 1960, p. vii.
58. W.H. McNeill, *The Pursuit of Power: Technology, Armed Forces and Society since AD 1000*, Oxford: Basil Blackwell, 1983, p. 373; Handel, 'Clausewitz in the Age of Technology', p. 88; Arnett, 'Beyond Threat Perception: Assessing Military Capacity and Reducing the Risk of War in Southern Asia', in Arnett, ed., *Military Capacity*, p. 9.
59. J.R. Isaac, S.C. Gupta, F.M.F. and S. Ramani, 'Computer Aids to Command and Control', *JUSII*, vol. CVI, no. 44 (1976), pp. 135, 145–6.
60. Venkateswaran, *Defence Organization*, p. 293.
61. M. Saint-Setiers, 'Science and Defence: Civil and Military Research are in Tune', *Defence and Armament*, no. 86 (July–August 1989), p. 52; Jasjit Singh, *India's Defence Spending*, pp. 150–1, 155–6.
62. Eric Arnett, 'Military Research and Development', in *SIPRI Yearbook 1998*, pp. 267–8; P.K. Ghosh, 'Economic Dimension of the Strategic Nuclear Triad', *Strategic Analysis*, vol. 26, no. 2 (2002), p. 287.
63. Kukreja, *Civil-Military Relations*, p. 199.
64. Paul H.B. Godwin, 'Military Technology and Doctrine in Chinese Military Planning: Compensating for Obsolescence', in Arnett, ed., *Military Capacity*, p. 40.
65. McNeill, *Pursuit of Power*, pp. 364–5, 369.
66. *IDR*, vol. 22, no. 2, 1989, p. 132.
67. Sapir, *Soviet Military System*, pp. 266, 286.
68. Vice-Admiral K.K. Nayyar, Air Marshal B.D. Tayal, Lieutenant-General V.K. Singh, Vice-Admiral R.B. Suri and Major-General Afsir Karim, *National Security: Military Aspects* (New Delhi: Rupa, 2003), p. 249.
69. Ramesh Thakur, 'India in the World: Neither Rich, Powerful, nor Principled', *Foreign Affairs*, vol. 76, no. 4 (1997), pp. 16–17.

70. Ruchita Beri, 'South Africa: An Overview of the Defence Industry', *Strategic Analysis*, vol. 25, no. 4, 2001, pp. 579, 581.
71. *54th Report of the Indian Estimates Committee*, p. 63.
72. Ibid., pp. 62–3.
73. Air Vice Marshal Samir K. Sen, *Military Technology and Defence Industrialization: The Indian Experience*, New Delhi: Manas, 2000, p. 130; Baranwal, ed., *SP's Military Yearbook: 1992–93*, p. 612.
74. Anand, *Joint Vision*, p. 77; Jasjit Singh, *India's Defence Spending*, p. 155; *Ayudh*, vol. 27, nos. 3–4, 2000, p. 8; Baranwal, ed., *SP's Military Yearbook: 1992–93*, p. 613.
75. *Strategic Digest*, vol. 32, no. 11 (2002), p. 1372.
76. Deba R. Mohanty, 'Trends in European Defence Industry: Fortress Europe or Atlantic Defence Industry?', *Strategic Analysis*, vol. 25, no. 4, 2001, pp. 586–7; Elisabeth Skons et al., 'Military Expenditure and Arms Production', in *SIPRI Yearbook 1998*, p. 186; Yezid Sayigh, 'Arms Production in Pakistan and Iran: The Limits of Self-Reliance', in Arnett, ed., *Military Capacity*, p. 194.
77. Deba R. Mohanty, 'Defence Industry Conversion in China', *Strategic Analysis*, vol. 22, no. 12 (1999), pp. 1837, 1843.
78. Richard Bernstein and Ross H. Munro, 'The Coming Conflict with America', *Foreign Affairs*, March–April (1999), p. 24.
79. Siddiqa-Agha, *Pakistan's Arms Procurement*, p. 111.
80. Naaz, 'Israel's Arms Industry', p. 2086.
81. *MODAR: 1993–94*, p. 20.
82. Avinash Nigudker and D. Wikrammanayake, 'India to Actively Export Arms', *Defence*, vol. 20, no. 4, 1989, p. 244.
83. *MODAR: 1994–95*, p. 24; *MODAR: 2000–2001*, p. 48.
84. The concept is adapted from Alex Roland's, 'Technology and War: The Historiographical Revolution of the 1980s', *Technology and Culture*, vol. 34, no. 1 (1993), p. 132.
85. Eric Arnett, 'Military Research and Development in Southern Asia: Limited Capalitities despite Impressive Resources', in Arnett, ed. *Military Capacity*, pp. 253–4; Roy-Chaudhury, *Sea Power and Indian Security*, p. 149.

Conclusion

> There is a race between offensive weapons and those meant for defence, and the pace of development of either is rapid today, so that even relatively new weapons tend to become obsolescent or out of date. The whole conception of war, on land or sea, has virtually changed with the coming of the atom bomb and, even more so, of the hydrogen bomb. In a country like India, that is, a country whose resources are limited, we have to take particular care that we do not buy something which may not be of much use soon after. . . . Personally, I think that with this entirely changed conception of warfare, all of us will have to think on wholly new lines. There is far too much a tendency to think on the lines of the previous war, ignoring subsequent developments.
>
> JAWAHARLAL NEHRU[1]

It is unfortunate that India's powerful first Prime Minister regarded himself a master of all trades. He also considered himself an expert on warfare. And he had little time or the right attitude for absorbing professional opinions from uniformed men. Jawaharlal Nehru's views on armed forces and warfare were at best warped and at worse ridiculous. To an extent, Nehru's ideas about statecraft were responsible for India's defeat in 1962.

Post-independent India has been lucky in not experiencing a great war on its soil till this date. The impact of war in independent India has been negligible. The Korean War resulted in the death of more than two million people (civilian and military). The Vietnam War resulted in the death of 2.2 million Vietnamese, Cambodians and Laotians. Another 3.2 million were wounded. Further, 14 million people became homeless. Only as regards unconventional warfare, the Indian Army's experience can be compared with the experiences of the big armies of the world. Nevertheless, the Indian Army's engagements in Kashmir and in the north-east were much limited in scale and scope compared to the US Army's experience in Vietnam and the Soviet Army's odyssey in Afghanistan. In Vietnam where 5,00,000 soldiers of the US Army fought both a conventional-cum-unconventional war for a decade, they lost 50,000 soldiers apart from the 1,50,000 wounded.

The counter-insurgency campaigns fought by the Indian Army were never as brutal as conducted by some foreign armies. About 10 per cent of the Algerians died because of the French Army's pacification campaign before Algeria obtained Independence from France.[2] Counter-insurgency campaign is especially relevant because probably in the future, policing role of the Indian Army is going to be its principal duty. The loss of one soldier becomes a headline news in the West.[3] This is not so in India, Pakistan and China. So, the Indian Army could afford to continue man-power intensive pacification campaigns both in the north-east and in Kashmir.

The air–land battles fought by the Indian Air Force and the armoured elements were slow moving and quite small. In the post-World War II era, the Israeli Defence Force could be taken as the model army as regards battlefield effectiveness. Neither the Pakistani nor the Indian militaries could reach the level of military efficiency displayed by the IDF. The Indian Army and the Air Force remain incapable of mobile operations in a fluid context. On the other hand, the Indian and the Pakistani forces were not as incompetent as the Egyptian and the Iraqi forces. The Indian Army can be categorized as the most well-behaved army as far as political control is concerned. In the spectrum of rogue armies, the IDF will occupy a moderate position and the Pakistan Army will occupy the most extreme position. The Pakistan Army is probably one of the most disobedient armies in Asia. One scholar writes that with the erosion of the prestige and legitimacy of the nation state, much of the army's former charisma is evaporating.[4] This is a global trend, and a trend that will continue in the near future. This development will probably impede the intrusion of army into the political sphere, at least in India.

In the light of recent developments, John Keegan's assertion, 'But the suspicion grows that battle has already abolished itself,'[5] appears a bit hollow. The reasons behind analysing conventional operations are Operation Desert Storm, the second war against Iraq and the experience of Kargil nearer home. Hence, it would be erroneous to agree with some Western analysts that the age of conventional war is over.[6] Somewhat echoing Keegan's approach, a civilian strategic analyst Kanti Bajpai writes 'that conventional war, which the Indian Armed Forces have concentrated on all these years, is, in fact, the least probable type of warfare in our future'.[7] None the less the possibility of conventional battles in the near future remains a reality. On the other hand it is too early to say that the Indian Army would face cyborgs, nano-technology and genetically engineered soldiers in the so called post-human age.[8] China and Pakistan's combined nuclear and conventional superiority, might lead to New Delhi, suffering diplomatic blackmails in the near future at best and defeats in the limited war at worse. To avert this, a paradigm shift is necessary in the Indian security policy. The ruling elite and the intelligentsia must accept that China and not Pakistan is India's principal threat. China due to its very size and the resource it commands constitutes a threat for India. From 1947 till 2003, very few senior Indian politicians accepted China being the principal threat for India. The Indian ruling class' ostrich-like refusal to realistically assess the threat posed by China is probably because at the subconscious level they have already accepted defeat.

Once it is accepted that the principal duty of the Indian military is to deter China, then radical restructuring of the armed forces is necessary. India could not afford to miss the Digital Revolution. A shift is occurring in global economy: from industrial to information-based economies. And this transformation is resulting in the emergence of Information Warfare in place of Industrial Warfare.[9] With the Revolution in Military Affairs (henceforth RMA) occurring at full thrust, it is apt to quote Stanislav Andreski who writes: 'Now we reach the stage when a sickly woman scientist suffering from an anxiety neurosis may constitute a far greater military asset than thousands of tough and fearless soldiers.'[10] The argument that China unlike India could afford to go for the RMA because the former's economy is booming as a result of Deng Xiaoping's reform but India's economy is growing very slowly after half-hearted liberalization, is only partly true.

India's defence budget is not elastic but New Delhi can afford to change the priorities. The government could save a lot of money by cutting the number of infantry divisions. At present, the government has to spend a lot on pay, pension and welfare of the infantry. Most of the infantry units are engaged in counter-insurgency duties. Such tasks sap the soldiers' morale and their training for conventional war and limited war in nuclear environment suffers. These tasks should be handed over to the restructured paramilitary forces. The Indian Army is also preparing for the wrong war at the wrong time and in the wrong place. The army maintains large number of tanks and armoured personnel carriers that are designed to launch deep penetration thrusts from Punjab in case of war with Pakistan. However, Pakistan's strategy is to conduct Kargil type low-intensity operations instead of large scale conventional operations. Against Pakistan's war by proxy, an armoured thrust is impracticable. Then due to nuclearization of the subcontinent, any deep penetration by Indian armoured forces into Pakistani territory is out of question. Due to the nature of terrain, armour cannot be used against China as well. So, India needs to drastically reduce armour along with infantry, though this move will hurt the vested interest of large number of regimental colonels. With the funds that will be available, the Indian armed force can integrate the tools of Information War. Whether India would be able to manufacture the information warfare gadgets remains to be seen.

New warfare requires combined arms operation. This will require further changes in the organizational format of the army. The separate infantry divisions and artillery plus armoured brigades need to be amalgamated into Combined All Arms Brigades for greater flexibility and manoeuvreability. Latest telecommunications satellite has increased the 'speed of battle'. For survival, quick response is necessary. So, the combined brigades should be air-mobile and these sorts of units could be rushed at breakneck speed to plug any gaps. In Kargil due to lack of adequate air transportation, the foot slogging infantry reached the battle zone too late when the enemy had already digged in. The result was heavy casualties. Air-mobile units have another advantage. At present India maintains separate plain infantry and mountain divisions to meet the (Pakistani) and the (Chinese) threats. The air-mobile units could be used in both the theatres.

For firepower support, the air-mobile units need to be supported by air-assault units. The latter units should include fighter-bombers equipped with precision guided munitions for giving firepower support to the light infantry of the air-mobile units operating at high altitudes. For avoiding the hand held stinger missiles as used by the *mujahideen* guerrillas in Kargil, these aircraft need to be fitted with electronic counter-measures. Then, for precision strikes, the aircraft as well as the ground units require aid of the Global Positioning System navigational satellites. For functioning as force-multipliers, these air-assault units would require to procure 'real time' intelligence about the enemy through sophisticated microprocessors and computer switching gears. Maintenance of these gadgets will require induction of large number of civilian technicians including females who will gradually replace the semi-literate peasant infantry. However, a word of caution is necessary. The spokesmen of the RMA are putting too much emphasis on intelligence acquisition. John Keegan's historical analysis rightly shows that intelligence in war, however good, does not always result in victory.[11] Generating surplus military power will aid India because increasing military capabilities open up new avenues of policymaking for the policymakers.[12] If India has to survive in the present state-system characterized by brutal power-politics, then harnessing the recent technological changes within an innovative managerial format is a must.

NOTES

1. Ravinder Kumar and H.Y. Sharada Prasad (eds.), *Selected Works of Jawaharlal Nehru*, Second Series, vol. 28, New Delhi: Jawaharlal Nehru Memorial Fund, 2001, p. 530.
2. David P. Barash and Charles P. Webel, *Peace and Conflict Studies*, New Delhi: Sage, 2002, pp. 11, 57, 300, 357.
3. Julian Thompson, 'The World in 1945: The View from May 2001', in Thompson, ed., *The Imperial War Museum Book of Modern Warfare: British and Commonwealth Forces at War, 1945-2000*, 2002, rpt., London: Pan Macmillan, 2003, p. 9.

4. Bernard Boene, 'Trends in the Political Control of Post-Cold War Armed Forces', in Stuart A. Cohen, ed., *Democratic Societies and Their Armed Forces: Israel in Comparative Context*, London: Frank Cass, 2000, p. 75.
5. John Keegan, *The Face of Battle: A Study of Agincourt, Waterloo and the Somme*, 1976; rpt., Harmondsworth: Penguin, 1978, p. 343.
6. Robert L.O'Connell in his *Ride of the Second Horseman: The Birth and Death of War*, Oxford: Oxford University Press, 1995, pp. 223–43 proclaims the death of war in the near future. Martin Van Creveld claims that the age of conventional war is *passe* and the post-modern age will witness merely low-intensity operations. Martin Van Creveld, 'Technology and War II: Postmodern War?', in Charles Townshend, ed., *The Oxford Illustrated History of Modern War*, New York: Oxford University Press, 1997, pp. 298–314. I agree with Martin Van Creveld that the age of *der totale krieg* had passed. But, limited conventional war under a nuclear umbrella remains a viable option.
7. Kanti Bajpai, 'India's Future Wars', in Jasjit Singh, ed., *Air Power and Joint Operations*, New Delhi: Knowledge World, 2003, p. 120.
8. For the nature of warfare in the post-modern age, see Christopher Coker, *Waging War Without Warriors? The Changing Culture of Military Conflict*, Boulder, Colorado: Lynne Rienner, 2002, pp. 192–4.
9. Charles Moskos, 'Towards a Postmodern Military?', in Cohen, ed., *Democratic Societies and Their Armed Forces*, p. 5.
10. Stanislav Andreski, *Military Organization and Society*, 1954, rpt., Berkeley/Los Angeles: University of California Press, 1968, p. 222.
11. John Keegan, *Intelligence in War: Knowledge of the Enemy From Napolean to Al-Qaeda*, New York: Alfred A. Knopf, 2003, p. 6.
12. Robert H. Bruce, 'Implications for International Security: Observations on the Security Dilemma and the Nature of Concerns provoked by Indian Naval Expansion', in idem., ed., *The Modern Indian Navy and the Indian Ocean: Developments and Implications*, Perth: Centre for Indian Ocean Regional Studies, 1989, p. 112.

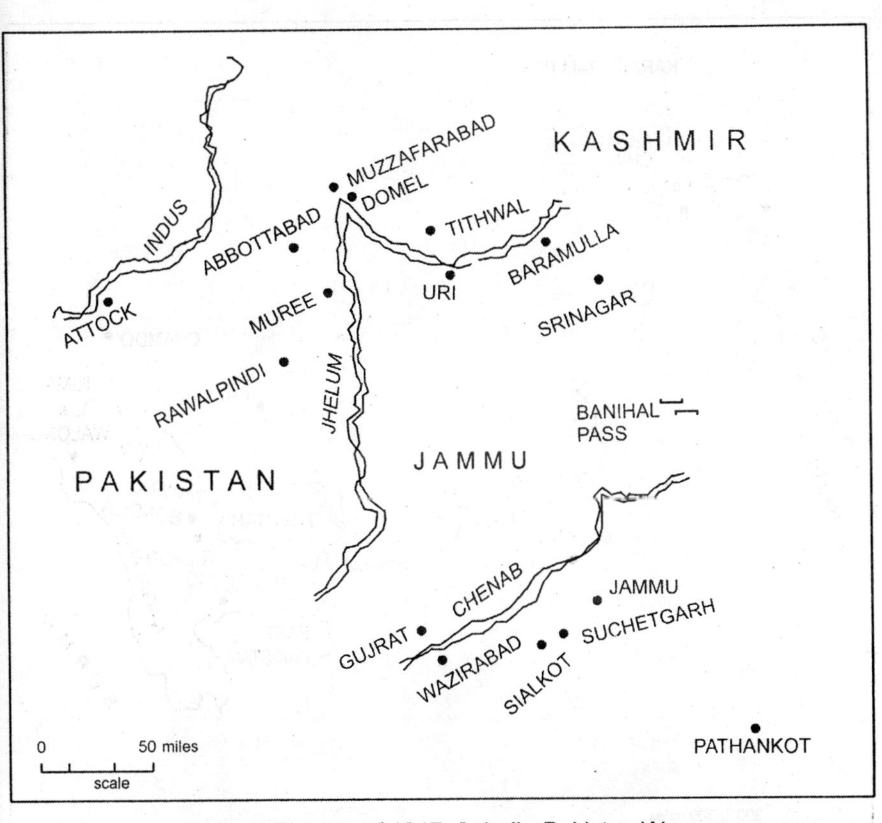

Map 1: Theatre of 1947–8, India-Pakistan War

Map 2: Theatre of 1962, India-China War

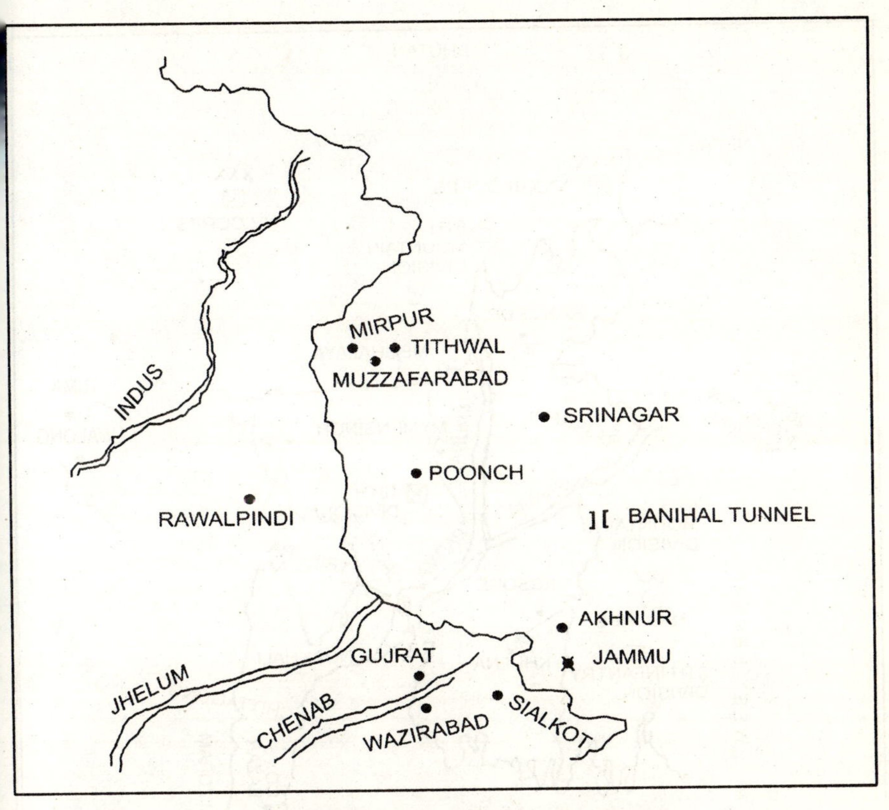

Map 3: 1965, Jammu and Kashmir Theatre

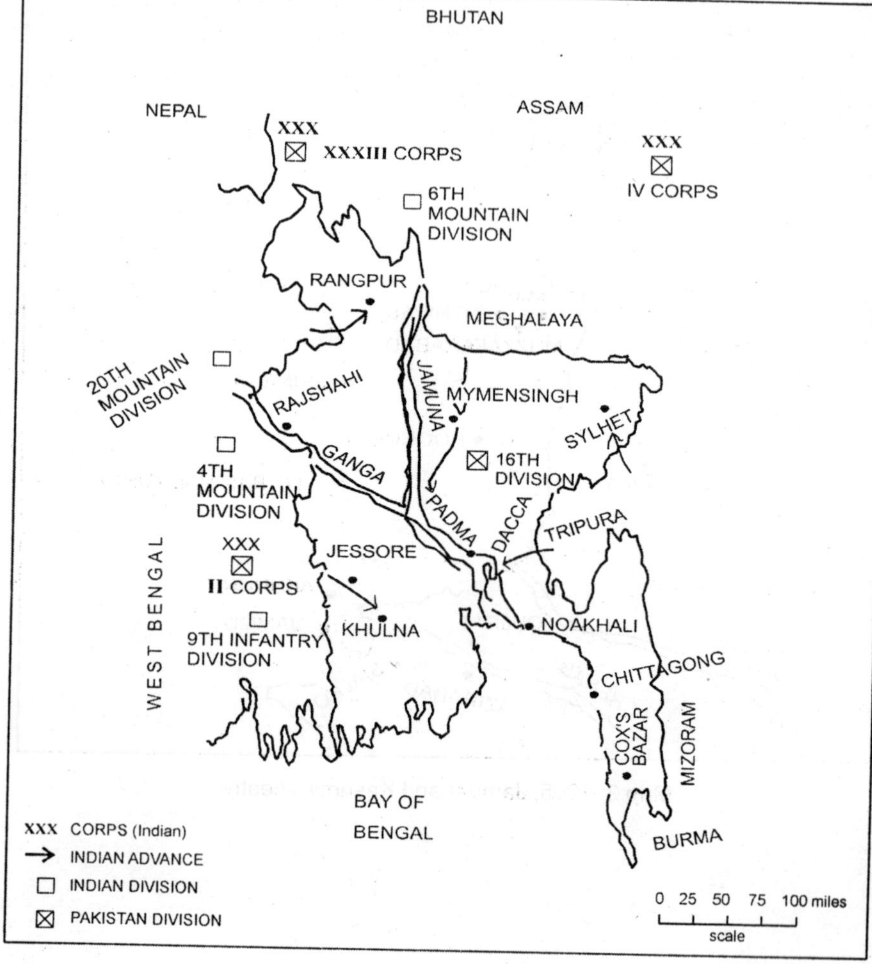

Map 4: 1971, East Pakistan

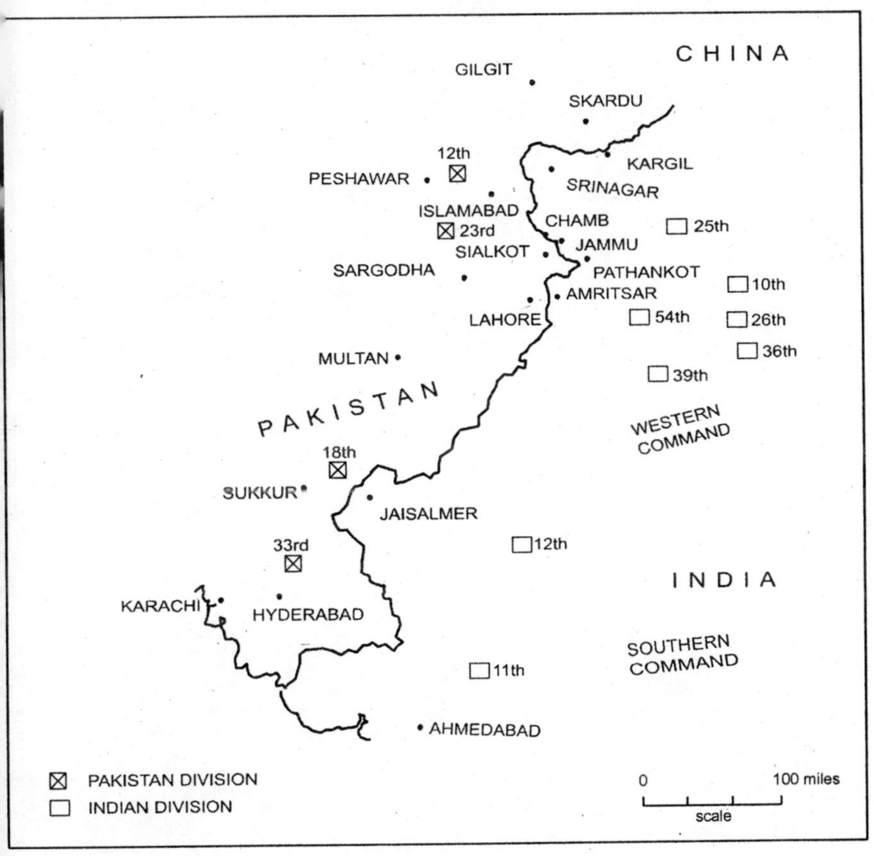

Map 5: 1971, Western Theatre

Bibliography

UNPUBLISHED TRANSCRIPTS OF ORAL INTERVIEWS, NEHRU MEMORIAL MUSEUM & LIBRARY, MANUSCRIPT SECTION, NEW DELHI

Oral Transcript of Interview with General Roy Bucher, 2 March 1970, Accession no. 59.

——— Prof. P.M.S. Blackett, Accession no. 284.

——— Lord Mountbatten, London, 26 July 1967, by B.R. Nanda, Accession no. 351.

——— General J.N. Chaudhuri, New Delhi, 23 March 1973, by B.R. Nanda, Accession no. 426.

PUBLISHED PAPERS

Gopal, S. and Uma Iyengar, eds., *The Essential Writings of Jawaharlal Nehru,* vols. 1 & 2, New Delhi: Oxford University Press, 2003.

Kumar, Ravinder and H.Y. Sharada Prasad, eds., *Selected Works of Jawaharlal Nehru,* Second Series, vol. 25 (1 February 1954–31 May 1954), New Delhi: Jawaharlal Nehru Memorial Fund, 1999; vol. 27 (1 October 1954–31 January 1955), New Delhi: Jawaharlal Nehru Memorial Fund, 2000; vol. 28, New Delhi: Jawaharlal Nehru Memorial Fund, 2001.

Mansergh, Nicholas, Editor in Chief and Penderel Moon, Asst. Editor, *The Transfer of Power,* vol. 5, *The Simla Conference: Background and Proceedings,* 1 September 1944–28 July 1945, London: Her Majesty's Stationery Office, 1974.

REPORTS

Amnesty International Report 1983, London: Amnesty International Publications, 1983.

Annual Accounts of the Ordnance and Clothing Factories and Agency Factories for the Year 1958–59, Calcutta: Govt. of India Press, 1960.

Eleventh Report, Ministry of Defence, Demands for Grants, Standing Committee on Defence, Thirteenth Lok Sabha, New Delhi: Lok Sabha Secretariat, 2001.

From Surprise to Reckoning: The Kargil Review Committee Report, New Delhi, 15 December 1999, New Delhi: Sage, 2000.

Joint Warfare of the Armed Forces of the United States, Joint Pub.–1, 10 January 1995.

Ministry of Defence Government of India, Annual Report: 1970–71; 1971–72; 1974–75; 1975–76; 1977–78; 1979–80; 1989–90; 1993–94; 1994–95; 1997–98; 1999–2000; 2000–01.

Notes, Memoranda and Letters Exchanged between the Governments of India and China, July 1962–October 1962, White Paper No. VII, Ministry of External Affairs Government of India, New Delhi: General Manager Government of India Press, 1962.

Recruitment into the Officer Corps of the Armed Forces, Report of a Seminar held at the United Service Institution of India on 14 February 1977, New Delhi: United Service Institution, 1977.

Reorganization of the Army and Air Forces in India, Report of a Committee set up by the Commander-in-Chief in India, vol. 1, 355.00954, IND, 31893, New Delhi: Institute of Defence Studies and Analyses Library.

Report for the Month of March 1944 for the Dominions, India, Burma and the Colonies and Mandated Territories to the War Cabinet, Report by the Secretary of State for India, 27 April 1944, WP(44) 229, New Delhi: Nehru Memorial Museum & Library.

Report of the Indian Parliament Estimates Committee: 1958–59, New Delhi: Ministry of Defence, 1959.

Tenth Report, Ministry of Defence, Standing Committee on Defence, Thirteenth Lok Sabha, New Delhi: Lok Sabha Secretariat, 2001.

The Imposition of a Manpower Ceiling on the Army: Report of a Seminar at the Untied Service Institution of India, Chairman General M.L. Thapan, 12 January 1976 , New Delhi: United Service Institution of India, 1977.

54th Report of the India Estimates Committee: 1956–57, New Delhi: Lok Sabha Secretariat, 1957.

PUBLISHED RESEARCH PAPERS, REGIMENTAL HISTORIES AND GOVERNMENT SPONSORED MONOGRAPHS

Anand, Vinod, *Joint Vision for the Indian Armed Forces*, Delhi Papers, no. 16, New Delhi: The Institute for Defence Studies and Analyses, 2001.

Bellers, Brigadier E.V.R., *The History of the 1st King George V's own Gurkha Rifles (The Malaun Regiment), 1920–47*, vol. 2, Aldershot: Gale & Polden, 1956.

Chibber, Lieutenant-General M.L., *Para Military Forces*, United Service Institution Paper, no. 4, New Delhi: United Service Institution of India, 1979.

———, *Leadership in the Indian Army During Eighties and Nineties*, United Service Institution Papers, no. 8, New Delhi: United Service Institution of India.

Cohen, Stuart A., *Studying the Israel Defense Forces: A Changing Contract with Israeli Society*, Ramat Gan, The BESA Centre: Bar-Ilan University, 1995.

———, *Towards a New Portrait of the (New) Israeli Soldier*, Ramat Gan, The BESA Centre: Bar Ilan University, 1997.

Field-Marshal K.M. Cariappa Memorial Lectures: 1995–2000, New Delhi: Directorate General of Infantry in association with Lancer Publishers & Distributors, 2001.

Heller, Mark A., *Continuity and Change in Israeli Security Policy*, Adelphi Paper, no. 335, Oxford: Oxford University Press, 2000.

Indo-Pakistan War—1965: A Flash-Back, 1966, ISPR Publications: Rawalpindi, 2002.

9 Gurkha Rifles: A Regimental History, 1817–1947, New Delhi: Vision Books, 1984.

Prasad, S.N. (Chief Editor), P.B. Sinha and Colonel A.A. Athale, *History of the Conflict with China: 1962*, New Delhi: History Division, Ministry of Defence, Government of India, 1992, www.timesofindia.com.

Praval, K.C., *Valour Triumphs: A History of the Kumaon Regiment*, Faridabad: Thomson Press, 1976.

Proudfoot, Lieutenant-Colonel C.L., *We Lead: 7th Light Cavalry, 1784–1990*, New Delhi: Lancer International, 1991.

Sandhu, Major-General Gurcharn Singh, *69th Armoured Regiment: 1968–1993*, New Delhi: Lancer, 1996.

Sharma, Lieutenant-Colonel Gautam, *Path of Glory: Exploits of 11 Gorkha Rifles*, Ahmedabad: Allied, 1988.

Sharma, General V.N., *India's Defence Forces: Building the Sinews of a Nation*, United Service Institution National Security Paper, no. 13, New Delhi: United Service Institution of India, 1994.

Singh, Pushpindar, *The Battle Axes: No. 7 Squadron Indian Air Force, 1942–1992*, New Delhi: The Society for Aerospace Studies, 1993.

Sinha, P.B., *Armed Forces of Bangladesh*, Occasional Paper, no. 1, New Delhi: Institute for Defence Studies and Analyses, 1979.

The Army in India and Its Evolution Including an Account of the Establishment of the Royal Air Force in India, 1924; rpt., Delhi: Anmol Publications, 1985.

The Indian Air Force and Its Aircraft: IAF Golden Jubilee, 1932–82, London: Ducimus Books, n.d.

The Indian Army, New Delhi: Lancer & Army Headquarter, 1990.

The Story of the Pakistan Air Force: 1988–1998, A Battle Against Odd, Islamabad: Shaheen Foundation in collaboration with Oxford University Press, 2000.

BOOKS AND ARTICLES

Abraham, Itty, 'India's "Strategic Enclave": Civilian Scientists and Military Technologies', *Armed Forces and Society*, vol. 18, no. 2 (1992).

———, *The Making of the Indian Atomic Bomb: Science, Secrecy and the Postcolonial State*, 1998; rpt., New Delhi: Orient Longman, 1999.

Abrahamsson, Bengt, 'Military Professionalization and Estimates on the Probability of War', in Jacques Van Doorn, ed., *Military Profession and Military Regimes: Commitments and Conflicts*, The Hague/Paris: Mouton, 1969.

Adas, Michael, *Machines as the Measure of Men: Science, Technology, and Ideologies of Western Dominance*, 1989; rpt., New Delhi: Oxford University Press, 1990.

Addy, Premen, 'Historical Problems in Sino-Indian Relations', in Ian Nish, ed., *South Asia in International Affairs: 1947–56, International Studies*, London: London School of Economics and Political Science, 1987.

Advani, Wing-Commander K., 'Why the Army Should Not Have an Air Corps', *Journal of the United Service Institution of India*, vol. 69, no. 417 (1969).

Afroz, Sultana, 'Pakistan and the 1951 Middle East Defence Plan—The US and UK Positions', *Asian Affairs*, vol. 19 (1988).

———, 'The Cold War and United States Military Aid to Pakistan 1947–60: A Reassessment', *South Asia*, vol. 17, no. 1 (1994).

Ahluwalia, Major A.S., 'Junior Leadership Training', *Journal of the United Service Institution of India*, vol. CII, no. 427 (1972).

Ahluwalia, Air-Commodore P.S., 'Application of Military Power in Space', *Indian Defence Review*, vol. 16, no. 3 (2001).

Ahmed, Major Akhtar, *Advance to Contact: A Soldier's Account of Bangladesh Liberation War*, Dhaka: The University Press Limited, 2000.

Albright, David E., 'Democratization and Civil–Military Relations in Russia and Ukraine', in John P. Lovell and David E. Albright, eds., *To Sheathe the Sword: Civil-Military Relations in the Quest for Democracy*, Westport: Greenwood Press, 1997.

Allen, Colonel Patrick D. and Lieutenant-Colonel Chris C. Demchak, 'The Palestinian-Israeli Cyberwar', *Military Review*, vol. LXXXIII, no. 2 (2003).

Anand, Vinod, 'Warfare in Transition and the Indian Subcontinent', *Strategic Analysis*, vol. 23, no. 5 (1999).

———, 'India's Military Response to the Kargil Aggression', *Strategic Analysis*, vol. 23, no. 7 (1999).

———, 'Joint and Integrated Logistics System for the Defence Services', *Strategic Analysis*, vol. 25, no. 1 (2001).

———, 'Pakistan and Taliban Role in Chechnya Conflict', in V.D. Chopra, ed., *Rise of Terrorism and Secessionism in Eurasia*, New Delhi: Gyan Publishing House, 2001.

Andreski, Stanislav, *Military Organization and Society*, 1954; rpt., Berkeley/ Los Angeles: University of California Press, 1968.

Anthony, Ian, 'Arms Exports to Southern Asia: Policies of Technology Transfer and Denial in Supplier Countries', in Eric Arnett, ed., *Military Capacity and the Risk of War: China, India, Pakistan and Iran*, Oxford: Oxford University Press, 1997.

Antia, Major-General S.N., 'Indo-Pak War 1971: Some Reflections', *Journal of the United Service Institution of India*, vol. CII, no. 427 (1972).

Arnett, Eric, 'Beyond Threat Perception: Assessing Military Capacity and Reducing the Risk of War in Southern Asia', in idem ed., *Military Capacity and the Risk of War: China, India, Pakistan and Iran*, Oxford: Oxford University Press, 1997.

———, 'Military Research and Development in Southern Asia: Limited Capabilities despite Impressive Resources', in idem ed., *Military Capacity and the Risk of War: China, India, Pakistan and Iran*, Oxford: Oxford University Press, 1997.

Arora, J.S.B., *War with Pakistan: 1971*, Delhi: Army Educational Stores, n.d.

Auchinleck, General Claude, 'The Future of India's Armed Forces', *Journal of the United Service Institution of India*, vol. LXXVI, no. 322 (1946).

Austin, Gregory, 'Soviet Perspectives on India's Developing Security Posture', in Sandy Gordon and Ross Babbage, eds., *India's Strategic Future: Regional State or Global Power?*, Delhi: Oxford University Press, 1992.

Babbage, Ross, 'Introduction', in idem and Sandy Gordon, eds., *India's Strategic Future: Regional State or Global Power?*, New Delhi: Oxford University Press, 1992.

———, 'India's Strategic Development: Issues for the Western Powers', in idem and Sandy Gordon, eds., *India's Strategic Future: Regional State or Global Power?*, New Delhi: Oxford University Press, 1992.

Babu, D. Shyam, 'India's National Security Council: Stuck in the Cradle?', *Security Dialogue*, vol. 34, no. 2 (2003).

Badri Maharaj, Sanjay, 'The Nuclear Battlefield: India vs Pakistan', *Indian Defence Review*, vol. 14, no. 2 (1999).

———, *The Armageddon Factor: Nuclear Weapons in the India-Pakistan Context*, New Delhi: Lancer, 2000.

Badrinath, Major Praveen, 'Psychological Impact of Protracted Service in Low Intensity Conflict Operations (LICO) on Armed Forces Personnel: Causes and Remedies', *Journal of the United Service Institution of India*, vol. CXXXIII, no. 551 (2003).

Baev, Pavel K., *The Russian Army in a Time of Troubles*, London: Sage, 1996.

Bahadur, Kalim, 'Pakistan and Afghanistan as Sources of Terrorism in the Region', in V.D. Chopra, ed., *Rise of Terrorism and Secessionism in Eurasia*, New Delhi: Gyan Publishing House, 2001.

Bajpai, Kanti P., 'State, Society, Strategy', in K.P. Bajpai and Amitabh Mattoo, eds., *Securing India: Strategic Thought and Practice*, New Delhi: Manohar, 1996.

———, *Roots of Terrorism*, New Delhi: Penguin, 2002.

———, 'India's Future Wars', in Jasjit Singh, eds., *Air Power and Joint Operations*, New Delhi: Knowledge World, 2003.

———, 'The War in Afghanistan and US Policy Towards India and Pakistan', in Salman Haidar, ed., *The Afghan War and Its Geopolitical Implications for India*, New Delhi: Manohar, 2004.

Bakshi, G.D., 'Yugoslavia: Air Strikes Test of the Air War Doctrine', *Strategic Analysis*, vol. 23, no. 5 (1999).

———, 'Operational Art in the Indian Context: An Open Sources Analysis', *Strategic Analysis*, vol. 25, no. 6 (2001).

Bammi, Lieutenant-General Y.M., 'The Evolution of the Pakistani Army and the Likely Military Strategy Against India from 2002 to 2010', in Vice-Admiral K.K. Nayyar, ed., *Pakistan at the Crossroads*, New Delhi: Rupa, 2003.

Barash, David P. and Charles P. Webel, *Peace and Conflict Studies*, New Delhi: Sage, 2002.

Bartov, Omer, *The Eastern Front, 1941–45: German Troops and the Barbarisation of Warfare*, 1985; rpt., Basingstoke: Palgrave, 2001.

Barua, Pradeep P., *The State at War in South Asia*, Lincoln/London: University of Nebraska Press, 2005.

Barzilai, Gad and Efraim Inbar, 'The Use of Force: Israeli Public Opinion on Military Options', *Armed Forces and Society*, vol. 23, no. 1 (1996).

Basu, Bikash B., 'Military Evaluation of Jungle Terrain: A Study in Military Geography', *Journal of the United Service Institution of India*, vol. 99, no. 416 (1969).

Battistelli, Fabrizio, 'The Postmodern Military: Conscription or Professionalism?', in Stuart A. Cohen, ed., *Democratic Societies and Their Armed Forces: Israel in Comparative Context*, London: Frank Cass, 2000.

Beckett, Ian F.W., 'The Rhodesian Army: Counter-Insurgency, 1972–79', in idem and John Pimlott, eds., *Armed Forces and Modern Counter-Insurgency*, London/Sydney: Croom Helm, 1985.

———, 'The Portuguese Army: The Campaign in Mozambique, 1964–74', in idem and John Pimlott, eds., *Armed Forces and Modern Counter-Insurgency*, London/Sydney: Croom Helm, 1985.

———, 'Command in the Late Victorian Army', in G.D. Sheffield, ed., *Leadership and Command: The Anglo-American Military Experience since 1861*, London/Washington: Brassey's, 1997.

Beckett, Ian F.W. and John Pimlott, 'Introduction', in Ian F.W. Beckett and John Pimlott, eds., *Armed Forces and Modern Counter-Insurgency*, London/Sydney: Croom Helm, 1985.

Beevor, Antony, *Stalingrad: The Fateful Siege, 1942–43*, 1998; rpt., Harmondsworth, Middlesex: Penguin, 1999.

Behera, Ajay Darshan, 'The Supporting Structures for Pakistan's Proxy War in Jammu and Kashmir', *Strategic Analysis*, vol. 25, no. 3 (2001).

Ben-Ari, Eyal, *Mastering Soldiers: Conflict, Emotions, and the Enemy in an Israeli Military Unit*, New York/Oxford: Berghahn Books, 1998.

Berghahn, Volker R., *Militarism: The History of an International Debate: 1861–1979*, Leamington Spa Berg, 1981.

Beri, Ruchita, 'South Africa: An Overview of the Defence Industry', *Strategic Analysis*, vol. 25, no. 4 (2001).

Berlin, Donald L., 'The Indian Ocean and the Second Nuclear Age', *Orbis*, vol. 48, no. 1 (2004).

Bernstein, Richard and Ross H. Munro, 'The Coming Conflict with America', *Foreign Affairs*, March–April (1999).

Bewoor, Lieutenant General G.G., 'High Altitude Mountain Warfare', *Institute for Defence Studies and Analysis Journal*, vol. 1 (1968–9).

Bhagat, Lieutenant-General P.S., *Forging the Shield: The Defence of India and South East Asia*, Calcutta: The Statesman, *c*.1965.

———, *The Shield and the Sword*, 1966; rpt., Delhi: Vikas, 1974.

———, 'The Divisional Commander', *Journal of the United Service Institution of India*, vol. CI, no. 425 (1971).

Bhatia, Major Harish, 'Re-Organization of the Armed Forces', *Journal of the United Service Institution of India*, vol. LXXXXV, no. 401 (1965).

Bhattacharya, Lieutenant-Colonel S.C., 'Motivating Forces for Combat', *Journal of the United Service Institution of India*, vol. LXXXXIII, no. 392 (1963).

Bidwell, Shelford and Dominick Graham, *Fire-Power: British Army Weapons and Theories of War, 1904–1945*, 1982; rpt., Boston: George Allen & Unwin, 1985.

Birth of an Air Force: The Memoirs of Air Vice-Marshal Harjinder Singh, ed. Air Commodore A.L. Saigal, New Delhi: Palit & Palit, 1977.

Biswas, Sukumar, *Three Weeks' War*, Calcutta: M.C. Sarkar & Sons Pvt. Ltd., 1966.

Blom, Amelie, 'The "Multi-Vocal State": The Policy of Pakistan on Kashmir', in Christophe Jaffrelot, ed., *Pakistan: Nationalism without a Nation?*, New Delhi: Manohar, 2002.

Blomberg, S. Brock and Gregory D. Hess, 'The Temporal Links Between Conflict and Economic Activity', *Journal of Conflict Resolution*, vol. 46, no. 1 (2002).

Boene, Bernard, 'Trends in the Political Control of Post-Cold War Armed Forces', in Stuart A. Cohen, ed., *Democratic Societies and Their Armed Forces: Israel in Comparative Context*, London: Frank Cass, 2000.

Bopegamage, A., 'Caste, Class and the Indian Military: A Study of the Social Origins of Indian Army Personnel', in Jacques Van Doorn, ed., *Military Profession and Military Regimes: Commitments and Conflicts*, The Hague/Paris: Mouton, 1969.

Bourne, J.M., 'British Generals in the First World War', in G.D. Sheffield, ed., *Leadership and Command: The Anglo-American Military Experience since 1861*, London/Washington: Brassey's, 1997.

Bracken, Paul, *Fire in the East: The Rise of Asian Military Power and the Second Nuclear Age*, New Delhi and San Francisco: HarperCollins, 1999.

Browning, Peter, *The Changing Nature of Warfare: The Development of Land Warfare from 1792 to 1945*, Cambridge: Cambridge University Press, 2002.

Bruce, Robert H., 'Implications for International Security: Observations on the Security Dilemma and the Nature of Concerns Provoked by Indian Naval Expansion', in idem ed., *The Modern Indian Navy and the Indian Ocean: Developments and Implications*, Perth: Centre for Indian Ocean Regional Studies, 1989.

Brzezinski, Zbigniew, 'A Geostrategy for Eurasia', *Foreign Affairs*, September–October 1997.

Bucher, General Roy, 'Advice to Gentlemen Cadets', *Journal of the United Service Institution of India*, vol. LXXVIII, no. 333 (1948).

Campbell, Christy, *Air War Pacific: The Fight for Supremacy in the Far East, 1937–45*, London: Hamlyn, 1991.

Candeth, Lieutenant-General K.P., *The Western Front: The Indo-Pakistan War 1971*, New Delhi: Allied, 1984.

Canton, Dario, 'Military Interventions in Argentina: 1900–1966', in Jacques Van Doorn, ed., *Military Profession and Military Regimes: Commitments and Conflicts*, The Hague/Paris: Mouton, 1969.

Cariappa, Lieutenant-General K.M., 'A Call to Duty', *Journal of the United Service Institution of India*, vol. LXXVIII, no. 330 (1948).

Carlton, Eric, *Militarism: Rule Without Law*, Aldershot: Ashgate, 2001.

Carr, William, *Hitler: A Study in Personality and Politics*, London: Edward Arnold, 1978.

Carver, Field Marshal Lord, 'Field Marshal Erich Von Manstein', in Correlli Barnett, ed., *Hitler's Generals*, London: Weidenfeld & Nicolson, 1989.

Chadha, Major Vivek, *Company Commander in Low Intensity Conflict*, New Delhi: Lancer, 1997.

Chand, Vice-Admiral S.K., 'The Naval Traditions of India', in K.K.N. Kurup, ed., *India's Naval Traditions*, New Delhi: Northern Book Centre, 1997.

Chant, Christopher, 'Yalu River: 1952', in idem, Richard Holmes and William Koenig, *Two Centuries of Warfare*, London: Octopus Books, 1978.

———, 'Six Day War: 1967', in idem, Richard Holmes and William Koenig, *Two Centuries of Warfare*, London: Octopus Books, 1978.

Chaphekar, Lieutenant S.G., 'A Frank Survey of India's Defence Problems–I', *Journal of the United Service Institution of India*, vol. LXXVIII, no. 327 (1947).

———,'A Frank Survey of India's Defence Problems–II', *Journal of the United Service Institution of India*, vol. LXXVIII, no. 328 (1947).

Chari, P.R., 'Civil-Military Relations in India', *Armed Forces and Society*, vol. 4, no. 1 (1977).

Chatterji, A.K., *Indian Navy's Submarine Arm*, New Delhi: Birla Institute of Scientific Research, 1982.

Chaturvedi, Air Marshal M.S., *History of the Indian Air Force*, New Delhi: Vikas, 1978.

Chaudhuri, General J.N., *Arms, Aims and Aspects*, Bombay: Manaktalas, 1966.

———, *An Autobiography* (as narrated to B.K. Narayan), New Delhi: Vikas, 1978.

———, 'A Chief of Defence Staff?', in Ravi Kaul, ed., *The Chanakya Defence Annual: 1973–74*, Allahabad: Chanakya Publishing House, n.d.

Chauhan, Brigadier D.S., 'Low Intensity Conflict in India: Organization and Resources Required to Meet this Threat', *National Defence College Journal*, vol. 15 (1993).

Cheema, Pervaiz Iqbal, 'Arms Procurement in Pakistan: Balancing the Needs for Quality, Self-Reliance and Diversity of Supply', in Eric Arnett, ed., *Military Capacity and the Risk of War: China, India, Pakistan and Iran*, Oxford: Oxford University Press, 1997.

———, *The Armed Forces of Pakistan*, 2002; rpt., Karachi: Oxford University Press, 2003.

Chenoy, Anuradha M., *Militarism and Women in South Asia*, New Delhi: Kali, 2002.

Chitnis, Captain A.H. and Commodore C.S. Patham, 'The Naval Tactics of Kunhali Marakkars and Their Relevance to Modern Warfare at Sea', in K.K.N. Kurup, ed., *India's Naval Traditions*, New Delhi: Northern Book Centre, 1997.

Chodoff, Elliot P., 'Ideology and Primary Groups', *Armed Forces and Society*, vol. 9, no. 4 (1983).

Chopra, Wing-Commander M.K., 'The Indian Air Force and the Nation', *Journal of the United Service Institution of India*, vol. LXXXVII, no. 366 (1957).

Chopra, V.D., 'Introduction', in idem, ed., *Rise of Terrorism and Secessionism in Eurasia*, New Delhi: Gyan Publishing House, 2001.

Clausewitz, Carl Von, *On War*, ed. and tr. Michael Howard and Peter Paret, 1976; rpt., Princeton, New Jersey: Princeton University Press, 1989.

Cohen, Eliot A., 'A Revolution in Warfare', *Foreign Affairs*, vol. 75, no. 2 (1996).

———,'Technology and Supreme Command', in Stuart A. Cohen, ed., *Democratic Societies and their Armed Forces: Israel in Comparative Context*, London: Frank Cass, 2000.

———, *Supreme Command: Soldiers, Statesmen and Leadership in Wartime*, London: Simon & Schuster, 2003.

Cohen, Stephen P., 'Issue, Role, and Personality: The Kitchener-Curzon Dispute', *Comparative Studies in Society and History*, vol. 10 (1967–8).

———, 'The Indian Military and Social Change', *Institute of Defence and Strategic Analysis Journal*, vol. 2 (1969–70).

———, *The Indian Army: Its Contribution to the Development of a Nation*, 1971; rpt., New Delhi: Oxford University Press, 1991.

———, 'US Weapons and South Asia: A Policy Analysis', *Pacific Affairs*, vol. 49, no. 1 (1976).

———, *The Pakistan Army*, 1984; rpt., Karachi: Oxford University Press, 1993.

———, 'The Military and Indian Democracy', in Atul Kohli, ed., *India's Democracy: An Analysis of Changing State-Society Relations*, 1988; rpt., New Delhi: Orient Longman, 1991.

———, *Emerging Power India*, 2001; rpt., New Delhi: Oxford University Press, 2003.

Cohen, Stuart A., 'The Israel Defence Forces (IDF) : From a "People's Army" to a "Professional Military"—Causes and Implications', *Armed Forces and Society*, vol. 21, no. 2 (1995).

———, 'The Scroll or the Sword? Tensions between Judaism and Military Service in Israel', in idem, ed., *Democratic Societies and their Armed Forces: Israel in Comparative Context*, London: Frank Cass, 2000.

Coker, Christopher, *Waging War Without Warriors? The Changing Culture of Military Conflict*, Boulder: Lynne Rienner, 2002.

Collier, Paul and Anke Hoeffler, 'On the Incidence of Civil War in Africa', *Journal of Conflict Resolution*, vol. 46, no. 1 (2002).

Collier, Paul and Nicholas Sambanis, 'Understanding Civil War: A New Agenda', *Journal of Conflict Resolution*, vol. 46, no. 1 (2002).

Coox, Alvin D., 'The Effectiveness of the Japanese Military Establishment in the Second World War', in Allan R. Millett and Williamson Murray, eds., *Military Effectiveness: The Second World War*, vol. 3, Boston: Allen & Unwin, 1988.

Corbett, Julian S., *Some Principles of Maritime Strategy* (With an Introduction and Notes by Eric J. Grove , 1911; rpt., London: Brassey's, 1988.

Coughlan, Elizabeth P., 'The Changing Structure and Values of the Polish Military', in John P. Lovell and David E. Albright, eds., *To Sheathe the Sword: Civil-Military Relations in the Quest for Democracy*, Westport, Greenwood Press, 1997.

Creveld, Martin Van, *Command in War*, Cambridge/Massachusetts: Harvard University Press, 1985.

———,'Technology and War II: Postmodern War?', in Charles Townshend, ed., *The Oxford Illustrated History of Modern War*, Oxford/New York: Oxford University Press, 1997.

Cushman, Lieutenant-General John H., 'Challenge and Response at the Operational and Tactical Levels: 1914–45', in Allan R. Millett and Williamson Murray, eds., *Military Effectiveness: The Second World War*, vol. 3, Boston: Allen & Unwin, 1988.

Dahiya, M.S., 'Indo-Soviet Relations under Kosygin', *Journal of the United Service Institution of India*, vol. CII, no. 427 (1972).

Dalvi, Brigadier J.P., *Himalayan Blunder: The Curtain Raiser to the Sino-Indian War of 1962*, Bombay: Thacker & Co., 1969.

Danopoulos, Constantine P. and Adem Chopani, 'Departyizing and Democratizing Civil-Military Relations in Albania', in John P. Lovell and David E. Albright, eds., *To Sheathe the Sword: Civil-Military Relations in the Quest for Democracy*, Westport: Greenwood Press, 1997.

Das, Gurudas, 'India's North East Soft Underbelly: Strategic Vulnerability and Security', *Strategic Analysis*, vol. 26, no. 4 (2002).

Das, Vice-Admiral P.S., 'A View from the Sea', in Air Commodore Jasjit Singh, ed., *Air Power and Joint Operations*, New Delhi: Knowledge World, 2003.

Das, S.T., *Indian Military: Its History and Development*, New Delhi: Sagar Publications, 1969.

Dasgupta, C., *War and Diplomacy in Kashmir: 1947–48*, New Delhi: Sage, 2002.

Dasgupta, Sanjay, 'Of Military Men and Politicians', in Maroof Raza, ed., *Generals and Governments in India and Pakistan*, New Delhi: Har-Anand, 2001.

———, 'Command and Control in the Nuclear Era', in Maroof Raza, ed., *Generals and Governments in India and Pakistan*, New Delhi: Har-Anand, 2001.

Dasgupta, Sumona, 'Militarization of the Indian State since the 1980s', in Maroof Raza, ed., *Generals and Governments in India and Pakistan*, New Delhi: Har-Anand, 2001.

Datta, Lieutenant-Commander Narapati, 'An Aircraft Carrier for the Indian Navy', *Journal of the United Service Institution of India*, vol. LXXXVII, no. 367 (1957).

Datta, Rakesh, 'From Henderson to Subrahmanyam: Army to be Blamed.

And Political Leaders?', *Indian Defence Review*, vol. 15, no. 1 (January 2000).

Datta, Sreeradha, 'What Ails the North-east: An Enquiry into the Economic Factors', *Strategic Analysis*, vol. 25, no. 1 (2001).

Davenport, Brian A., 'The Ogarkov Ouster: The Development of Soviet Military Doctrine and Civil/Military Relations in the 1980s', *Journal of Strategic Studies*, vol. 14, no. 2 (1991).

David, Steven R., 'Democracy, Internal War and Israeli Security', in Stuart A. Cohen, ed., *Democratic Societies and their Armed Forces: Israel in Comparative Context*, London: Frank Cass, 2000.

Davidson, Lieutenant-General Phillip B., *Vietnam at War, The History: 1946–75*, 1988; rpt., London: Sidgwick & Jackson, 1989.

Davis, Vincent, 'Foreword', in John P. Lovell and David E. Albright, eds., *To Sheathe the Sword: Civil-Military Relations in the Quest for Democracy*, Westport: Greenwood Press, 1997.

Dawson III, Joseph G., '"Zealous for Annexation": Volunteer Soldiering, Military Government, and the Service of Colonel Alexander Doniphan in the Mexican–American War', *Journal of Strategic Studies*, vol. 19, no. 4 (1996).

DeBlois, Major Bruce M., 'Ascendant Realms: Characteristics of Airpower and Space Power', in Colonel Phillip S. Meilinger, ed., *The Paths of Heaven: The Evolution of Airpower Theory*, New Delhi: Lancer, 2000.

Deighton, Len, *Blitzkrieg: From the Rise of Hitler to the Fall of Dunkirk*, 1979; rpt., London: Granada, 1981.

Demchak, Chris C., 'Numbers or Networks: Social Constructions of Technology and Organizational Dilemmas in IDF Modernization', *Armed Forces and Society*, vol. 23, no. 2 (1996).

Deo, Arvind R., 'Pakistan's Unending Search for a Viable Foreign Policy', in Vice-Admiral K.K. Nayyar, ed., *Pakistan at the Crossroads*, New Delhi: Rupa, 2003.

Dewey, Clive, 'Some Consequences of Military Expenditure in British India: The Case of the Upper Sind Sagar Doab, 1849–1947', in idem, ed., *Arrested Development in India: The Historical Dimension*, New Delhi: Manohar, 1988.

Doorn, Jacques Van, 'Political Changes and the Control of the Military: Some General Remarks', in idem, ed., *Military Profession and Military Regimes: Commitments and Conflicts*, The Hague/Paris: Mouton, 1969.

Dreze, Jean, 'Militarism, Development and Democracy', in M.V. Ramana and C. Rammanohar Reddy, eds., *Prisoners of the Nuclear Dream*, New Delhi: Orient Longman, 2003.

Dudley, B.J., 'The Military and Politics in Nigeria', in Jacques Van Doorn, ed., *Military Profession and Military Regimes: Commitments and Conflicts*, The Hague/Paris: Mouton, 1969.

Dunn, Peter M., 'The American Army: The Vietnam War, 1965–73', Ian F.W. Beckett and John Pimlott, eds., *Armed Forces & Modern Counter-Insurgency*, London/Sydney: Croom Helm, 1985.

Ekovich, Steven, 'Iran and New Threats in the Persian Gulf and Middle East', *Orbis*, vol. 48, no. 1 (2004).

Elkin, Jerrold F. and W. Andrew Ritzel, 'Military Role Expansion in India', *Armed Forces and Society*, vol. 11, no. 4 (1985).

Elron, Efrat, Boas Shamir and Eyal Ben-Ari, 'Why Don't They Fight Each Other? Cultural Diversity and Operational Unity in Multinational Forces', in Stuart A. Cohen, ed., *Democratic Societies and Their Armed Forces: Israel in Comparative Context*, London: Frank Cass, 2000.

English, John A., *On Infantry*, 1981; rpt., New York: Praeger, 1984.

Enloe, Cynthia H., *Ethnic Soldiers: State Security in Divided Societies*, Harmondsworth, Middlesex: Penguin, 1980.

———, 'Ethnicity in the Evolution of Asia's Armed Bureaucracies', in idem and DeWitt Ellinwood, eds., *Ethnicity and the Military in Asia*, New Brunswick/London: Transaction Books, 1981.

Etzioni-Halevy, Eva, 'Civil-Military Relations and Democracy: The Case of the Military-Political Elites' Connection in Israel', *Armed Forces and Society*, vol. 22, no. 3 (1996).

Fadok, Lieutenant-Colonel David S., 'John Boyd and John Warden: Airpower's Quest for Strategic Paralysis', in Colonel Phillip S. Meilinger, ed., *The Paths of Heaven: The Evolution of Airpower Theory*, New Delhi: Lancer, 2000.

Farrar-Hockley, Anthony, 'Korea, 1950–53: The Far Side of the World', in Major-General Julian Thompson, ed., *The Imperial War Museum Book of Modern Warfare: British and Commonwealth Forces at War, 1945–2000*, 2002; rpt., London: Pan Macmillan, 2003.

Farrell, Theo, 'Figuring Out Fighting Organizations: The New Organizational Analysis in Strategic Studies', *Journal of Strategic Studies*, vol. 19, no. 1 (1996).

Felker, Lieutenant-Colonel Edward J., 'Soviet Military Doctrine and Air Theory: Change Through the Light of a Storm', in Colonel Phillip S. Meilinger, ed., *The Paths of Heaven: The Evolution of Airpower Theory*, New Delhi: Lancer, 2000.

Ford, Roger, *The World's Great Tanks*, Kent: Grange, 1997.

Forster, Jurgen E., 'The Dynamics of *Volksgemeinschaft*: The Effectiveness of the German Military Establishment in the Second World War', in Allan R. Millett and Williamson Murray, eds., *Military Effectiveness: The Second World War*, vol. 3, Boston: Allen & Unwin, 1988.

Forward, 'Control of Reinforcements', *Journal of the United Service Institution and India*, vol. LXXVIII, no. 330 (1948).

Fox, Robert, 'Bosnia and Kosovo, 1992–2000: Balkan Stew', in Major-

General Julian Thompson, ed., *The Imperial War Museum Book of Modern Warfare: British and Commonwealth Forces at War, 1945–2000,* 2002; rpt., London: Pan Macmillan, 2003.

Francillon, Rene J., *Vietnam Air Wars,* London: Temple Press, 1987.

Freedman, Lawrence, *Atlas of Global Strategy: War and Peace in the Nuclear Age,* London: Macmillan, 1985.

———,'Strategic Studies and the Problem of Power', in idem, Paul Hayes and Robert O'Neill, eds., *War, Strategy and International Politics: Essays in Honour of Michael Howard,* Oxford: Clarendon Press, 1992.

———, ed., *War,* Oxford: Oxford University Press, 1994.

Freemantle, Major-General F.L., *Fred's Foibles,* New Delhi: Lancer, 2000.

French, David, '"Tommy is no Soldier": The Morale of the Second British Army in Normandy, June–August 1944', *Journal of Strategic Studies,* vol. 19, no. 4 (1996).

Fuller, Major-General J.F.C., *The Second World War: 1939–45, A Strategical and Tactical History,* 1954; rpt., New York: Da Capo, 1993.

Furlan, Lieutenant Andrew W., 'The Effects of Cold Weather on Infantry Weapons', *Journal of the United Service Institution in India,* vol. LXXXXIII, no. 393 (1963).

Gabriel, Richard A. and Paul L. Savage, *Crisis in Command: Mismanagement in the Army,* 1981; rpt., New Delhi: Himalayan Books, 1986.

Ganguly, Sumit, 'Explaining the Indian Nuclear Tests of 1998', in Raju G.C. Thomas and Amit Gupta, eds., *India's Nuclear Security,* New Delhi: Vistaar Publications, 2000.

———, 'The Islamic Dimensions of the Kashmir Insurgency', in Christophe Jaffrelot, ed., *Pakistan: Nationalism without a Nation?* New Delhi: Manohar, 2002.

———, *Conflict Unending: India-Pakistan Tensions since 1947,* New Delhi: Oxford University Press, 2002.

Garlinski, Jozef, *Hitler's Last Weapons: The Underground War Against the V1 and V2,* London: Julian Friedman Publishers, 1978.

Garver, John W., *Protracted Contest: Sino-Indian Rivalry in the Twentieth Century,* New Delhi: Oxford University Press, 2001.

Gat, Azar, *The Development of Military Thought: The Nineteenth Century,* 1992; rpt., Oxford: Clarendon Press, 1999.

———, *A History of Military Thought: From the Enlightenment to the Cold War,* Oxford: Oxford University Press, 2001.

Gates, Scott, 'Recruitment and Allegiance: The Microfoundations of Rebellion', *Journal of Conflict Resolution,* vol. 46, no. 1 (2002).

Gaylor, John, *Sons of John Company: The Indian and Pakistan Armies, 1903–91,* 1992; rpt., New Delhi: Lancer, 1993.

Ghosh, C.N., 'EMP Weapons', *Strategic Analysis,* vol. 24, no. 7 (2000).

———, 'Directed Energy Weapons', *Strategic Analysis*, vol. 24, no. 11 (2001).

———, 'Application of Unmanned Combat Aerial Vehicles in Future Battles of the Subcontinent', *Strategic Analysis*, vol. 25, no. 4 (2001).

Ghosh, P.K., 'Revisiting Gunboat Diplomacy: An Instrument of Threat or Use of Limited Naval Force', *Strategic Analysis*, vol. 24, no. 11 (2001).

———, 'Economic Dimension of the Strategic Nuclear Triad', *Strategic Analysis*, vol. 26, no. 2 (2002).

Ghosh, S.K., 'Insurgency in South and South-East Asia, 1969–70', *Institute of Defence Studies and Analyses Journal*, vol. 3 (1970–1).

Gittings, John, *The Role of the Chinese Army*, London: Oxford University Press, 1967.

Glantz, David M., *The Soviet Conduct of Tactical Maneuver: Spearhead of the Offensive*, London: Frank Cass, 1991.

Godfrey, F.A., 'The Latin American Experience: The Tupamaros Campaign in Uruguay, 1963–73', in Ian F.W. Beckett and John Pimlott, eds., *Armed Forces and Modern Counter-Insurgency*, London/Sydney: Croom Helm, 1985.

Godwin, Paul H.B., 'Military Technology and Doctrine in Chinese Military Planning: Compensating for Obsolescence', in Eric Arnett, ed., *Military Capacity and the Risk of War: China, India, Pakistan and Iran*, Oxford: Oxford University Press, 1997.

Gooch, John, 'Clausewitz Disregarded: Italian Military Thought and Doctrine, 1815–1943', *Journal of Strategic Studies*, vol. 9, nos. 2–3 (1986).

———, 'Addition, Multiplication and Step-Function: The "Arithmetic" of Military Revolution in Modern Times', in Stuart A. Cohen, ed., *Democratic Societies and Their Armed Forces: Israel in Comparative Context*, London: Frank Cass, 2000.

Gopal, Sarvepalli, *Jawaharlal Nehru: A Biography, 1947–56*, vol. 2, 1979; rpt., New Delhi: Oxford University Press, 1988.

Gordon, Andrew, 'Ratcatchers and Regulators at the Battle of Jutland', in Gary Sheffield and Geoffrey Till, eds., *The Challenges of High Command: The British Experience*, Basingstoke: Palgrave Macmillan, 2003.

Gordon, Sandy, 'Domestic Foundations of India's Security Policy', in idem and Ross Babbage, eds., *India's Strategic Future: Regional State or Global Power?*, New Delhi: Oxford University Press, 1992.

———, 'Conclusion', in idem and Ross Babbage, eds., *India's Strategic Future: Regional State or Global Power?*, New Delhi: Oxford University Press, 1992.

———, 'Indian Security Policy and the Rise of the Hindu Right', *South Asia*, vol. 17, Special Issue, 1994.

Graczyk, Jozef, 'Social Promotion in the Polish People's Army', in Jacques Van Doorn, ed., *Military Profession and Military Regimes: Commitments and Conflicts*, The Hague/Paris: Mouton, 1969.

Grant, Brigadier N.B., 'Inter-Services Cooperation and All That: A Top Defence Management Myth', *Journal of the United Service Institution of India*, vol. CVI, no. 455 (1976).

Gray, Colin S., 'History for Strategists: British Seapower as a Relevant Past', *Journal of Strategic Studies*, vol. 17, no. 1 (1994).

Gray, Colin S. and Roger W. Barnett, 'Reflections', in idem and Barnett, eds., *Seapower and Strategy*, London: Tri Service Press, 1989.

Gregorian, Raffi, *The British Army, the Gurkhas and the Cold War Strategy in the Far East: 1947–54*, Basingstoke: Palgrave, 2002.

Grewal, Brigadier R.S., 'Ethno Nationalism in North Eastern India', *Journal of the United Service Institution of India*, vol. CXXXIII, no. 552 (2003).

Grey, Jeffrey, *A Military History of Australia*, Cambridge: Cambridge University Press, 1990.

———, 'Malaya, 1948–60: Defeating Communist Insurgency', in Major-General Julian Thompson, ed., *The Imperial War Museum Book of Modern Warfare: British and Commonwealth Forces at War, 1945–2000*, 2002; rpt., London: Pan Macmillan, 2003.

Griffith, Brigadier-General Samuel B., 'Introduction', in idem and Major Harris-Clichy Peterson, eds., *Guerrilla Warfare*, 1962; rpt., London: Cassell, 1963.

Groove, Eric, 'Maritime Forces and Stability in South Asia', in Eric Arnett, ed., *Military Capacity and the Risk of War: China, India, Pakistan and Iran*, Oxford: Oxford University Press, 1997

Guderian, General Heinz, *Panzer Leader* , tr. from German by Constantine Fitzgibbon, 1952; rpt., London: Arrow Books, 1990.

Gulati, Brigadier Y.B., 'A Size and a Shape for the Army', *Journal of the United Service Institution of India*, vol. LXXXXIII, no. 391 (1963).

Gunaratna, Rohan, 'Transnational Terrorism: Support Networks and Trends', in K.P.S. Gill and Ajai Sahni, eds., *Faultlines: Writings on Conflict and Resolution*, vol. 7, New Delhi: Bulwark Books and the Institute for Conflict Management, 2000.

Gunston, Bill, 'Soviet Warships', in Ray Bonds, ed., *The Soviet War Machine: An Encyclopedia of Russian Military Equipment and Strategy*, 1976; rpt., London: Salamander, 1977.

———, 'Soviet Aircraft', in Ray Bonds, ed., *The Soviet War Machine: An Encyclopedia of Russian Military Equipment and Strategy*, 1976; rpt., London: Salamander, 1977.

Gupta, Amit, 'Indian Security Planning in the 1990s: Learning to Live in a New World Order', in Marvin G. Weinbaum and Chetan Kumar, eds., *South Asia Approaches the Millennium: Reexamining National Security*, Boulder: Westview, 1995.

Gupta, Rakesh, 'Towards a Political Economy of Intra State Conflicts', in V.D. Chopra, ed., *Rise of Terrorism and Secessionism in Eurasia*, New Delhi: Gyan Publishing House, 2001.

Gutteridge, William, *Military Institutions and Power in the New States*, New York: Federick A. Praeger Publishers, 1965.

Habib, A. Hasnan, 'South-East Asian Perceptions of India's Strategic Development: An Indonesian View', in Sandy Gordon and Ross Babbage, eds., *India's Strategic Future: Regional State or Global Power?* New Delhi: Oxford University Press, 1992.

Hacker, Barton C., 'Engineering a New Order: Military Institutions, Technical Education, and the Rise of the Industrial State', *Technology and Culture*, vol. 34, no. 1 (1993).

Haider, Ejaz, 'Managing Nuclear Weapons in South Asia: In Search of a Model', in M.V. Ramana and C. Rammanohar Reddy, eds., *Prisoners of the Nuclear Dream*, New Delhi: Orient Longman, 2003.

Hall, David, 'Lessons not Learned: The Struggle Between the Royal Air Force and Army for the Tactical Control of Aircraft, and the Post-Mortem on the Defeat of the British Expeditionary Force in France in 1940', in Gary Sheffield and Geoffrey Till, eds., *The Challenges of High Command: The British Experience*, Basingstoke: Macmillan, 2003.

Hamilton, Nigel, *The Full Monty: Montgomery of Alamein, 1887–1942*, London: Penguin, 2001.

Hammond, Grant T., 'The Perils of Peacekeeping for the US: Relearning Lessons from Beirut for Bosnia', in Edward Moxon-Browne, ed., *A Future for Peacekeeping?*, Basingstoke: Macmillan, 1998.

Handel, Michael I., 'Clausewitz in the Age of Technology', *Journal of Strategic Studies*, vol. 9, nos. 2–3 (1986).

Harris, Jerry, 'The US Military in the era of Globalization', *Race and Class*, vol. 44, no. 2 (2002).

Harsh, Brigadier Surinder, 'Desert Warfare in Our Context', *Journal of the United Service Institution of India*, vol. 99, no. 417 (1969).

Hart, Stephen, 'Montgomery, Morale, Casualty Conservation and "Colossal Cracks": 21st Army Group's Operational Technique in North West Europe, 1944–45', *Journal of Strategic Studies*, vol. 19, no. 4 (1996).

Hayde, Brigadier Desmond E., *The Battle of Dograi*, 1984; rpt., Dehra Dun: Natraj, 1991.

Heathcote, T.A., *The Indian Army: The Garrison of British Imperial India, 1822–1922*, Newton Abbot: David & Charles, 1974.

Herzog, Chaim, *The War of Atonement*, 1975; rpt., New Delhi: Dehra Dun: Natraj, 1984.

Hewitt, Vernon, *Towards the Future: Jammu and Kashmir in the 21st Century*, Cambridge: Granta, 2001.

Hiranandani, Vice-Admiral G.M., *Transition to Triumph: History of the Indian Navy, 1965–75*, New Delhi: Lancer, 2000.

Holcomb, Major James F., 'Developments in Soviet Helicopter Tactics', in Air Commodore E.S. Williams, ed., *Soviet Air Power: Prospects for the Future, Perestroyka and the Soviet Air Forces*, London: Tri Service Press, 1990.

Holmes, Richard, 'Kursk: 1943', in idem, Christopher Chant and William Koenig, *Two Centuries of Warfare*, London: Octopus Books, 1978.

Hood III, Ronald Chalmers, 'Bitter Victory: French Military Effectiveness during the Second World War', in Allan R. Millett and Williamson Murray, eds., *Military Effectiveness: The Second World War*, vol. 3, Boston: Allen & Unwin, 1988.

Howard, Michael, 'Leadership in the British Army in the Second World War: Some Personal Observations', in G.D. Sheffield, ed., *Leadership and Command: The Anglo-American Military Experience since 1861*, London/Washington: Brassey's, 1997.

Huntington, Samuel P., *The Soldier and the State: The Theory and Politics of Civil-Military Relations*, 1957; rpt., Cambridge, Massachusetts: The Belknap Press, 1981.

Husain, Ross Masood, 'Threat Perception and Military Planning in Pakistan: The Impact of Technology, Doctrine and Arms Control', in Eric Arnett, ed., *Military Capacity and the Risk of War: China, India, Pakistan and Iran*, Oxford: Oxford University Press, 1997.

Hussain, Wasbir, 'Cross-Border Human Traffic in South Asia: Demographic Invasion, Anxiety and Anger in India's North-east', in K.P.S. Gill and Ajai Sahni, eds., *Faultlines: Writings on Conflict and Resolution*, vol. 7, New Delhi: Bulwark Books and the Institute for Conflict Management, 2000.

Indian Defence Review Research Team, 'OP TOPAC: The Kashmir Imbroglio', *Indian Defence Review*, vol. 14, no. 2 (1999).

Irwin, Alistair, 'The Buffalo Thorn: The Nature of the Future Battlefield', *Journal of Strategic Studies*, vol. 19, no. 4 (1996).

Isaac, J.R., S.C. Gupta, E.M.E. and S. Ramani, 'Computer Aids to Command and Control', *Journal of the United Service Institution of India*, vol. CVI, no. 44 (1976).

Jackson, Robert, 'The Strategic Outlook for the Indian Subcontinent: Some Lessons of Recent History', *Asian Affairs*, vol. 3 (1972).

Jacob, Lieutenant-General J.F.R., *Surrender at Dacca: Birth of a Nation*, New Delhi: Manohar, 1997.

Jacobs, G., 'Unmanned Aerial Vehicles: Some Programmes', *Indian Defence Review*, vol. 14, no. 2 (1999).

Jaffrelot, Christophe, 'Introduction, Nationalism Without a Nation: Pakistan Searching for its Identity', in idem ed., *Pakistan: Nationalism Without a Nation?* New Delhi: Manohar, 2002.

Jain, Satish K, 'Operation Safed Sagar', *Indian Defence Review*, vol. 16, no. 3 (2001).

Jalal, Ayesha, *The State of Martial Rule: The Origins of Pakistan's Political Economy of Defence*, 1990; rpt., New Delhi: Foundation Books, 1992.

Jamwal, N.S., 'Counter Terrorism Strategy', *Strategic Analysis*, vol. 27, no. 1 (2003).

———, 'Terrorists' Modus Operandi in Jammu and Kashmir', *Strategic Analysis*, vol. 27, no. 3 (2003).

Jessup, Jr., John E. and Robert W. Coakley, eds., *A Guide to the Study and Use of Military History*, Washington D.C.: Centre of Military History US Army, 1982.

Jha, Prem Shankar, *Kashmir 1947: The Origins of a Dispute*, New Delhi: Oxford University Press, 2003.

Jingmin, Colonel Xiao and Major Bao Bin, '21st Century Land Operations', in Michael Pillsbury, ed., *Chinese Views of Future Warfare*, 1997; rpt., New Delhi: Lancer, 1998.

Joe, 'The Assam Rifles–An Appraisal', *Journal of the United Service Institution of India*, vol. LXXXXV, no. 398 (1965).

Johnson, Air Vice-Marshal J.E. 'Johnnie', *The Story of Air Fighting*, London: Hutchinson, 1985.

Johnson, Wray R., *Vietnam and American Doctrine for Small Wars*, Bangkok: White Lotus Press, 2001.

Jolly, Group Captain R.K., 'Joint Operations: The Way Forward', in Air Commodore Jasjit Singh, ed., *Air Power and Joint Operations*, New Delhi: Knowledge World, 2003.

Jones, Owen Bennet, *Pakistan: Eye of the Storm*, New Delhi: Viking, 2002.

Joshi, Akshay, 'A Holistic View of the Revolution in Military Affairs (RMA)', *Strategic Analysis*, vol. 22, no. 11 (1999).

———, 'Information Warfare in Kargil Operations', *Indian Defence Review*, vol. 14, no. 2 (1999).

Joshi, Manoj K., 'Directions in India's Defence and Security Policies', in Sandy Gordon and Ross Babbage, eds., *India's Strategic Future: Regional State or Global Power?*, New Delhi: Oxford University Press, 1992.

Juneja, V.P., *Indo-Pak War: 1971*, New Delhi: New Light Publishers, 1971.

Kadian, Rajesh, *India and Its Army*, New Delhi: Vision Books, 1990.

Kaegi, Walter E., 'The Crisis in Military Historiography', *Armed Forces and Society*, vol. 7, no. 2 (1981).

Kain, Philip K., 'Kant's Political Theory and Philosophy of History', *CLIO*, vol. 18, no. 4 (1989).

Kak, Kapil, 'India's Conventional Defence: Problems and Prospects', *Strategic Analysis*, vol. 22, no. 11 (1999).

———, 'A Century of Air Power: Lessons and Pointers', *Strategic Analysis*, vol. 24, no. 12 (2001).

Kalkat, Major-General O.S., 'Approach to Tactics', *Journal of the United Service Institution of India*, vol. CII, no. 427 (1972).

Kalyanaraman, S., 'Operation Parakram: An Indian Exercise in Coercive Diplomacy', *Strategic Analysis*, vol. 26, no. 4 (2002).

———, 'Conceptualizations of Guerrilla Warfare', *Strategic Analysis*, vol. 27, no. 2 (2003).

Kamath, Commander V.A., 'India and Sea Power', *Journal of the United Service Institution of India*, vol. LXXXIV, no. 354 (1954).

Kantha, Pramod K., 'Transition from Authoritarianism in South Asia: Pakistan's Renewed Hope for Democracy', in John P. Lovell and David E. Albright, eds., *To Sheathe the Sword: Civil-Military Relations in the Quest for Democracy*, Westport, Connecticut: Greenwood Press, 1997.

Kanwal, Colonel Gurmeet, 'Weapons and Technology', *Indian Defence Review*, vol. 2, no. 1 (1987).

———,'NATO's Kosovo Intervention: Threat to Stable World Order', *Indian Defence Review*, vol. 14, no. 2 (1999).

———, 'Safety and Security of India's N-Weapons', *Strategic Analysis*, vol. 25, no. 1 (2001).

———, 'Airland Operations', in Air-Commodore Jasjit Singh, ed., *Air Power and Joint Operations*, New Delhi: Knowledge World, 2003.

Kapila, Air Vice Marshal Viney, *The Indian Air Force: A Balanced Strategic and Tactical Application*, New Delhi: Ocean Books Pvt. Ltd., 2002.

Karim, Major-General Afsir, 'Confrontation in Kargil', *Indian Defence Review*, vol. 14, no. 2 (1999).

———, 'Terrorism and National Security Concerns', in V.D. Chopra, ed., *Rise of Terrorism and Secessionism in Eurasia*, New Delhi: Gyan Publishing House, 2001.

———, 'War on Terrorism: Strategic Diversion', *Aakrosh*, vol. 6, no. 18 (2003).

———, 'Pak-Sponsored Terrorism', in Vice-Admiral K.K. Nayyar, ed., *Pakistan at the Crossroads*, New Delhi: Rupa, 2003.

Karnad, Bharat, *Nuclear Weapons and Indian Security: The Realist Foundations of Strategys*, Delhi: Macmillan, 2002.

Karsten, Peter, 'The Coup d'État and Civilian Control of the Military in Competitive Democracies', in John P. Lovell and David E. Albright, eds., *To Sheathe the Sword: Civil-Military Relations in the Quest for Democracy*, Westport, Connecticut: Greenwood Press, 1997.

Katari, Admiral R.D., *A Sailor Remembers*, New Delhi: Vikas, 1982.

Kathpalia, Lieutenant-General P.N., 'Internal Security and CI Operations in Urban Areas', *Indian Defence Review*, vol. 2, no. 1 (1987).

Katsouris, Andreas and Daniel Goure, 'Strategic Crossroads in South Asia: The Potential Roles for Missile Defence', *Comparative Strategy: An International Journal*, vol. 18, no. 2 (1999).

Katyal, K.K. 'Big Brother Comes Calling', *The Hindu*, 11 February 2002.

Katz, Samuel M., *Israeli Tank Battles: Yom Kippur to Lebanon*, London: Arms and Armour Press, 1988.

Kaul, Lieutenant-General B.M., *The Untold Story*, Bombay: Allied, 1967.

Kavic, Lorne J., *India's Quest for Security: Defence Policies, 1947–65*, Berkeley/Los Angeles: University of California Press, 1967.

Keegan, John, *The Face of Battle: A Study of Agincourt, Waterloo and the Somme*, 1976; rpt., Harmondsworth, Middlesex: Penguin, 1978.

———, *The Mask of Command*, 1987; rpt., New York: Penguin, 1988.

———, *The Price of Admiralty: The Evolution of Naval Warfare*, 1988; rpt., New York: Penguin, 1989.

———, *The Second World War*, 1989; rpt., London: Pimlico, 1997.

———, 'Introduction', in idem, ed., *Churchill's Generals*, 1991; rpt., London: Weidenfeld & Nicholson, 1993.

———, *Intelligence in War: Knowledge of the Enemy from Napoleon to Al-Qaeda*, New York: Alfred A. Knopf, 2003.

Kennedy, Paul, 'Grand Strategies and Less-than-Grand Strategies: A Twentieth Century Critique', in Lawrence Freedman, Paul Hayes and Robert O'Neill, eds., *War, Strategy and International Politics: Essays in Honour of Michael Howard*, Oxford: Clarendon Press, 1992.

Kennedy, Colonel William V., 'The Defence of China's Homeland', in Ray Bonds, ed., *The Chinese War Machine: A Technical Analysis of the Strategy and Weapons of the People's Republic of China*, London: Salamander, 1979.

Khan, Mohammad Asghar, *The First Round: Indo-Pakistan War 1965*, Ghaziabad: Vikas, 1979.

———, *Generals in Politics: Pakistan 1958–1982*, London/Canberra: Croom Helm, 1983.

Khan, Mohammad Ayub, *Friends not Masters: A Political Autobiography*, Lahore: Oxford University Press, 1967.

Khanduri, Brigadier Chandra B., *Field Marshal K.M. Cariappa: A Biographical Sketch*, Delhi: Dev Publications, 2000.

———, 'The Face of Proxy War in J&K', in V.D. Chopra, ed., *Rise of Terrorism and Secessionism in Eurasia*, New Delhi: Gyan Publishing House, 2001.

Killen, John, *The Luftwaffe: A History*, 1967; rpt., London: Sphere, 1970.

Kimmerling, Baruch, 'The Social Construction of Israel's "National Security"', in Stuart A. Cohen, ed., *Democratic Societies and their Armed Forces: Israel in Comparative Context*, London: Frank Cass, 2000.

Kiralfy, Alexander, 'Japanese Naval Strategy', in Edward Mead Earle, ed., *Makers of Modern Strategy: Military Thought from Machiavelli to Hitler*, 1943; rpt., Princeton: Princeton University Press, 1971.

Kiszely, John, 'The British Army and Approaches to Warfare since 1945', *Journal of Strategic Studies*, vol. 19, no. 4 (1996).

Klintworth, Gary, 'Chinese Perspectives on India as a Great Power', in Sandy Gordon and Ross Babbage, eds., *India's Strategic Future: Regional State or Global Power?*, New Delhi: Oxford University Press, 1992.

Knox, MacGregor, 'The Italian Armed Forces, 1940–43', in Allan R. Millett and Williamson Murray, eds., *Military Effectiveness: The Second World War*, vol. 3, Boston: Allen & Unwin, 1988.

Kohli, S.N., *Sea Power and the Indian Ocean*, New Delhi: Tata McGraw-Hill, 1978.

Kohn, Richard H., 'The Social History of the American Soldier: A Review and Prospectus for Research', *American Historical Review*, vol. 86, no. 3 (1981).

Kondapalli, Srikanth, *China's Military: The PLA in Transition*, New Delhi: Knowledge World, 1999.

———, 'China's Naval Structure and Dynamics', *Strategic Analysis*, vol. 23, no. 7 (1999).

———, 'China's Naval Training Programme', *Strategic Analysis*, vol. 23, no. 8 (1999).

———, 'China's Naval Equipment Acquisition', *Strategic Analysis*, vol. 23, no. 9 (1999).

———, 'Chinese Navy's Political Work and Personnel', *Strategic Analysis*, vol. 23, no. 10 (2000).

———, 'China's Naval Strategy', *Strategic Analysis*, vol. 23, no. 12 (2000).

———, 'Towards a Lean and Mean Army: Aspects of China's Ground Forces Modernisation', *Strategic Analysis*, vol. 26, no. 4 (2002).

Kotwal, Dinesh, 'The Contours of Assam Insurgency', *Strategic Analysis*, vol. 24, no. 12 (2001).

Krishnaswamy, Air Chief-Marshal S., 'Air Power Today and Tomorrow', in Air Commodore Jasjit Singh, ed., *Air Power and Joint Operations*, New Delhi: Knowledge World, 2003.

Kudryashov, Sergei, 'The Soviet Perspective', in Paul Addison and Jeremy A. Crang, eds., *The Burning Blue: A New History of the Battle of Britain*, London: Pimlico, 2000.

Kukreja, Veena, *Civil-Military Relations in South Asia: Pakistan, Bangladesh and India*, New Delhi: Sage, 1991.

Kumar, Ashwini, 'Latest Trend in Terrorism', in V.D. Chopra, ed., *Rise of Terrorism and Secessionism in Eurasia*, New Delhi: Gyan Publishing House, 2001.

Kumar, Air Marshal Bharat, 'The Pakistani Air Force—An Asessment', in Vice-Admiral K.K. Nayyar, ed., *Pakistan at the Crossroads*, New Delhi: Rupa, 2003.

Kumar, Satish, 'Sources of Democracy and Pluralism in India', in Vice-Admiral K.K. Nayyar and Jorg Schultz, eds., *South Asia Post 9/11: Searching for Stability*, New Delhi: Rupa, 2003.

Kumar, Sumita, 'The Role of Islamic Parties in Pakistani Politics', *Strategic Analysis*, vol. 25, no. 2 (2001).

Kundu, Apurba, 'The Indian Armed Forces' Sikh and Non-Sikh Officers' Opinion of Operation Blue Star', *Pacific Affairs*, vol. 67, no. 1 (1994).

———, *Militarism in India: The Army and Civil Society in Consensus*, New Delhi: Viva Books, 1998.

La, Fang, 'What's Wrong with our Army-Air Cooperation', *Journal of the United Service Institution of India*, vol. CI, no. 425 (1971).

Lahiri, Major Dipes, 'Mountains and Armed Helicopters', *Journal of the United Service Institution of India*, vol. CII, no. 428 (1972).

Lal, Air Chief Marshal P.C., *My Years with the IAF*, 1986; rpt., New Delhi: Lancer International, 1987.

Lambah, S.K., 'Democracy in Pakistan', in Vice-Admiral K.K. Nayyar and Jorg Schultz, eds., *South Asia Post 9/11: Searching for Stability*, New Delhi: Rupa, 2003.

Lansford, Tom, *All for One: Terrorism, NATO and the United States*, Aldershot, Hamphire: Ashgate, 2002.

Larus, Joel, 'India and Its Ocean: The Atypical Relationship Ends', in Robert H. Bruce, ed., *The Modern Indian Navy and the Indian Ocean*, Perth: Centre for Indian Ocean Regional Studies Curtin University of Technology, 1989.

Leach, Barry A., *German Strategy Against Russia: 1939–41*, Oxford: Clarendon Press, 1973.

Lee, Nigel de, '"A Brigadier is only a Coordinator": British Command at Brigade Level in North-West Europe, 1944, A Case Study', in G.D. Sheffield, ed., *Leadership and Command: The Anglo-American Military Experience since 1861*, London/Washington: Brassey's, 1997.

———, 'Scandinavian Disaster: Allied Failure in Norway in 1940', in Gary Sheffield and Geoffrey Till, eds., *The Challenges of High Command: The British Experience*, Houndmills, Basingstoke: Palgrave Macmillan, 2003.

Lele, A.V., 'China as a Space Power', *Strategic Analysis*, vol. 26, no. 2 (2002).

Lennox, Jane, 'Morale Implications of *Perestroyka*', in Air Commodore E.S. Williams, ed., *Soviet Air Power: Prospects for the Future, Perestroyka and the Soviet Air Forces*, London: Tri-Service Press, 1990.

Liddell Hart, B.H., *Thoughts on War*, 1944; rpt., Staplehurst: Spellmount, 1999.

Lind, William S., *Maneuver Warfare Handbook*, Boulder/London: Westview, 1985.

Longer, V., *Red Coats to Olive Green: A History of the Indian Army, 1600–1974*, Bombay: Allied, 1974.

Loveman, Brian, '"Protected Democracy" in Latin America', in John P. Lovell and David E. Albright, eds., *To Sheathe the Sword: Civil-Military Relations in the Quest for Democracy*, Westport: Greenwood Press, 1997.

Lucas, James, *Hitler's Enforcers: Leaders of the German War Machine, 1939–45*, 1996; rpt., London: Brockhampton Press, 1999.

Lyman, Robert, 'The Art of Maneuver at the Operational Level of War:

Lieutenant-General W.J. Slim and Fourteenth Army, 1944–45', in Gary Sheffield and Geoffrey Till, eds., *The Challenges of High Command: The British Experience*, Basingstoke: Palgrave Macmillan, 2003.

Lyon, Hugh, 'China's Navy for Coastal Defence Only', in Ray Bonds, ed., *The Chinese War Machine: A Technical Analysis of the Strategy and Weapons of the People's Republic of China*, London: Salamander, 1979.

Macgregor, Douglas A., *Breaking the Phalanx: A New Design for Landpower in the 21st Century*, Westport: Praeger, 1997.

Mackenzie, S.P., *Revolutionary Armies in the Modern Era: A Revisionist Approach*, London/New York: Routledge, 1997.

Mackintosh, Malcolm, 'Continuity and Change in Soviet Strategic Thought', in Lawrence Freedman, Paul Hayes and Robert O'Neill, eds., *War, Strategy and International Politics: Essays in Honour of Michael Howard*, Oxford: Clarendon Press, 1992.

Macksey, Kenneth, *Tank versus Tank: The Illustrated Story of Armoured Battlefield Conflict in the Twentieth Century*, 1999; rpt., London: Chancellor Press, 2001.

Macmillan, Margaret, 'The Indian Army Since Independence', *South Asian Review*, vol. 3, no. 1 (1969).

Malhotra, Brigadier Ashok, *Trishul: Ladakh and Kargil, 1947–93*, New Delhi: Lancer, 2003.

Malhotra, Air Vice Marshal S.S. and Group Captain Ranbir Singh, eds., *In and Out of Cockpit: Yadein Memories*, New Delhi: Ebouz Classics, 1993.

Malkasian, Carter, *A History of Modern Wars of Attrition*, Westport: Praeger, 2002.

Mamik, M.S., 'Tactical Nuclear Weapons at Sea: Are Thev Essential?', *Strategic Analysis*, vol. 13, no. 3 (1990).

Mande, Lieutenant-Colonel Y.A., 'Commanding Officer and His Team', *Journal of the United Service Institution of India*, vol. CII, no. 427 (1972).

———, 'Our Training Establishments', *Journal of the United Service Institution of India*, vol. CVI, no. 442 (1976).

Mankekar, D.R., *Pakistan Cut to Size*, Delhi: Hind Pocket Books, 1972.

Martin, Michel Louis, 'Operational Weakness and Political Activism: The Military in Sub-Saharan Africa', in John P. Lovell and David E. Albright, eds., *To Sheathe the Sword: Civil-Military Relations in the Quest for Democracy*, Westport, Connecticut: Greenwood Press, 1997.

Martz, John D., 'Contrasting Military Roles in Democratization: Colombia and Venezuela', in John P. Lovell and David E. Albright, eds., *To Sheathe the Sword: Civil-Military Relations in the Quest for Democracy*, Westport: Greenwood Press, 1997.

Marwah, Ved, *Uncivil Wars: Pathology of Terrorism in India*, 1995; rpt., New Delhi: Indus, 1996.

Mason, Francis K., *Hawker Aircraft since 1920*, London: Putnam & Company, 1961.

Mason, Philip, *A Matter of Honour: An Account of the Indian Army, Its Officers and Men*, 1974; rpt., Dehra Dun: Educational Private Limited, 1988.

Mathew, Dean, 'Aircraft Carriers: An Indian Introspection', *Strategic Analysis*, vol. 23, no. 12 (2000).

Mathur, Anand, 'Growing Importance of the Indian Ocean in the Post-Cold War Era and Its Implication for India', *Strategic Analysis*, vol. 26, no. 4 (2002).

'Matters of Moment', *Journal of the United Service Institution of India*, vol. LXXVII, no. 326 (1947).

'Matters of Moment', *Journal of the United Service Institution of India*, vol. LXXVIII, no. 328 (1947).

'Matters of Moment', *Journal of the United Service Institution of India*, vol. LXXVIII, no. 329 (1947).

Matthews, Rupert, *Hitler: Military Commander*, London: Arcturus, 2003.

Matthews, R.G., 'The Development of India's Defence-Industrial Base', *Journal of Strategic Studies*, vol. 12, no. 4 (1989).

Mattoo, Amitabh, 'Raison *d'état* or Adhocism', in idem and Kanti Bajpai, eds., *Securing India: Strategic Thought and Practice, Essays by George K. Tanham with Commentaries*, New Delhi: Manohar 1996.

Maxwell, Neville, *India's China War*, 1970; rpt., Bombay: Jaico, 1971.

Mazrui, Ali A., 'Anti-Militarism and Political Militancy in Tanzania', in Jacques Van Doorn, ed., *Military Profession and Military Regimes: Commitments and Conflicts*, The Hague/Paris: Mouton, 1969.

McCrabb, Colonel Maris 'Buster', 'The Evolution of NATO Air Doctrine', in Colonel Phillip S. Meilinger, ed., *The Paths of Heaven: The Evolution of Airpower Theory*, New Delhi: Lancer, 2000.

McKinley, Michael, 'Indian Naval Developments and Australian Strategy in the Indian Ocean', in Robert H. Bruce, ed., *The Modern Indian Navy and the Indian Ocean*, Perth: Centre for Indian Ocean Regional Studies, 1989.

McNeill, W.H., *The Pursuit of Power: Technology, Armed Forces and Society since AD 1000*, Oxford: Basil Blackwell, 1983.

McNeilly, Mark, *Sun Tzu and the Art of Modern Warfare*, Oxford/New York: Oxford University Press, 2001.

McPherson, Kenneth and Peter Reeves, 'The Foundations of Indian Naval Power', in Robert H. Bruce, ed., *The Modern Indian Navy and the Indian Ocean*, Perth: Centre for Indian Ocean Regional Studies, 1989.

Mehta, Jagat S., 'The Geopolitical and Socio-Economic Implications of USA's Involvement in Afghanistan', in Salman Haidar, ed., *The Afghan War and its Geopolitical Implications for India*, New Delhi: Manohar, 2004.

Meilinger, Colonel Phillip S., 'Introduction', in idem ed., *The Paths of Heaven: The Evolution of Airpower Theory*, New Delhi: Lancer, 2000.

Melvin, Mungo and Stuart Peach, 'Reaching for the End of the Rainbow: Command and the Revolution in Military Affairs', in Gary Sheffield and Geoffrey Till, eds., *The Challenges of High Command: The British Experience*, Basingstoke: Palgrave Macmillan, 2003.

Menezes, Lieutenant-General S.L., *Fidelity and Honour: The Indian Army from the Seventeenth to the Twenty-First Century*, New Delhi: Viking, 1993.

Mengxiong, Chang, 'Weapons of the 21st Century', in Michael Pillsbury, ed., *Chinese Views of Future Warfare*, 1997; rpt., New Delhi: Lancer, 1998.

Menon, Rear-Admiral Raja, *Maritime Strategy and Continental Wars*, London: Frank Cass, 1998.

———, 'Reflections on India's Nuclear Doctrine and Command and Control', *Journal of the United Service Institution of India*, vol. CXXXIII, no. 552 (2003).

Millett, Allan R., 'The United States Armed Forces in the Second World War', in idem and Williamson Murray, eds., *Military Effectiveness: The Second World War*, vol. 3, Boston: Allen & Unwin, 1988.

Mintz, Alex and Randolph T. Stevenson, 'Defence Expenditures, Economic Growth, and the "Peace Dividend"', *Journal of Conflict Resolution*, vol. 39, no. 2 (1995).

Misra, K.P., 'Paramilitary Forces in India', *Armed Forces and Society*, vol. 6, no. 3 (1980).

Mishra, Shitanshu, 'Exploitation of Information and Communication Technology by Terrorist Organisations', *Strategic Analysis*, vol. 27, no. 3 (2003).

Mitcham, Jr., Samuel W., *The Desert Fox in Normandy: Rommel's Defence of Fortress Europe*, London: Praeger, 1997.

Moghadam, Assaf, 'Palestinian Suicide Terrorism in the Second Intifada: Motivations and Organizational Aspects', *Studies in Conflict & Terrorism*, vol. 26, no. 2 (2003).

Mohan, P.V.S. Jagan and Samir Chopra, *The India-Pakistan Air War of 1965*, New Delhi: Manohar, 2005.

Mohanty, Deba R., 'Defence Industry Conversion in China', *Strategic Analysis*, vol. 22, no. 12 (1999).

———, 'Trends in European Defence Industry: Fortress Europe or Atlantic Defence Industry?' *Strategic Analysis*, vol. 25, no. 4 (2001).

Moore, Captain John E., *The Soviet Navy Today*, London: Macdonald and Jane's, 1975.

Moreman, T.R., '"Small Wars" and "Imperial Policing": The British Army and the Theory and Practice of Colonial Warfare in the British Empire, 1919–39', *Journal of Strategic Studies*, vol. 19, no. 4 (1996).

Morton, Colonel C.W., 'Man Management', *Journal of the United Service Institution of India*, vol. LXXVIII, no. 330 (1948).

Moskos, Charles, 'Towards a Postmodern Military?', in Stuart A. Cohen, ed., *Democratic Societies and Their Armed Forces: Israel in Comparative Context*, London: Frank Cass, 2000.

Mullik, B.N., *My Years with Nehru: The Chinese Betrayal*, Bombay: Allied Publishers, 1971.

———, *My Years with Nehru: 1948-64*, Bombay: Allied, 1972.

Murray, Williamson, 'Clausewitz: Some Thoughts on What the Germans got Right', *Journal of Strategic Studies*, vol. 9, nos. 2–3 (1986).

———, 'British Military Effectiveness in the Second World War', in idem and Alan R. Millett, eds., *Military Effectiveness: The Second World War*, vol. 3, Boston: Allen & Unwin, 1988.

———, 'Armoured Warfare: The British, French, and German Experiences', in idem and Alan R. Millett, eds., *Military Innovation in the Interwar Period*, Cambridge: Cambridge University Press, 1996.

Naaz, Farah, 'Israel's Arms Industry', *Strategic Analysis*, vol. 23, no. 12 (2000).

Naidu, G.V.C., 'India and Southeast Asia—An Activist Role for Indian Navy', *Journal of Indian Ocean Studies*, vol. 11, no. 2 (2003).

Nair, D.D., 'Russian Defence Industry in Asia', *Indian Defence Review*, vol. 18, no. 4 (2003).

Nair, Brigadier Vijay K., 'Employment of Military Helicopters, Part II: The Indian Experiences and Compulsions', *Indian Defence Review*, January 1992.

Nanavatty, Brigadier R.K., 'Use of Armed Forces in Aid to Civil Authority: Problems and Prospects', *National Defence College Journal*, vol. 15 (1993).

Nanda, Admiral S.M., *The Man Who Bombed Karachi: A Memoir*, New Delhi: HarperCollins, 2004.

Narain, Major-General Pratap, *Indian Arms Bazaar*, Delhi: Shipra Publications, 1994.

Nasr, S.V.R., 'Islam, the State and the Rise of Sectarian Militancy in Pakistan', in Christophe Jaffrelot, ed., *Pakistan: Nationalism Without a Nation?* New Delhi: Manohar, 2002.

Naveh, Shimon, *In Pursuit of Military Excellence: The Evolution of Operational Theory*, 1997, reprint, London: Frank Cass, 2000.

Nayyar, Vice-Admiral K.K, Air-Marshal B.D. Tayal, Lieutenant-General V.K. Singh, Vice-Admiral R.B. Suri and Major-General Afsir Karim, *National Security: Military Aspects*, New Delhi: Rupa, 2003.

Neillands, Robin, 'Cyprus, 1955–59: EOKA', in Major-General Julian Thompson, ed., *The Imperial War Museum Book of Modern Warfare: British and Commonwealth Forces at War, 1945–2000*, 2002; rpt., London: Pan Macmillan, 2003.

Nelson, Harold W., 'Space and Time in *On War*', *Journal of Strategic Studies*, vol. 9, nos. 2–3 (1986).

Nepram, Binalakshmi, *South Asia's Fractured Frontier: Armed Conflicts, Narcotics and Small Arms Proliferation in India's North-east*, New Delhi: Mittal, 2002.

Newall, Clayton R., *The Framework of Operational Warfare*, London: Routledge, 1991.

Niazi, Lieutenant-General A.A.K., *The Betrayal of East Pakistan*, 1998; rpt., Karachi: Oxford University Press, 2000.

O'Ballance, Edgar, *The Red Army of China*, London: Faber & Faber, 1962.

———, *Korea: 1950–53*, Dehra Dun: Natraj, 1969.

———, *Afghan Wars, Battles in a Hostile Land: 1839 to the Present*, 2002; rpt., Karachi: Oxford University Press, 2003.

Oberoi, Lieutenant-General Vijay, 'Doctrinal Challenges', in Air Commodore Jasjit Singh, ed., *Air Power and Joint Operations*, New Delhi: Knowledge World, 2003.

O'Connell, Robert L., *Ride of the Second Horseman: The Birth and Death of War*, Oxford: Oxford University Press, 1995.

Omissi, David, *The Sepoy and the Raj: The Indian Army, 1860–1940*, Houndmills, Basingstoke: Macmillan, 1994.

O'Neill, Mark A., 'Air Combat on the Periphery: The Soviet Air Force in Action during the Cold War, 1945–89', in Robin Higham, John T. Greenwood and Von Hardesty, eds., *Russian Aviation and Air Power in the Twentieth Century*, London: Frank Cass, 1998.

O'Neill, Robert, 'Problems of Command in Limited Warfare: Thoughts from Korea and Vietnam', in Lawrence Freedman, Paul Hayes and Robert O'Neill, eds., *War, Strategy and International Politics: Essays in Honour of Michael Howard*, Oxford: Clarendon Press, 1992.

Overy, Richard, *Russia's War*, 1997; rpt., London: Penguin, 1998.

Pahwa, Pran Krishan, 'The Nuclear Dimension', in Rear-Admiral Raja Menon, ed., *Weapons of Mass Destruction*, New Delhi: Sage, 2004.

Palit, Major-General D.K., *The Lightning Campaign: Indo-Pakistan War 1971*, New Delhi: Thomson Press, 1972.

———, *War in High Himalaya: The Indian Army in Crisis, 1962*, New Delhi: Lancer, 1991.

Palsokar, R.D., 'The Essentials of Guerrilla War', *Journal of the United Service Institution of India*, vol. 103, no. 429 (1973).

Panikkar, K.M. *Problems of Indian Defence*, Bombay: Asia Publishing House, 1960.

Panwar, Sango, 'The Shape of India's Emerging Missile Shield', *Strategic Analysis*, vol. 13, no. 3 (1990).

Parab, Major S.D., 'Cold Weather Operations in Mountainous Terrain', *Journal of the United Service Institution of India*, vol. LXXXXIII, no. 392 (1963).

Parik, Squadron-Leader K.N., 'Fear and Its Effect on Combat Efficiency', *Journal of the United Service Institution of India*, vol. CII, no. 427 (1972).

Parmar, Leena, 'Resettlement of Ex-Servicemen in India: Problems of Army Socialization and Adjustment', in idem, ed., *Military Sociology: Global Perspectives*, Jaipur/New Delhi: Rawat, 1999.

Pattanaik, Smruti S., 'Military Decision-Making in Pakistan', in Maroof Raza, ed., *Generals and Governments in India and Pakistan*, New Delhi: Har-Anand, 2001.

———, 'Pakistan's Kashmir Policy: Objectives and Approaches', *Strategic Analysis*, vol. 26, no. 2 (2002).

———, 'Pakistan's Nuclear Strategy', *Strategic Analysis*, vol. 27, no. 1 (2003).

Pawloski, Richard A., 'Lanes, Trains and Technology', in Air Commodore E.S. Williams, ed., *Soviet Air Power: Prospects for the Future, Perestroyka and the Soviet Air Forces*, London: Tri Service Press, 1990.

Pay, John, 'Full Circle: The US Navy and Its Carriers, 1974–93', *Journal of Strategic Studies*, vol. 17, no. 1 (1994).

Perry, William J., 'Defence in an Age of Hope', *Foreign Affairs*, November–December (1996).

Peters, Jimi, *The Nigerian Military and the State*, London/New York: I.B. Tauris, 1997.

Petersen, P.A., '*Perestroyka* and Planning Soviet Air Power', in Air Commodore E.S. Williams, ed., *Soviet Air Power: Prospects for the Future, Perestroyka and the Soviet Air Forces*, London: Tri-Service Press, 1990.

Phadke, R.V., 'Response Options: Future of Indian Air Power Vision 2020', *Strategic Analysis*, vol. 24, no. 10 (2001).

Pillai, Sushil K., 'The Invisible Country: Ethnicity & Conflict Management in Myanmar', in K.P.S. Gill and Ajai Sahni, eds., *Faultlines: Writings on Conflict & Resolution*, vol. 7, New Delhi: Bulwark Books and the Institute for Conflict Management, 200[illegible].

Pimlott, John, 'The French Army: From Indo-China to Chad, 1946–84', in idem and Ian F.W. Beckett, eds., *Armed Forces and Modern Counter-Insurgency*, London/Sydney: Croom Helm, 1985.

———, 'The British Army: The Dhofar Campaign, 1970–75', in idem and Ian F.W. Beckett, eds., *Armed Forces and Modern Counter-Insurgency*, London/Sydney: Croom Helm, 1985.

———, 'The Theory and Practice of Strategic Bombing', in Colin McInnes and G.D. Sheffield, eds., *Warfare in the Twentieth Century: Theory and Practice*, London: Unwin Hyman, 1988.

Ponnappa, Lieutenant-Colonel C.B., 'My Views on Post-War Forces', *Journal of the United Service Institution of India*, vol. LXXVII, no. 326 (1947).

Pope-Atkins, G., 'Redefining Security in Inter-American Relations: Implications for Democracy and Civil-Military Relations', in John P. Lovell and David E. Albright, eds., *To Sheathe the Sword: Civil-Military Relations in the Quest for Democracy*, Westport, Greenwood, 1997.

Porch, Douglas, *The Portuguese Armed Forces and the Revolution*, London: Croom Helm, 1977.

Posen, Barry R., *The Sources of Military Doctrine: France, Britain, and Germany Between the World Wars*, Ithaca/London: Cornell University Press, 1984.

Prabha, Kshitij, 'Narco-Terrorism and India's Security', *Strategic Analysis*, vol. 24, no. 10 (2001).

Prakash, Vice-Admiral Arun, 'Evolution of the Joint Andaman and Nicobar Command (ANC) and Defence of Our Island Territories', Part II, *Journal of the United Service Institution of India*, vol. CXXXIII, no. 551 (2003).

Prakash, Squadron Leader B.G., 'Indigenous Cryogenic Engine', *Indian Defence Review*, vol. 14, no. 2 (1999).

Prakash, Shri, 'Transformation of Pakistan and Afghanistan as Spring-Boards of Cross-Country Terrorism in the Region', in V.D. Chopra, ed., *Rise of Terrorism and Secessionism in Eurasia*, New Delhi: Gyan Publishing House, 2001.

———, 'USA's Involvement in Afghanistan: Third World Perceptions with Special Reference to South Asian Countries and Implications for the Future', in Salman Haidar, ed., *The Afghan War and Its Geopolitical Implications for India*, New Delhi: Manohar, 2004.

Prasad, Major-General Har, 'The Principles of War: Lessons from History to Remember', *Journal of the United Service Institution of India*, vol. LXXXXVI, no. 404 (1966).

Proudfoot, Lieutenant-Colonel C.L., 'The Indian Soldier: Cornerstone of Democracy', *Journal of the United Service Institution of India*, vol. C, no. 421 (1970).

Puri, Major M.K., 'Local Counter Attacks in the Mountains', *Journal of the United Service Institution of India*, vol. LXXXXVI, no. 402 (1966).

Racine, Jean-Luc, 'Pakistan and the "India Syndrome": Between Kashmir and the Nuclear Predicament', in Christophe Jaffrelot, ed., *Pakistan: Nationalism without a Nation?* New Delhi: Manohar, 2002.

Radbruch, Hans Eberhard, 'From Scharnhorst to Schmidt: The System of Education and Training in the German Bundeswehr', *Armed Forces and Society*, vol. 5, no. 4 (1979).

Raghavan, Lieutenant-General V.R., *Siachen: Conflict without End*, New Delhi: Viking, 2002.

Raina, Lieutenant-Colonel B.L., 'Military Medical Services in War', *Journal of the United Service Institution of India*, vol. LXXVIII, no. 330 (1948).

Rais, Rasul B., 'Indian Naval Developments: Implications for Pakistan', in Robert H. Bruce, ed., *The Modern Indian Navy and the Indian Ocean: Developments and Implications*, Perth: Centre for Indian Ocean Regional Studies, 1989.

Rajagopalan, Rajesh, *Second Strike: Arguments about Nuclear War in South Asia*, New Delhi: Penguin, 2005.

———, *Fighting Like a Guerrilla: The Indian Army and Counterinsurgency*, New Delhi: Routledge, 2008.

Rama Rao, R., 'Peace in the Subcontinent', *Journal of the United Service Institution of India*, vol. 103, no. 429 (1973).

Raman, B., 'The Kosovo Tragedy and NATO's Psywar', *Indian Defence Review*, vol. 14, no. 2 (1999).

Ramdas, Colonel B.S., 'The Computer and Its Effect on Management from Army to Regimental Level', *Indian Defence Review*, vol. 2, no. 1 (1987).

Ramdas, Admiral L., 'Nuclear Weapons and National Security', in M.V. Ramana and C. Rammanohar Reddy, eds., *Prisoners of the Nuclear Dream*, New Delhi: Orient Longman, 2003.

Rammohan, E.N., 'Disquiet on the Western Front', *Journal of the United Service Institution of India*, vol. CXXXIII, no. 552 (2003).

Rao, General K.V. Krishna, *In the Service of the Nation: Reminiscences*, New Delhi: Viking, 2001.

Rao, P.V.R., 'Governmental Machinery for the Evolution of National Defence Policy and the Higher Direction of War', *Institute of Defence Studies and Analyses Journal*, vol. 1 (1969).

Rappai, M.V., 'China's Nuclear Arsenal and Missile Defence', *Strategic Analysis*, vol. 26, no. 1 (2002).

Rastogi, Air Vice-Marshal S.C., 'Indian Air Force and Technology', *Journal of the United Service Institution of India*, vol. CXXXIII, no. 552 (2003).

Raza, Maroof, *Low Intensity Conflicts: The New Dimension to India's Military Commitments*, Meerut: Kartikeya Publications, 1995.

———, 'Generals and Government in India and Pakistan', in idem, ed., *Generals and Governments in India and Pakistan*, New Delhi: Har-Anand, 2001.

Reddy, K., 'India's Defence Expenditure: 1872–1967', *Indian Economic and Social History Review*, vol. 7 (1970).

Regan, Patrick M., 'Third-Party Interventions and the Duration of Intra-State Conflicts', *Journal of Conflict Resolution*, vol. 46, no. 1 (2002).

Reid, Brian Holden, 'Introduction', in *Journal of Strategic Studies*, vol. 19, no. 4 (1996).

———, 'The Commander and his Chief of Staff: Ulysses S. Grant and John A. Rawlins, 1861–65', in G.D. Sheffield, ed., *Leadership and Command: The Anglo-American Military Experience since 1861*, London/Washington: Brassey's, 1997.

Reid, Julian, 'Foucault on Clausewitz: Conceptualizing the Relationship between War and Power', *Alternatives*, vol. 28, no. 1 (2003).

Reynal-Querol, Marta, 'Ethnicity, Political Systems, and Civil Wars', *Journal of Conflict Resolution*, vol. 46, no. 1 (2002).

Rikhye, Ravi, 'Tricap: New Dimensions in Armoured Warfare', *Journal of the United Service Institution of India*, vol. CI, no. 425 (1971).

———, 'Rethinking Mechanized Infantry Concepts', *Journal of the United Service Institution of India*, vol. CII, no. 427 (1972).

Ripley, Tim, *The New Illustrated Guide to Modern US Army*, London: Salamander, 1992.

Robson, Brian, *The Road to Kabul: The Second Afghan War, 1878–81*, London: Arms and Armour, 1986.

Rogers, Colonel H.C.B., *Tanks in Battle*, 1965; rpt., London: Sphere, 1972.

Roland, Alex, 'Technology and War: The Historiographical Revolution of the 1980s', *Technology and Culture*, vol. 34, no. 1 (1993).

Ropp, Theodore, 'Continental Doctrines of Sea Power', in Edward Mead Earle, ed., *Makers of Modern Strategy: Military Thought from Machiavelli to Hitler*, 1943; rpt., Princeton: Princeton University Press, 1971.

Rosen, Stephen Peter, *Societies and Military Power: India and Its Armies*, New Delhi: Oxford University Press, 1996.

Rothenberg, Gunther E., *The Anatomy of the Israeli Army*, London: B.T. Batsford Ltd., 1979.

Roy, Kaushik, 'Good Governance Versus Bad Governance in South Asia: Civil-Military Relations in India and Pakistan (1947–2000)', *Asian Studies*, vol. 18, nos. 1–2 (2000).

———, 'The Battle for Kargil: Post-Modern War in South Asia', in Kanti Bajpai, Afsir Karim and Amitabh Matoo, eds., *Kargil and After: Challenges for Indian Policy*, New Delhi: Har-Anand, 2001.

———, 'Limitations of Western Warfare: American Military Operations in Afghanistan, 2001', in *The Afghanistan Crisis: Problems Perspectives*, New Delhi: Nehru Memorial Museum & Library, 2002.

———, 'Goliath Against David: Militaries against the Militias', *Contemporary India*, vol. 1, no. 2 (2002).

———, 'Mars Defeated? Conventional Militaries in Unconventional Warfare', *Contemporary India*, vol. 2, no. 2 (2003).

Roy, Vice-Admiral Mihir K., 'Asymmetry of India's Defence Forces: A Sailor's View', *Strategic Analysis*, vol. 13, no. 3 (1990).

———, *War in the Indian Ocean*, New Delhi: Lancer, 1995.

Roy-Chaudhury, Rahul, *Sea Power and Indian Security*, London/Washington: Brassey's, 1995.

———, 'The Limits to Naval Expansion', in Kanti P. Bajpai and Amitabh Mattoo, eds., *Securing India: Strategic Thought and Practice*, New Delhi: Manohar, 1996.

———, 'India's Maritime Challenges in the early 21st Century', *Indian Defence Review*, vol. 14, no. 2 (1999).

———, *India's Maritime Security*, New Delhi: Knowledge World, 2000.

Roychowdhury, Shankar, 'Management of Power in the Defence Services: Values Dimensions', in S.K. Chakraborty and Pradip Bhattacharya, eds., *Leadership and Power: Ethical Explorations*, New Delhi: Oxford University Press, 2001.

———, *Officially at Peace: Reflections on the Army and Its Role in Troubled Times*, New Delhi: Viking, 2002.

Ryan, Cornelius, *A Bridge Too Far*, 1974; rpt., London: Coronet Books, 1975.

Sachdev, A.K., 'Modernization of the Chinese Air Force', *Strategic Analysis*, vol. 23, no. 6 (1999).

Saikal, Amin, 'The Future of India and South-west Asia', in Sandy Gordon and Ross Babbage, eds., *India's Strategic Future: Regional State or Global Power?*, New Delhi: Oxford University Press, 1992.

Saikia, Jaideep, 'Autumn in Springtime: The ULFA Battles for Survival', in K.P.S. Gill and Ajai Sahni, eds., *Faultlines: Writings on Conflict and Resolution*, vol. 7, New Delhi: Bulwark Books and the Institute for Conflict Management, 2000.

———, 'Changing Contours of Separatism', *Aakrosh*, vol. 6, no. 18 (2003).

Sakhuja, Vijay, 'Dragon's Dragonfly: The Chinese Aircraft Carrier', *Strategic Analysis*, vol. 24, no. 7 (2000).

———, 'Maritime Power of People's Republic of China: The Economic Dimension', *Strategic Analysis*, vol. 24, no. 11 (2001).

———, 'Sea Based Deterrence and Indian Security', *Strategic Analysis*, vol. 25, no. 1 (2001).

———, 'Indian Ocean and the Safety of Sea Lines of Communication', *Strategic Analysis*, vol. 25, no. 5 (2001).

———, 'Pakistan's Naval Strategy: Past and Future', *Strategic Analysis*, vol. 26, no. 4 (2002).

Saksena, Commodore N.M.L., 'Major Naval Operations since Independence', in Major-General Afsir Karim, ed., *The Indian Armed Forces: A Basic Guide*, New Delhi: Lancer, 1995.

———, 'Role of the Indian Navy', in Major-General Afsir Karim, ed., *The Indian Armed Forces: A Basic Guide*, New Delhi: Lancer, 1995.

———, 'Manpower Recruitment and Branches of the Navy', in Major-General Afsir Karim, ed., *The Indian Armed Forces: A Basic Guide*, New Delhi: Lancer, 1995.

Saksena, Major Yogi, 'Indo-Pak War, 1971: Some Lessons', *Journal of the United Service Institution of India*, vol. CII, no. 428 (1972).

Santhanam, K., Sreedhar, Sudhir Saxena and Manish, *Jihadis in Jammu and Kashmir*, New Delhi: Sage, 2003.

———, 'Sources of Terror: India', in Vice-Admiral K.K. Nayyar and Jorg Schultz, eds., *South Asia Post 9/11: Searching for Stability*, New Delhi: Rupa, 2003.

Sanyal, S., 'One Man's Terrorist: Victims' Perspectives on Terrorism', in K.P.S. Gill and Ajai Sahni, eds., *Faultlines: Writings on Conflict and Resolution*, vol. 7, New Delhi: Bulwark Books and the Institute for Conflict Management, 2000.

Sapir, Jacques, *The Soviet Military System*, 1987; tr. by David Macey, Cambridge: Polity Press, 1991.

Sarkar, Colonel Bhaskar, *Tackling Insurgency and Terrorism: Blueprint for Action*, Delhi: Vision Books, 1998.

———, *Outstanding Victories of the Indian Army: 1947–71*, New Delhi: Lancer, 2000.

Sarkar, Jagadish Narayan, 'Aspects of Military Policy in Medieval India', *Proceedings of the Indian History Congress*, 36th Session, Aligarh (1975).

Sarkesian, Sam C., 'Combat Effectiveness', in idem, ed., *Combat Effectiveness: Cohesion, Stress, and the Volunteer Military*, Beverly Hills/London: Sage, 1980.

Sarma, Group-Captain S. D., 'Meteorology and War', *Journal of the United Service Institution of India*, vol. CII, no. 427 (1972).

Sarma, Vice-Admiral S.H., *My Years at Sea*, New Delhi: Lancer, 2001.

Sawhney, Pravin, 'Operation Rhino: A Case Study', *Indian Defence Review*, January 1992.

———, *The Defence Makeover: 10 Myths that Shape India's Image*, New Delhi: Sage, 2002.

Sawhny, Rathy, 'Decision Making for Defence', *Institute of Defence Studies and Analyses Journal*, vol. 2 (1969–70).

Sayigh, Yezid, 'Arms Production in Pakistan and Iran: The Limits of Self-Reliance', in Eric Arnett, ed., *Military Capacity and the Risk of War: China, India, Pakistan and Iran*, Oxford: Oxford University Press, 1997.

Schiff, Rebecca L., 'Concordance Theory: A Response to Recent Criticism', *Armed Forces and Society*, vol. 23, no. 2 (1996).

———, 'The Indian Military and Nation Building: Institutional and Cultural Concordance', in John P. Lovell and David E. Albright, eds., *To Sheathe the Sword: Civil-Military Relations in the Quest for Democracy*, Westport: Greenwood, 1997.

Schmidl, Erwin A., 'The Evolution of Peace Operations', in idem, ed., *Peace Operations Between War and Peace*, London: Frank Cass, 2000.

Scobell, Andrew, *China's Use of Military Force: Beyond the Great Wall and the Long March*, Cambridge: Cambridge University Press, 2003.

Sen, Lieutenant-General L.P., *Slender Was the Thread: Kashmir Confrontation, 1947–48*, 1969; rpt., New Delhi: Orient Longman, 1994.

Sen, Air Vice-Marshal Samir K., *Military Technology and Defence Industrialization: The Indian Experience*, New Delhi: Manas, 2000.

Sen, Air-Commodore T.K., 'Development of the IAF after Independence', in Major-General Afsir Karim, ed., *The Indian Armed Forces: A Basic Guide*, New Delhi: Lancer, 1995.

———, 'Operational Role and Evolution 1962–1977', in Afsir Karim, ed., *The Indian Armed Forces, A Basic Guide*, New Delhi: Lancer, 1995.

Seth, Vijay, *The Flying Machines: Indian Air Force 1933 to 1999*, New Delhi: Seth Communications, 2000.

Sethi, M.L., 'Higher Defence Organization in India', *Journal of the United Service Institution of India*, vol. CXX, no. 499 (1990).

Shafqat, Saeed, 'From Official Islam to Islamism: The Rise of *Dawat-ul-Irshad* and *Lashkar-e-Toiba*', in Christophe Jaffrelot, ed., *Pakistan: Nationalism Without a Nation?*, New Delhi: Manohar, 2002.

Shah, G.M., 'Militancy in Jammu and Kashmir: A Study of Actors, Objectives and Strategies', in V.D. Chopra, ed., *Rise of Terrorism and Secessionism in Eurasia*, New Delhi: Gyan Publishing House, 2001.

Shankar, Lieutenant-General Vinay, 'Dynamics of Future Wars', in Air Commodore Jasjit Singh, ed., *Air Power and Joint Operations*, New Delhi: Knowledge World, 2003.

Sharma, Lieutenant-Colonel Gautam, *Nationalization of the Indian Army: 1885–1947*, New Delhi: Allied, 1996.

Sheffield, G.D., 'Introduction: Command, Leadership and the Anglo-American Experience', in idem, ed., *Leadership and Command: The Anglo-American Military Experience Since 1861*, London/Washington: Brassey's, 1997.

Shekhar, Chandra, 'Indigenization of the Defence Industry', *Journal of the United Service Institution of India*, vol. 131, no. 544 (2001).

Shrivastava, B.K., 'The United States and International Terrorism in South Asia', in V.D. Chopra, ed., *Rise of Terrorism and Secessionism in Eurasia*, New Delhi: Gyan Publishing House, 2001.

Shrivastava, V.K., 'Indian Army: The Challenge Ahead', *Strategic Analysis*, vol. 25, no. 4 (2001).

———, 'Indian Army 2020: A Vision Statement on Strategy and Capability', *Strategic Analysis*, vol. 25, no. 6 (2001).

———, 'Indian Air Force in the Years Ahead: An Army View', *Strategic Analysis*, vol. 25, no. 8 (2001).

Siddiqa-Agha, Ayesha, *Pakistan's Arms Procurement and Military Buildup, 1979–1999: In Search of a Policy*, Basingstoke: Palgrave, 2001.

Simonsen, Sven Gunnar, 'Marching to a Different Drum? Political Orientations and Nationalism in Russia's Armed Forces', *The Journal of Communist Studies and Transition Politics*, vol. 17, no. 1 (2001).

Simpkin, Richard E., *Race to the Swift: Thoughts on Twenty-First Century Warfare*, London: Brassey's, 1985.

Singh, Anil Kumar, 'India's Maritime Security: Challenges Ahead', *Contemporary India*, vol. 2, no. 2 (2003).

Singh, Anita Inder, 'Containment Through Diplomacy: Britain, India and the Cold War in Indochina, 1954–56', in Ian Nish, ed., *South Asia in International Affairs: 1947–56, International Studies*, London: London School of Economics and Political Science, 1987.

Singh, Marshal of the Indian Air Force, Arjan, 'Presidential Address', in Air-Commodore Jasjit Singh, ed., *Air Power and Joint Operations*, New Delhi: Knowledge World, 2003.

Singh, Major K. Brahma, 'Infantry Mobility in the Jungle', *Journal of the United Service Institution of India*, vol. LXXXXV, no. 401 (1965).

———,'Military Leadership–An Objective Study', *Journal of the United Service Institution of India*, vol. 99, no. 416 (1969).

Singh, Daljit, 'Military Education in India: Changes from the British Tradition', *Journal of the United Service Institution of India*, vol. CIV, no. 436 (1974).

Singh, Daljit and Katherine Singh, 'Decentralization and Diffusion of Power in the Army Structures of India and Pakistan', *Journal of the United Service Institution of India*, vol. CI, no. 425 (1971).

Singh, Lieutenant-General Depinder, *Field-Marshal Sam Manekshaw: Soldiering with Dignity*, Dehra Dun: Natraj, 2002.

Singh, Major Gurbachan, 'The Right Type and Some Thoughts on Indianization', *Journal of the United Service Institution of India*, vol. LXXVI, no. 324 (1946).

Singh, Lieutenant-General Harbakhsh, *War Despatches: Indo-Pak Conflict, 1965*, New Delhi: Lancer, 1991.

———, *In the Line of Duty: A Soldier Remembers*, New Delhi: Lancer, 2000.

Singh, Colonel Harjeet, *Doda: An Insurgency in the Wilderness*, New Delhi: Lancer, 1999.

Singh, Major-General Jagjit, *With Honour and Glory: Wars fought by India, 1947–99*, New Delhi: Lancer, 2001.

Singh, Air Commodore Jasjit, 'Strategic Framework for Defence Planners· Air Power in the 21st Century', *Strategic Analysis*, vol. 22, no. 12 (1999).

———,'Pakistan's Fourth War', *Strategic Analysis*, vol. 23, no. 5 (1999).

———, 'Dynamics of Limited War', *Strategic Analysis*, vol. 24, no. 7 (2000).

———, *India's Defence Spending: Assessing Future Needs*, New Delhi: Knowledge World, 2000.

———, 'Nuclear Command and Control', *Strategic Analysis*, vol. 25, no. 2 (2001).

———, 'Our Wars in Future', in idem, ed., *Air Power and Joint Operations*, New Delhi: Knowledge World, 2003.

Singh, Jaswant, *Defending India*, Chennai: Macmillan, 1999.

———, ed., *Indian Armed Forces Year Book: 1981–82*, Bombay: n.d.

Singh, K.R., 'Regional Cooperation in the Bay of Bengal: Non-Conventional Threats-Maritime Dimension', *Strategic Analysis*, vol. 24, no. 12 (2001).

———, *Navies of South Asia*, New Delhi: Rupa, 2002.

Singh, Major-General Lachhman, *Victory in Bangladesh*, 1981; rpt., Dehra Dun: Natraj, 1991.

Singh, Air-Commodore N.B., *Air Power in the New Millennium*, New Delhi: Manas, 2000.

Singh, Pushpindar, Ravi Rikhye and Peter Steinemann, *Fiza'ya: Psyche of the Pakistan Air Force*, New Delhi: Society for Aerospace Studies, 1991.

Singh, Lieutenant-Colonel Rajendra, 'Is Scientific Selection Successful?', *Journal of the United Service Institution of India*, vol. LXXVI, no. 325 (1946).

———, *Soldier and Soldiering in India: Twelve Essays*, 1954; rpt., New Delhi: Army Educational Stores, 1964.

Singh, Group-Captain Ranbir, *Indian Air Force: In the Footsteps of Our Legends*, Noida: Book Mates, 1998.

Singh, Leintenant-General R.K. Jasbir, *Marshal Arjan Singh: Life and Times*, New Delhi: Ocean Books, 2002.

Singh, Roopinder, *Arjan Singh: Marshal of the Indian Air Force*, New Delhi: Rupa, 2002.

Singh, Commander Sanjay J., 'The Nature of War in the 21st Century', in Air Commodore Jasjit Singh, ed., *Air Power and Joint Operations*, New Delhi: Knowledge World, 2003.

———, 'The Past Five Decades', in Air Commodore Jasjit Singh, ed., *Air Power and Joint Operations*, New Delhi: Knowledge World, 2003.

Singh, Rear Admiral Satyindra, 'The Kargil Review Committee Report: Much to Learn and Implement', *Indian Defence Review*, vol. 15, no. 1 (2000).

Singh, Major-General Sukhwant, *India's War Since Independence: The Liberation of Bangladesh*, vol. 1, 1980; rpt., New Delhi: Lancer, 1998.

———, *India's War Since Independence: Defence of the Western Border*, vol. 2, 1981; rpt., New Delhi: Lancer, 1998.

———, *General Trends: India's War Since Independence*, vol. 3, 1982; rpt., New Delhi: Lancer, 1998.

Singh, Swaran, 'Continuity and Change in China's Maritime Strategy', *Strategic Analysis*, vol. 23, no. 9 (1999).

Sinha, P.M., 'Air Power in the Information Age', *Indian Defence Review*, vol. 16, no. 3 (2001).

Sinha, Lieutenant-General S.K., 'Counter Insurgency Operations', *Journal of the United Service Institution of India*, vol. 100, no. 420 (1970).

———, *Operation Rescue: Military Operations in Jammu and Kashmir 1947–9*, 1977; rpt., New Delhi: Vision Books, 2002.

Sinha, S.P., 'Prabhakaran as Leader of the LTTE', *Journal of the United Service Institution of India*, vol. 131, no. 544 (2001).

Smith, Chris, *India's Ad Hoc Arsenal: Direction or Drift in Defence Policy*, Oxford: Oxford University Press, 1994.

Smith, Joseph, 'The Spanish-American War: Land Battles in Cuba, 1895–98', *Journal of Strategic Studies*, vol. 19, no. 4 (1996).

Sodhi, Brigadier H.S., 'The Higher Echelons of Leadership', *Journal of the United Service Institution of India*, vol. CIV, no. 436 (1974).

———, *Top Brass: A Critical Appraisal of the Indian Military Leadership*, Noida: Trishul Publications, 1993.

Sohn, Jae Souk, 'Factionalism and Party Control of the Military in Communist North Korea', in Jacques Van Doorn, ed., *Military Profession and Military Regimes: Commitments and Conflicts*, The Hague/Paris: Mouton, 1969.

Solomon, David N., 'The Soldierly Self and the Peacekeeping Role: Canadian Officers in Peacekeeping Forces', in Jacques Van Doorn, ed., *Military Profession and Military Regimes: Commitments and Conflicts*, The Hague/Paris: Mouton, 1969.

Sood, Lieutenant-General (Retd). V.K. and Pravin Sawhney, *Operation Parakram: The War Unfinished*, New Delhi: Sage, 2003.

'Special Document: Mahmud Abbas' Call for a Halt to the Militarization of the Intifada', *Journal of Palestine Studies*, vol. 32, no. 2 (2003).

Sprout, Margaret Tuttle, '*Mahan*: Evangelist of Sea Power', in Edward Mead Earle, ed., *Makers of Modern Strategy: Military Thought from Machiavelli to Hitler*, 1943; rpt., Princeton: Princeton University Press, 1971.

Sreedhar, 'Defence Expenditure During the Rajiv Gandhi Years', *Strategic Analysis*, vol. 13, no. 3 (1990).

Sridharan, Commander K., *A Maritime History of India*, Delhi: Publications Division, 1965.

Stephens, Lieutenant-Colonel J. Wilson, 'The Question', *Journal of the United Service Institution of India*, vol. LXXVIII, no. 330 (1948).

Sterner, Eric R., 'You Say You Want a Revolution (in Military Affairs)', *Comparative Strategy: An International Journal*, vol. 18, no. 4 (1999).

Stroup, Jr., Theodore G. and Leonard Wong, 'Re-establishing the Force: The Revolution in Military Affairs in the Human Resource and Leadership Systems', in Stuart A. Cohen, ed., *Democratic Societies and Their Armed Forces: Israel in Comparative Context*, London: Frank Cass, 2000.

Strudwick, Brigadier M.J., 'A National Plan to Contain Militancy and Insurgency', *National Defence College Journal*, vol. 15 (1993).

Subrahmanyam, K., 'Decision Making in Defence', *Institute for Defence Studies and Analysis Journal*, vol. 2 (1969–70).

Sukumaran, R., 'The 1962 India–China War and Kargil 1999: Restrictions on the Use of Air Power', *Strategic Analysis*, vol. 27, no. 3 (2003).

Sundarji, General K., *Blind Men of Hindoostan: Indo-Pak Nuclear War*, 1993; rpt., New Delhi: UBSPD, 1996.

———, *Of Some Consequences: A Soldier Remembers*, New Delhi: HarperCollins, 2000.

———, *Vision 2100: A Strategy for the Twenty-First Century*, Delhi: Konark, 2003.

Suri, Vice-Admiral R.B., 'The Pakistani Navy—Force Level Options', in Vice-Admiral K.K. Nayyar, ed., *Pakistan at the Crossroads*, New Delhi: Rupa, 2003.

Suryanarayan, V., 'Ethnic Cauldron in Sri Lanka', in V.D. Chopra, ed., *Rise of Terrorism and Secessionism in Eurasia*, New Delhi: Gyan Publishing House, 2001.

Sweetman, Bill, 'The Modernization of China's Air Force', in Ray Bonds, ed., *The Chinese War Machine: A Technical Analysis of the Strategy and Weapons of the People's Republic of China*, London: Salamander Book, 1979.

Tahiliani, R.H., 'Corruption in Arms Trade', *Journal of the United Service Institution of India*, vol. 131, no. 544 (2001).

Tahir-Kheli, Shirin, 'The Military in Contemporary Pakistan', *Armed Forces and Society*, vol. 6, no. 4 (1980).

Talbot, Ian, 'Does the Army Shape Pakistan's Foreign Policy?', in Christophe Jaffrelot, ed., *Pakistan: Nationalism Without a Nation?*, New Delhi: Manohar, 2002.

———, 'The Punjabization of Pakistan: Myth or Reality', in Christophe Jaffrelot, ed., *Pakistan: Nationalism Without a Nation?*, New Delhi: Manohar, 2002.

Tanham, George K., 'Indian Strategic Thought: An Interpretative Essay', in Kanti P. Bajpai and Amitabh Mattoo, eds., *Securing India: Strategic Thought and Practice*, New Delhi: Manohar, 1996.

———, 'Indian Strategy in Flux', in Kanti P. Bajpai and Amitabh Mattoo, eds., *Securing India: Strategic Thought and Practice*, New Delhi: Manohar, 1996.

Tanham, George K. and Marcy Agmon, *The Indian Air Force: Trends and Prospects*, 1995; rpt., New Delhi: Vikas, 1996.

Tellis, Ashley J., 'Aircraft Carriers and the Indian Navy: Assessing the Present, Discerning the Future', *Journal of Strategic Studies*, vol. 10, no. 2 (1987).

———, 'Securing the Barrack: The Logic, Structure and Objectives of India's Naval Expansion', in Robert H. Bruce, ed., *The Modern Indian Navy and the Indian Ocean: Developments and Implications*, Perth: Centre for Indian Ocean Regional Studies, 1989.

———, *India's Emerging Nuclear Posture: Between Recessed Deterrent and Ready Arsenal*, New Delhi: Oxford University Press, 2001.

Thakur, Ramesh, 'India in the World: Neither Rich, Powerful, nor Principled', *Foreign Affairs*, vol. 76, no. 4 (1997).

The Military Art of People's War: Selected Writings of General Vo Nguyen Giap, edited and with an introduction by Russell Stetler, 1970; rpt., New York/London: Monthly Review Press, 1971.

The Military Maxims of Napoleon, tr. from the French by Lieutenant-General

George C.D'Aguilar, Introduction and Commentary by David G. Chandler, 1987; rpt., New York: Da Capo, 1995.

Thilo, Lieutenant-General Karl Wilhelm, 'A Perspective from the Army High Command (OKH)', in David M. Glantz, ed., *The Initial Period of War on the Eastern Front: 22 June–August 1941*, 1993; rpt., London: Frank Cass, 1997.

Thomas, Martin, 'Order Before Reform: The Spread of French Military Operations in Algeria, 1954–58', in David Killingray and David Omissi, eds., *Guardians of Empire: The Armed Forces of the Colonial Powers c. 1700–1964*, Manchester/New York: Manchester University Press, 1999.

Thomas, Raju G.C., 'The Sources of Indian Naval Expansion', in Robert H Bruce, ed., *The Modern Indian Navy and the Indian Ocean: Developments and Implications*, Perth: Centre for Indian Ocean Regional Studies, 1989.

———, 'The Growth of Indian Military Power: From Sufficient Defence to Nuclear Deterrence', in Sandy Gordon and Ross Babbage, eds., *India's Strategic Future: Regional State or Global Power?*, New Delhi: Oxford University Press, 1992.

———, 'Arms Procurement in India: Military Self-Reliance Versus Technological Self-Sufficiency', in Eric Arnett, ed., *Military Capacity and the Risk of War: China, India, Pakistan and Iran*, Oxford: Oxford University Press, 1997.

———, 'India's Nuclear and Missile Programs: Strategy, Intentions, Capabilities', in idem and Amit Gupta, eds., *India's Nuclear Security*, New Delhi: Vistaar Publications, 2000.

Thomas, Raju G.C. and Amit Gupta, 'Introduction', in Thomas and Gupta, eds., *India's Nuclear Security*, New Delhi: Vistaar Publications, 2000.

Thompson, Julian, 'The World in 1945: The View from May 2001', in idem, ed., *The Imperial War Museum Book of Modern Warfare: British and Commonwealth Forces at War, 1945–2000*, 2002; rpt., London: Pan Macmillan, 2003.

Thorat, Lieutenant-General S.P.P., *From Reveille to Retreat*, New Delhi: Allied, 1986.

Till, Geoffrey, 'Maritime Strategy and the Twenty-First Century', *Journal of Strategic Studies*, vol. 17, no. 1 (1994).

Tinker, Hugh, 'From British Indian Army to Indian Army', *Indo-British Review*, vol. 16, no. 1 (1989).

Tiwary, Air-Commodore A.K., 'Future of Fighter Operations', *Indian Defence Review*, vol. 15, no. 1 (2000).

Toase, Francis, 'The South African Army: The Campaign in South West Africa/Namibia since 1966', in idem and Ian F.W. Beckett, eds., *Armed Forces and Modern Counter-Insurgency*, London/Sydney: Croom Helm, 1985.

Towle, Philip, 'Air Power in Afghanistan', in Air Commodore E.S. Williams, ed., *Soviet Air Power: Prospects for the Future, Perestroyka and the Soviet Air Forces*, London: Tri Service Press, 1990.

Trehan, Jyoti, 'Violence in J&K: Complexities and Pathways', in K.P.S. Gill and Ajai Sahni, eds., *Faultlines: Writings on Conflict and Resolution*, vol. 7, New Delhi: Bulwark Books and the Institute for Conflict Management, 2000.

Tritten, James John, 'Is Naval Warfare Unique?', *Journal of Strategic Studies*, vol. 12, no. 4 (1989).

Tse-Tung, Mao, 'Guerrilla Warfare', in Brigadier-General Samuel B. Griffith and Major Harris-Clichy Peterson, eds., *Guerrilla Warfare*, 1962; rpt., London: Cassell, 1963.

Tully, Mark and Satish Jacob, *Amritsar: Mrs Gandhi's Last Battle*, Calcutta: Rupa, 1985.

Tzu, Sun, *The Art of War* (tr. with an Introduction by Samuel B. Griffith), Oxford: Oxford University Press, 1963.

Vallance, Andrew G.B., *The Air Weapon: Doctrines of Air Power Strategy and Operational Art*, London: Macmillan, 1996.

Valluri, S.R., 'Lest We Forget: The Futility and Irrelevance of Nuclear Weapons for India', in Raju G.C. Thomas and Amit Gupta, eds., *India's Nuclear Security*, New Delhi: Vistaar Publications, 2000.

Vandervort, Bruce, *Wars of Imperial Conquest in Africa: 1830–1914*, Bloomington/Indianapolis: Indiana University Press, 1998.

Vasudeva, P.K., 'The Role of the Chief of Defence Staff', *Journal of the United Service Institution of India*, vol. CXXX, no. 542 (2000).

Venkateswaran, A.L., *Defence Organization in India: A Study of Major Developments in Organization and Administration since Independence*, Delhi: Publications Division, 1967.

Verma, Bharat, 'Pakistan: The Counter Strategy', *Indian Defence Review*, vol. 14, no. 2 (1999).

———, 'MAKS 2001 International Air Show: Moscow on the Comeback Trail', *Indian Defence Review*, vol. 16, no. 3 (2001).

Vertzberger, Yacov, 'India's Strategic Posture and the Border War Defeat of 1962: A Case Study in Miscalculation', *Journal of Strategic Studies*, vol. 5, no. 3 (1982).

Vohra, A.M., 'A Military Policy for the Nineties', *Journal of the United Service Institution of India*, vol. CXX, no. 499 (1990).

Walker, Walter, 'Brunei and Borneo, 1962–66: An Efficient Use of Military Force', in Major-General Julian Thompson, ed., *The Imperial War Museum Book of Modern Warfare: British and Commonwealth Forces at War, 1945–2000*, 2002; rpt., London: Pan Macmillan, 2003.

Wallace, J.J.A., 'Maneuver Theory in Operations Other than War', *Journal of Strategic Studies*, vol. 19, no. 4 (1996).

Warikoo, K., 'Islamist Extremism and Terrorism in Kashmir', in V.D. Chopra, ed., *Rise of Terrorism and Secessionism in Eurasia*, New Delhi: Gyan Publishing House, 2001.

Watson, Bruce Allen, *When Soldiers Quit: Studies in Military Disintegration*, Westport: Praeger, 1997.

Watts, Anthony J., 'The Falklands War', in Antony Preston, ed., *History of the Royal Navy in the 20th Century*, London: Bison, 1987.

Weist, Andrew A., 'Haig, Gough and Passchendaele', in G.D. Sheffield, ed., *Leadership and Command: The Anglo-American Military Experience since 1861*, London/Washington: Brassey's, 1997.

Wettern, Desmond, 'The Navy Post-1945', in Antony Preston, ed., *History of the Royal Navy in the 20th Century*, London: Bison, 1987.

Wheeler, Geoffrey, 'The Indian Ocean Area: Soviet Aims and Interests', *Asian Affairs*, vol. 3 (1972).

Whistler, Major-General L.G., 'A Story of Honourable Dealing', *Journal of the United Service Institution of India*, vol. LXXVIII, no. 329 (1947).

Wiatr, Jerzy J., 'Social Prestige of the Military: A Comparative Approach', in Jacques Van Doorn, ed., *Military Profession and Military Regimes: Commitments and Conflicts*, The Hague/Paris: Mouton, 1969.

Wickremasuriya, Kingsley, 'Community Oriented Policing and Conflict Management: The Jaffna Experiment', in K.P.S. Gill and Ajai Sahni, eds., *Faultlines: Writings on Conflict and Resolution*, vol. 7, New Delhi: Bulwark Books and the Institute for Conflict Management, 2000.

Wilson, Major A.J., 'Officer Production and Nationalization', *Journal of the United Service Institution of India*, vol. LXXVI, no. 324 (1946).

Winton, Harold R., 'An Ambivalent Partnership: US Army and Air Force Perspectives on Air-Ground Operations, 1973–90', in Colonel Phillip S. Meilinger, ed., *The Paths of Heaven: The Evolution of Air Power Theory*, New Delhi: Lancer, 2000.

Woerner, General Fred, 'Foreword', in John P. Lovell and David E. Albright, eds., *To Sheathe the Sword: Civil-Military Relations in the Quest for Democracy*, Westport: Greenwood, 1997.

Woodward, Dennis, 'The People's Liberation Army: A Threat to India?', *Contemporary South Asia*, vol. 12, no. 2 (2003).

Yapp, M.A., 'British Perceptions of the Russian Threat to India', *Modern Asian Studies*, vol. 24, no. 4 (1987).

Young, Brigadier Peter, ed., *Strategy and Tactics of the Great Generals and Their Battles*, 1984; rpt., London: Bison Books, 1986.

Young, Peter Lewis, 'Australia's Military Doctrine: Outlines of Changes', *Indian Defence Review*, vol. 14, no. 2 (1999).

Zahab, Mariam Abou, 'The Regional Dimension of Sectarian Conflicts in Pakistan', in Christophe Jaffrelot, ed., *Pakistan: Nationalism without a Nation?* New Delhi: Manohar, 2002.

Zetterling, Niklas and Anders Frankson, *Kursk 1943: A Statistical Analysis*, London: Frank Cass, 2000.

Zhilin, P., 'The Armed Forces of the Soviet State: Fifty Years of Experience in Military Construction', in Jacques Van Doorn, ed., *Military Profession and Military Regimes: Commitments and Conflicts*, The Hague/Paris: Mouton, 1969.

———, 'Policy and Strategy of the Soviet Union in the Second World War', in Arthur L. Funk, ed., *Politics and Strategy in the Second World War: Germany, Great Britain, Japan, the Soviet Union and the United States*, Manhattan, Kansas: Kansas State University, 1976.

Zobel, Colonel Horst, '3rd Panzer Division Operations', in David M. Glantz, ed., *The Initial Period of War on the Eastern Front: 22 June–August 1941*, 1993; rpt., London: Frank Cass, 1997.

Zolberg, Astride R., 'Military Rule and Political Development in Tropical Africa', in Jacques Van Doorn, ed., *Military Profession and Military Regimes: Commitments and Conflicts*, The Hague/Paris: Mouton, 1969.

Zuberi, Matin, 'The Proposed Indian Nuclear Doctrine', *Contemporary India*, vol. 1, no. 1 (2002).

NEWSPAPERS

The Hindu, The Telegraph, Times of India.

MAGAZINES

Ayudh, Defence, Defence and Armament, Defence Watch, Frontline, India Today, Jane's Defence Weekly, Newsweek, Outlook, Strategic Digest, The Economist, The Week, Vayu 2000 Aerospace Review.

MISCELLANEOUS PUBLICATIONS

Baranwal, J., Editor-in-Chief, *SP's Military Yearbook: 1992–93*, New Delhi: Guide Publications, 1992.

———, *SP's Military Yearbook: 1993–94*, New Delhi: Guide Publications, 1993.

———, *SP's Military Yearbook: 1995*, New Delhi: Guide Publications, 1995.

———, *SP's Military Yearbook: 1998–99*, New Delhi: Guide Publications, 1998.

Kaul, Ravi, ed., *The Chanakya Defence Annual: 1973–74*, Allahabad: Chanakya Publishing House, n.d.

Peace and Conflict: An IPCS Bulletin, vol. 5, no. 9 (2002).

Singh, Jaswant, ed., *Indian Armed Forces Yearbook: 1981–82*, Bombay, n.d.

Singh, Lieutenant-General R.K. Jasbir, ed., *Indian Defence Yearbook: 2003*, Dehra Dun: Natraj, 2003.

SIPRI Yearbook 1998: Armaments, Disarmament and International Security, Oxford: Oxford University Press, 1998.

The Military Balance: 1990–91, London: The International Institute for Strategic Studies, 1990.

Index